D0866381

THE FROZEN CELL

THE FROZEN CELL

A Ciba Foundation Symposium

Edited by
G. E. W. WOLSTENHOLME
and
MAEVE O'CONNOR

J. & A. CHURCHILL
104 GLOUCESTER PLACE, LONDON
1970

First published 1970

With 115 illustrations

Standard Book Number 7000 1445 4

© *Longman Group Ltd 1970*

All rights reserved. No part of this publication may be reproduced, stored in a retrieval system, or transmitted, in any form or by any means, electronic, mechanical, photocopying, recording or otherwise, without the prior permission of the copyright owner.

Printed in Great Britain

Contents

Membership

Symposium on The Frozen Cell held 20th–22nd May 1969

P. Mazur (Chairman)	Biology Division, Oak Ridge National Laboratory, P.O. Box Y, Oak Ridge, Tenn. 37830, U.S.A.
J. F. Brandts	Department of Chemistry, University of Massachusetts, Amherst, Mass. 01002, U.S.A.
J. D. Davies	Department of Pathology, University of Cambridge, Tennis Court Road, Cambridge CB2 1QP, England
B. C. Elford	Clinical Research Centre Laboratories, National Institute for Medical Research, Mill Hill, London, N.W.7, England
J. Farrant	Clinical Research Centre Laboratories, National Institute for Medical Research, Mill Hill, London, N.W.7, England
R. I. N. Greaves	Department of Pathology, University of Cambridge, Tennis Court Road, Cambridge, CB2 1QP, England
U. Heber	Botanisches Institut der Universität, Ulenbergstrasse 127, 4 Düsseldorf, West Germany
C. Huggins	Surgical Low Temperature Unit, Massachusetts General Hospital, Boston, Mass. 02114, U.S.A.
I. M. Klotz	Biochemistry Division, Department of Chemistry, Northwestern University, Evanston, Ill. 60201, U.S.A.
D. Lee	Clinical Research Centre Laboratories, National Institute for Medical Research, Mill Hill, London, N.W.7, England
S. P. Leibo	Biology Division, Oak Ridge National Laboratory, P.O. Box Y, Oak Ridge, Tenn. 37830, U.S.A.
J. Levitt	Department of Botany, University of Missouri, Columbia, Mo. 65201, U.S.A.
B. Luyet	3837 Monona Drive, Madison, Wis. 53714, U.S.A.
A. P. MacKenzie	American Foundation for Biological Research, R.5, Box 137, Madison, Wis. 53704, U.S.A.
H. T. Meryman	Blood Research Laboratory, American National Red Cross, 9312 Old Georgetown Road, Bethesda, Maryland 20014, U.S.A.

T. Nei — Institute of Low Temperature Science, Hokkaido University, Sapporo, Japan

D. E. Pegg — Clinical Research Centre Laboratories, National Institute for Medical Research, Mill Hill, London, N.W.7, England

A. P. Rinfret — Union Carbide Corporation, Linde Division, 8th floor, 270 Park Avenue, New York, N.Y. 10017, U.S.A.

Denise Simatos — Laboratoire de Biologie Physico-chimique, Université de Dijon, Dijon, France

H. A. D. Walder — Department of Neurosurgery, Katholieke Universiteit, St. Annastraat 315, Nijmegen, Holland

C. A. Walter — National Institute for Medical Research, Mill Hill, London, N.W.7, England

A. E. Woolgar — National Institute for Medical Research, Mill Hill, London, N.W.7, England

The Ciba Foundation

The Ciba Foundation was opened in 1949 to promote international cooperation in medical and chemical research. It owes its existence to the generosity of CIBA Ltd, Basle, who, recognizing the obstacles to scientific communication created by war, man's natural secretiveness, disciplinary divisions, academic prejudices, distance, and differences of language, decided to set up a philanthropic institution whose aim would be to overcome such barriers. London was chosen as its site for reasons dictated by the special advantages of English charitable trust law (ensuring the independence of its actions), as well as those of language and geography.

The Foundation's house at 41 Portland Place, London, has become well known to workers in many fields of science. Every year the Foundation organizes six to ten three-day symposia and three to four shorter study groups, all of which are published in book form. Many other scientific meetings are held, organized either by the Foundation or by other groups in need of a meeting place. Accommodation is also provided for scientists visiting London, whether or not they are attending a meeting in the house.

The Foundation's many activities are controlled by a small group of distinguished trustees. Within the general framework of biological science, interpreted in its broadest sense, these activities are well summed up by the motto of the Ciba Foundation: *Consocient Gentes*—let the peoples come together.

Preface

THIS meeting originated from Dr John Farrant's suggestion of a symposium to discuss the precise effects of different freezing or cooling procedures and different concentrations of protective compounds on physical parameters in biological systems. The aim was to try to relate the physical changes to the biological properties necessary for cell viability and to learn more about the basic mechanism, or mechanisms, of freezing injury and protection.

Dr Farrant, Dr Peter Mazur, and their colleagues gave invaluable advice during the organization of the meeting, and the Ciba Foundation is most grateful to them for all their help. We are also indebted to Dr Farrant for much kind assistance with editorial difficulties. Dr Peter Mazur's firm and persuasive direction as chairman helped to throw many of the questions into sharper focus.

During the meeting the Clinical Research Centre Laboratories at the National Institute for Medical Research invited the participants, together with members of the British Cryogenic Council and other guests, to attend an afternoon of demonstrations and a garden party at Mill Hill. Dr Audrey Smith in this way contributed a valuable part to the symposium, although she could not manage to attend the actual proceedings. Also, very appropriately, Sir Alan and Lady Parkes acted as hosts for the Ciba Foundation at the main social function held at 41 Portland Place.

CHAIRMAN'S OPENING REMARKS

PETER MAZUR

ALL of us here are convinced, I am sure, that the responses of cells and macromolecules to subzero temperatures are matters of interest and of biological significance. Perhaps the best evidence that this view is not over-parochial is the fact that Dr Wolstenholme and the staff of the Ciba Foundation have organized this present symposium. The original impetus for the meeting came from Dr John Farrant, and we are especially indebted to him, to Dr Audrey Smith, and to Dr Wolstenholme for bringing it to fruition.

Whether by accident or design, our discussion is occurring on the 20th anniversary of the year that Polge, Smith and Parkes (1949) published their findings that glycerol enables mammalian sperm to survive slow freezing. The impact of that paper and its immediate successors on the field of low-temperature biology is difficult to exaggerate. It is perhaps equalled only by Lovelock's detailed and critical analysis of freezing injury in the human red cell, published some four years later. As you recall, Lovelock concluded that freezing and thawing cause haemolysis because freezing subjects cells to concentrations of electrolyte above mole fraction 0·014 (~ 0·8 M), and because thawing subjects them to dilution. He further concluded that glycerol and dimethylsulphoxide prevent haemolysis by preventing the electrolytes from concentrating to the critical value, and that this action is a consequence of the colligative properties of solutions. The total mole fraction of solute in a partially frozen solution is determined by temperature. If all the solutes are electrolytes, then the required mole fraction will consist entirely of electrolytes, but if a non-electrolyte such as glycerol is present, the concentration of electrolyte will be reduced, and the extent of reduction at a given temperature will be approximately proportional to the osmolar ratio of glycerol to electrolyte.

Lovelock stated that only low-molecular weight hydrophilic solutes with low eutectic points could protect cells, because only such solutes could yield solutions of high molar concentration at sufficiently low temperatures. He also stated that to be protective a solute must permeate a cell, for otherwise it could not prevent a rise in intracellular electrolyte during freezing.

So persuasive were the findings of Smith, Lovelock, and their colleagues that other investigators have tended to assume that the freezing procedures developed for sperm are also optimal for cells in general, and they have tended to assume that Lovelock's explanation of freezing injury and protection holds for cells in general.

The central question I would like to pose for our consideration is: to what extent are these assumptions valid ? Let me elaborate:

(1) Lovelock's explanation of freezing injury is detailed and specific. But slow freezing not only causes electrolytes to concentrate, it also causes the removal of water, the osmotic shrinkage of cells, and a decrease in the spatial separation of macromolecules. Some investigators, notably Meryman (1968) and Levitt (1966), believe that freezing injury is due more to these events than to the concomitant concentration of electrolytes.

(2) However, Lovelock's, Meryman's, and Levitt's theories agree in ascribing most freezing injury to a single cause, and in ascribing this single cause to changes in properties of solutions brought about by dehydration. Injury is not a result of ice formation *per se*. On the other hand, some of us believe that there are *at least* two factors responsible for injury, and that one of these is in fact ice formation within the cell. Furthermore, one of our speakers, Dr Brandts, has strong theoretical and experimental evidence for the existence of a third potentially injurious factor, namely, lowered temperature itself.

(3) These questions as to causes of injury have their counterpart in the protection of cells. Do all additives protect by a single mechanism, and, if so, is it by the colligative mechanism proposed by Lovelock?

(4) Related to questions of mechanisms of injury and of protection are questions as to the site and nature of the lesions. I believe that several speakers will present evidence to strengthen the growing suspicion that the chief targets of injury are membranes. If so, it then follows that additives probably protect cells by somehow protecting their semi-permeable membranes.

Cells contain both surface and internal membranes. Are they both susceptible to injury? Apparently so if additives must permeate cells in order to protect them. But is this long-held assumption correct? Is permeation by an additive obligatory for protection? The main affirmative evidence again appears to be that provided by Lovelock. He showed that fewer red cells survive freezing in non-permeating additives than survive in permeating additives. However, what has been generally ignored is that he also found that many more red cells survived freezing in non-permeating additives than survived in the absence of any additive. Furthermore, Rapatz and Luyet (1963) and Rapatz, Sullivan and Luyet (1968) have reported that human red cells and bovine red cells respond similarly to freezing in glycerol in spite of the fact that the latter are many times less permeable to glycerol. Some of these points are summarized in Table I. Other evidence on this matter with other cells will be presented by several speakers.

TABLE I

RELATION BETWEEN PERMEATION OF GLYCEROL AND RECOVERY OF RED CELLS AFTER FREEZING

Cell	*Time for permeation of glycerol at*		*Recovery (%) of cells frozen slowly to −20 °C or below after exposure to indicated concentrations of glycerol*				
						14%, + Cu++ at	
	20 °C	0 °C	0%	10%, 30 sec at 0 °C	10%, 10 min (?) at 20 °C	0 °C	40 °C
Human	1 min*	4 min*	2†	78‡	96†	72*	97*
Bovine	~30 min§	—	0‖	75‡	92‖	—	—

* Lovelock (1953).
† Rapatz, Sullivan and Luyet (1968).
‡ Lovelock and Bishop (1959).
§ Jacobs (1931).
‖ Rapatz and Luyet (1963).

The question of the relation between the ability of a solute to permeate and its ability to protect against freezing injury is, it seems to me, a vital one. The answer is essential to understanding the fundamental causes of injury and it may well be essential to the successful freezing of cellular aggregates.

REFERENCES

JACOBS, M. H. (1931). *Ergebn. Biol.*, **7**, 1. [Cited by Davson, H., and Danielli, J. D. (1952). In *Permeability of Natural Membranes*, p. 83. Cambridge University Press.]
MERYMAN, H. T. (1968). *Nature, Lond.*, **218**, 333–336.
LOVELOCK, J. E. (1953). *Biochim. biophys. Acta*, **11**, 28.
LOVELOCK, J. E., and BISHOP, M. W. H. (1959). *Nature, Lond.*, **183**, 1394.
LEVITT, J. (1966). In *Cryobiology*, pp. 495–563, ed. Meryman, H. T. London & New York: Academic Press.
POLGE, C., SMITH, A. U., and PARKES, A. S. (1949). *Nature, Lond.*, **164**, 666.
RAPATZ, G., and LUYET, B. (1963). *Biodynamica*, **9**, 125.
RAPATZ, G., SULLIVAN, J. J., and LUYET, B. (1968). *Cryobiology*, **5**, 18.

POLYHEDRAL CLATHRATE HYDRATES

IRVING M. KLOTZ

Biochemistry Division, Department of Chemistry, Northwestern University, Evanston, Illinois

WATER is a very versatile substance, both at the macroscopic and at the molecular levels. It will dissolve a wide variety of substances—ionic, polar and apolar solutes. Each solute interacts with solvent to produce a unique hydrated structure which is the resultant of the interplay of the specific character of the solute molecule and the cooperative structural adaptability of the solvent molecules.

The detailed molecular structures of hydrated solute species in liquid water are difficult to establish since available probes provide only indirect information. For many solutes, however, solid crystalline hydrates can be isolated. Their molecular structures can be definitely established by X-ray diffraction techniques, and have been described for a wide variety of solutes.

In ionic hydrates the intense centrosymmetric electrostatic field orients water dipoles very sharply towards the charged centre species. On the other hand, for apolar solutes direct pair-wise interactions are very much weaker. This difference is illustrated, for example, by the extremely large hydration energies (about 100 kcal/mole [$10^3 \times 418{\cdot}4$ joules/mole]) for even singly charged ions as contrasted to values of only about 3 kcal/mole ($10^3 \times 12{\cdot}552$ J/mole) for apolar molecules such as methane or argon.

In many hydrates of apolar solutes, therefore, H_2O–H_2O interactions dominate the structure. Thus water molecules form polyhedra of very different size and shape which can enclose an apolar solute molecule as in a closed box or cage. Within recent years a very large number of different substances have been found to form hydrates with such polyhedral clathrate structures. It seems appropriate therefore to describe some of these structures as a prelude to an examination of the character of the frozen cell.

STOICHEIOMETRY AND CHEMISTRY

On the basis of chemical stoicheiometry as well as molecular structure studies it is convenient to classify the polyhedral hydrates into three major groups. These are assembled in Table I.

TABLE I

POLYHEDRAL HYDRATES

	Class I		*Class II*	*Class III*
Guest Molecules	Ar	CH_4	$CHCl_3$	$(n\text{-}C_4H_9)_4N^+F^-$
	Kr	C_2H_2	CH_3CHCl_2	$(n\text{-}C_4H_9)_4N^{+-}O_2CC_6H_5$
	Cl_2	C_2H_4	$(CH_3)_2O$	$[(n\text{-}C_4H_9)_4N^+]_2WO_4^=$
	H_2S	C_2H_6	C_3H_8	$(i\text{-}C_5H_{11})_4N^+F^-$
	PH_3	CH_3Cl	$(CH_3)_3CH$	$(n\text{-}C_4H_9)_3S^+F^-$
	SO_2	CH_3SH	C_3H_7Br	$(n\text{-}C_4H_9)_4P^+Cl^-$
	$C_2H_5NH_2$	CH_3CHF_2	$(CH_3)_2CO$	$(CH_3)_3N$ $(CH_2)_6N_4$
	$(CH_3)_2NH$	CH_2CHF		$n\text{-}C_3H_7NH_2$ $(CH_3)_4N^+OH^-$
				$i\text{-}C_3H_7NH_2$
				$(C_2H_5)_2NH$
				$(CH_3)_3C\text{-}NH_2$
				C_4H_9OH
Stoicheiometry	$M \cdot 5\frac{3}{4}\ H_2O$ $M \cdot 7\frac{2}{3}\ H_2O$		$M \cdot 17H_2O$ $M \cdot M'_2 \cdot 17H_2O$	$M \cdot (5\text{-}40)H_2O$
Unit Cell	$46H_2O$		$136H_2O$	Variable number H_2O
	Polyhedra $2H_{12}$ (5 Å) $6H_{14}$ (6 Å)		Polyhedra $16H_{12}$ (5 Å) $8H_{16}$ (7 Å)	Polyhedra H_8, H_{12}, H_{14}, H_{15} H_{16}, H_{17}, H_{18}, H_{60}, etc.
	Faces Pentagons Hexagons		Faces Pentagons Hexagons	Faces Quadrilaterals Hexagons Heptagons
	$M_2 \cdot M_6 \cdot 46H_2O$		$M_8 \cdot M'_{16} \cdot 136H_2O$	

(H_n symbolizes a polyhedron with n faces)

Listed at the top of Table I are representative examples of guest solute molecules which can be enclosed within the cage-like lattice formed by host water molecules. The guest molecules of Class I include inorganic as well as organic molecules, but in all cases the molecular size is relatively small. The earliest hydrates known (Davy, 1811*a*, *b*; Faraday, 1823;

Villard, 1896; de Forcrand, 1925) were those of relatively inert solute molecules, such as argon or the hydrocarbons. More recently, however, it has also been found that even molecules such as small amines or small cyclic oxides, which in principle could form disruptive hydrogen bonds directly to the water, can nevertheless be enclosed within a polyhedral cage (Glew, Mak and Rath, 1967; Jeffrey and McMullan, 1967; McMullan, Jordan and Jeffrey, 1967). Guest molecules in Class II are larger in size than those in Class I, but span the same range of chemical properties, including relatively inert hydrocarbons as well as molecules such as tetrahydrofuran or acetone, which in principle could form direct hydrogen bonds to water molecules. As we shall see shortly, guest molecules of Classes I and II are enclosed totally within a single polyhedral water lattice. In contrast the molecules in Class III are enclosed in larger and less symmetric polyhedra which can generally be visualized as being constituted by the fusion of several of the small polyhedra (McMullan and Jeffrey, 1959; Jeffrey and McMullan, 1967).

Polyhedral hydrates are also known which contain two guest species. These double hydrates are particularly common for molecules of Class II in circumstances where a second solute such as H_2S, H_2Se, CO_2, or some of the rare gases are purposely added. These have been named "helper gases" by von Stackelberg and Meinhold (1954), because their presence within the hydrate lattices very significantly increases the stability of the structure. It seems likely that, even when helper gases are not added purposely, air (i.e. oxygen and nitrogen) also participates to some extent in stabilizing the structure by being trapped within some of the open cages. This conclusion has been reached by several observers (see, e.g., Jeffrey and McMullan, 1967) who have found that gases are evolved from most hydrate crystals as they are melted.

MOLECULAR STRUCTURE

The unique characteristic feature of the structure of the polyhedral hydrates is the presence of pentagonal faces (Claussen, 1951; Pauling and Marsh, 1952; von Stackelberg and Müller, 1954). As is very well known, water is a molecule which can form tetrahedral bonds, two being contributed by the bonded hydrogen atoms and two being accepted by the two pairs of free electrons on the oxygen atom. This disposition of hydrogen atoms and free electron pairs makes H_2O particularly suited for four-coordinated interactions in condensed phases (Bernal and Fowler, 1933). Each of the two protons can form a hydrogen bond to one of the two centres of negative charge on an adjacent oxygen atom. Since each oxygen atom can act as acceptor for two hydrogen bonds and as donor for two, each water molecule is surrounded tetrahedrally by four other water molecules.

In ice this tetrahedral orientation of neighbouring water molecules extends in all directions to build up the familiar hexagonal rings of oxygen atoms in the crystal. With slight distortions in the tetrahedral bond length water molecules can also be arranged in pentagonal faces, which in the simplest situation can be fitted together to produce the pentagonal dodecahedron (Fig. 1), one of the five regular platonic solids. Alternatively polyhedra with 14, 15, or 16 faces (Fig. 1) may be produced (see Jeffrey and McMullan, 1967).

HYDRATE POLYHEDRA

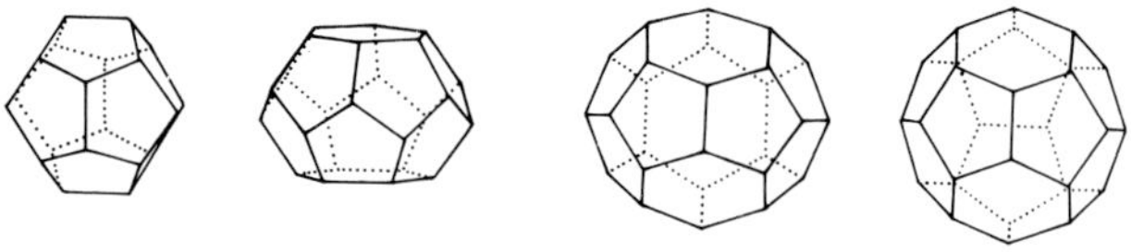

	Dodecahedron	Tetrakai decahedron	Pentakai decahedron	Hexakai decahedron
Faces	12	14	15	16
Vertices	20	24	26	28
Edges	30	36	39	42
Volume enclosed	160 Å³	230 Å³	260 Å³	290 Å³

FIG. 1. Simple polyhedra found in crystalline clathrate hydrates of Class I and Class II molecules.

The dodecahedron has 20 vertices and 30 edges. At each vertex is an oxygen atom, and each edge contains an OH···O bond. Since there are 20 oxygen atoms at the vertices these must be associated with 40 hydrogen atoms ($H_{40}O_{20}$ corresponds stoicheiometrically to H_2O). Of the 40 protons, 30 can be placed in the 30 edges. Consequently ten remain to form external hydrogen bonds extending from ten vertices; and the remaining ten oxygen atoms in the dodecahedron can act as acceptors of hydrogens from external water molecules. Thus by bringing polyhedra together so that they can share faces, one can build up a superstructure which fits well into a macroscopic crystalline array. The unit cell (Claussen, 1951; Pauling and Marsh, 1952; von Stackelberg and Müller,

1954) for the Class I polyhedral hydrates (see Table I) contains eight such polyhedra, two having 12 faces and six having 14 faces. In contrast the unit cell of Class II hydrates contains, as shown in Fig. 2, 16 dodecahedra and eight hexakaidecahedra.

The dodecahedra enclose a volume of about 160 $Å^3$ and have an internal diameter of about 5·2 Å (0·52 nm) (Fig. 1). The 14-sided polyhedron has an internal diameter of 5·9 Å and a 16-sided enclosure has an internal diameter of 6·9 Å. Those guest molecules which crystallize as Class I polyhedral hydrates may fill all the cavities in a unit

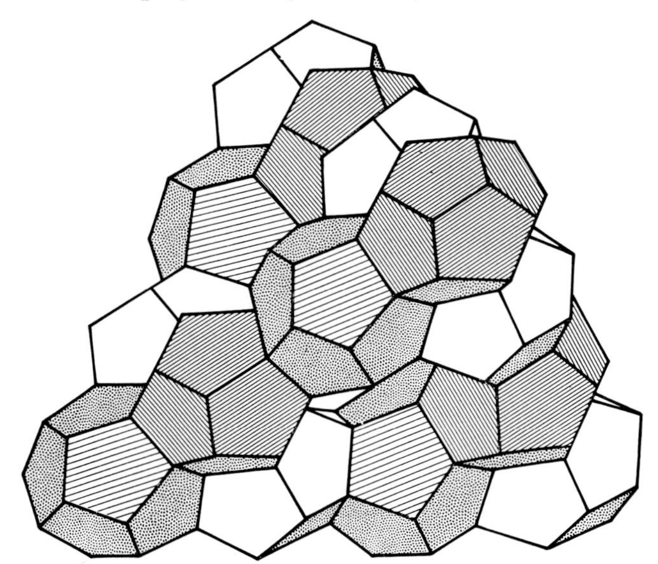

FIG. 2. Face-sharing arrangement of polyhedra in the unit cell of Class II hydrates. (See Jeffrey and McMullan, 1967.)

cell or may be inserted only in the larger enclosures. If only the larger cavities are filled by the inert molecule M then the formula for the unit cell turns out to be $6M \cdot 46H_2O$ or stoicheiometrically $M(H_2O)_{7 \cdot 67}$. If all the cavities are occupied by M the corresponding formulas are $8M \cdot 46H_2O$ and $M(H_2O)_{5 \cdot 75}$, respectively. In the Class II unit cell containing 24 cavities and 136 H_2O molecules, the guest molecules M are generally too large to occupy any of the smaller enclosures and hence fit only into the eight large cavities. Thus the formula for the unit cell is $8M \cdot 136H_2O$ or $M(H_2O)_{17}$ stoicheiometrically. It is in this group of hydrates that mixed hydrates are commonly found since a small molecule such as H_2S can fill the 16 smaller cavities. These mixed hydrates thus have the stoicheiometric formula, $M(H_2S)_2(H_2O)_{17}$.

In both Class I and Class II hydrates each guest molecule is completely enclosed in one polyhedron with 12 to 16 faces, that is, in one of the polyhedra illustrated in Fig. 1. A skeletal diagram of the relative positions of the guest molecule in one such polyhedron is illustrated in Fig. 3 (McMullan and Jeffrey, 1965) which shows ethylene oxide in two of its possible positions within a 14-hedron. Since the interaction between the host lattice molecules and the guest molecule enclosed is a very weak one,

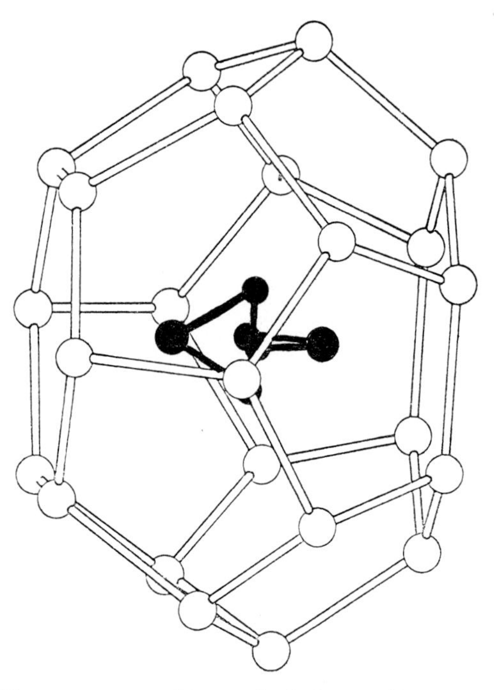

FIG. 3. Two average positions of ethylene oxide guest molecule (filled circles) within a 14-hedron (water oxygens shown as open circles). (See McMullan and Jeffrey, 1965.)

the guest molecules have a much greater degree of rotational freedom than one would expect to find in the solid state. This is also illustrated in Fig. 4, which shows the symmetrical electron density distribution of tetrahydrofuran within the 16-hedron of its polyhedral hydrate (Mak and McMullan, 1965).

Class III polyhedra are much more complicated than those of Class I or Class II (see Fig. 5). They demonstrate that it is not necessary to squeeze a guest molecule into a simple single small polyhedron in order to have a stable hydrate. The cages in Class III may be considered as constituted from the fusion of several 12–16 hedra into a single much larger poly-

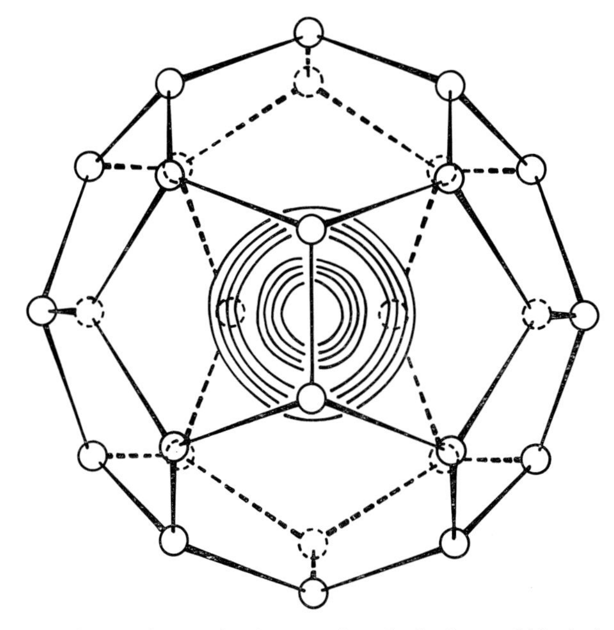

FIG. 4. Electron density distribution of tetrahydrofuran within the hexakaidecahedron of its clathrate double hydrate $8C_4H_8O \cdot 6 \cdot 4H_2S \cdot 136H_2O$. (See Mak and McMullan, 1965.)

MULTIPLE FUSED POLYHEDRA

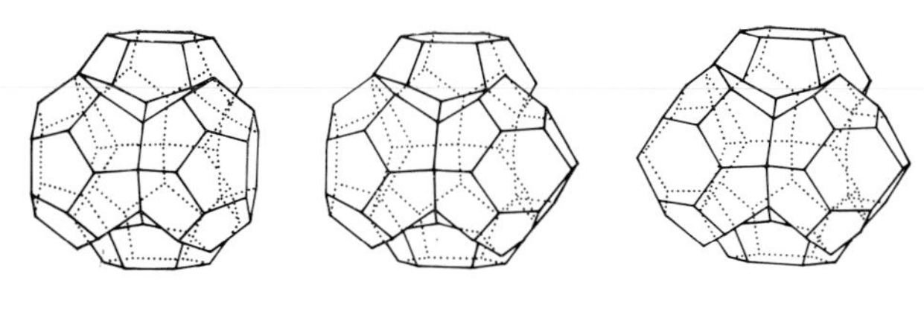

Fusion of	4(14-hedra)	3(14-hedra) 1(15-hedron)	2(14-hedra) 2(15-hedra)
Faces	44	45	46
Vertices	70	72	74
Edges	112	115	118
Volume enclosed	1000 Å^3	1000 Å^3	1000 Å^3

FIG. 5. Fused polyhedra found in crystalline clathrate hydrates of Class III molecules.

hedron (McMullan and Jeffrey, 1959; Jeffrey and McMullan, 1967). Some examples are shown in Fig. 5. As is evident in a comparison of Figs. 1 and 5, the fused polyhedra are composed of many more water molecules and enclose much larger volumes. Consequently guest molecules of much larger size may be contained within the cages. Fig. 6 shows the orientation of a tetraisoamylammonium ion within its polyhedral cage, which is actually constituted from two 14 hedra and two 15 hedra.

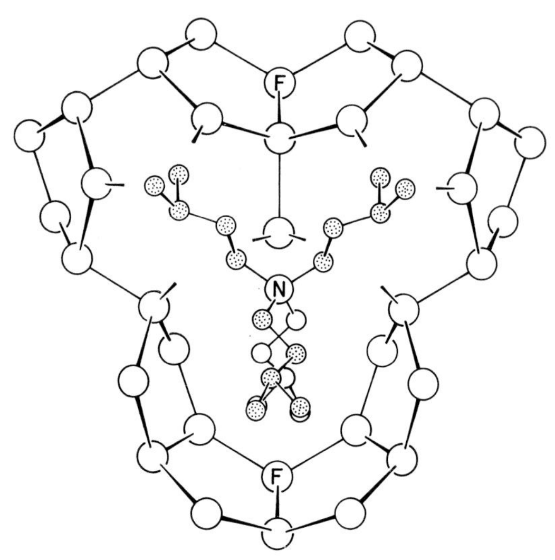

FIG. 6. Orientation of tetraisoamylammonium cation within lobes of fused (two 14- and two 15-) polyhedra of water molecules in $(i\text{-}C_5H_{11})_4N^+F^- \cdot 38H_2O$. (See McMullan and Jeffrey, 1959.)

All of the polyhedra shown so far are made up of faces containing pentagons or hexagons. Fig. 7 (Beurskens and Jeffrey, 1964) shows an even further departure from the relative symmetry illustrated so far. The cages in Fig. 7 are much more unsymmetrical and complex and are bounded by quadrilateral rhombuses as well as by pentagonal and hexagonal faces. The cavities are both large and irregular and the volume enclosed, 1600 Å^3, is substantially larger than that of any previously illustrated polyhedron. The hydrate structure shown in Fig. 7 has the formula $(n\text{-}C_4H_9)_3S^+F^- \cdot 23H_2O$. Only half of the complete water cage is shown in Fig. 7; the remaining half would be a mirror image of that illustrated. Within a full cage are two guest molecules:

$$(C_4H_9)_3S^+ \quad {}^+S(C_4H_9)_3$$

Particularly striking is the fact that the two triplets of apolar butyl groups of the two molecules face away from each other; they tend to be far apart and embedded in (three pairs of subsidiary) water lobes rather than adjacent to each other and excluding intervening water molecules. The six incomplete polyhedra forming the lobes enclosing each of the alkyl

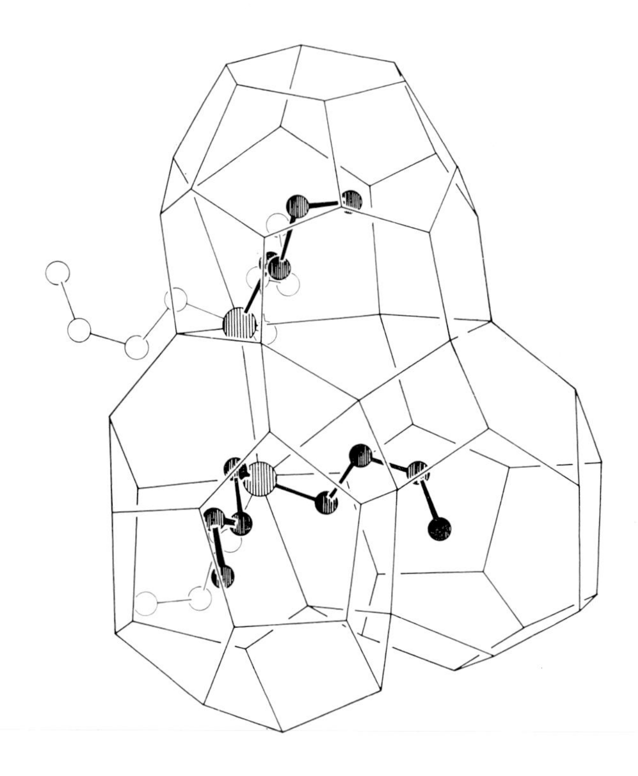

FIG. 7. Disposition of butyl side chains within lobes of polyhedra of water molecules in $(C_4H_9)_3S^+F^- \cdot 23H_2O$. (See Beurskens and Jeffrey, 1964.)

side chains are symmetrically arranged about the centre of symmetry at the mid-point of the $S^+ \cdots S^+$ line. There are three types of these incomplete compartments with different sizes and shapes made up of the following polygonal faces: (*a*) nine pentagons and two hexagons; (*b*) seven pentagons, two hexagons and one quadrilateral; and (*c*) eight pentagons and one hexagon.

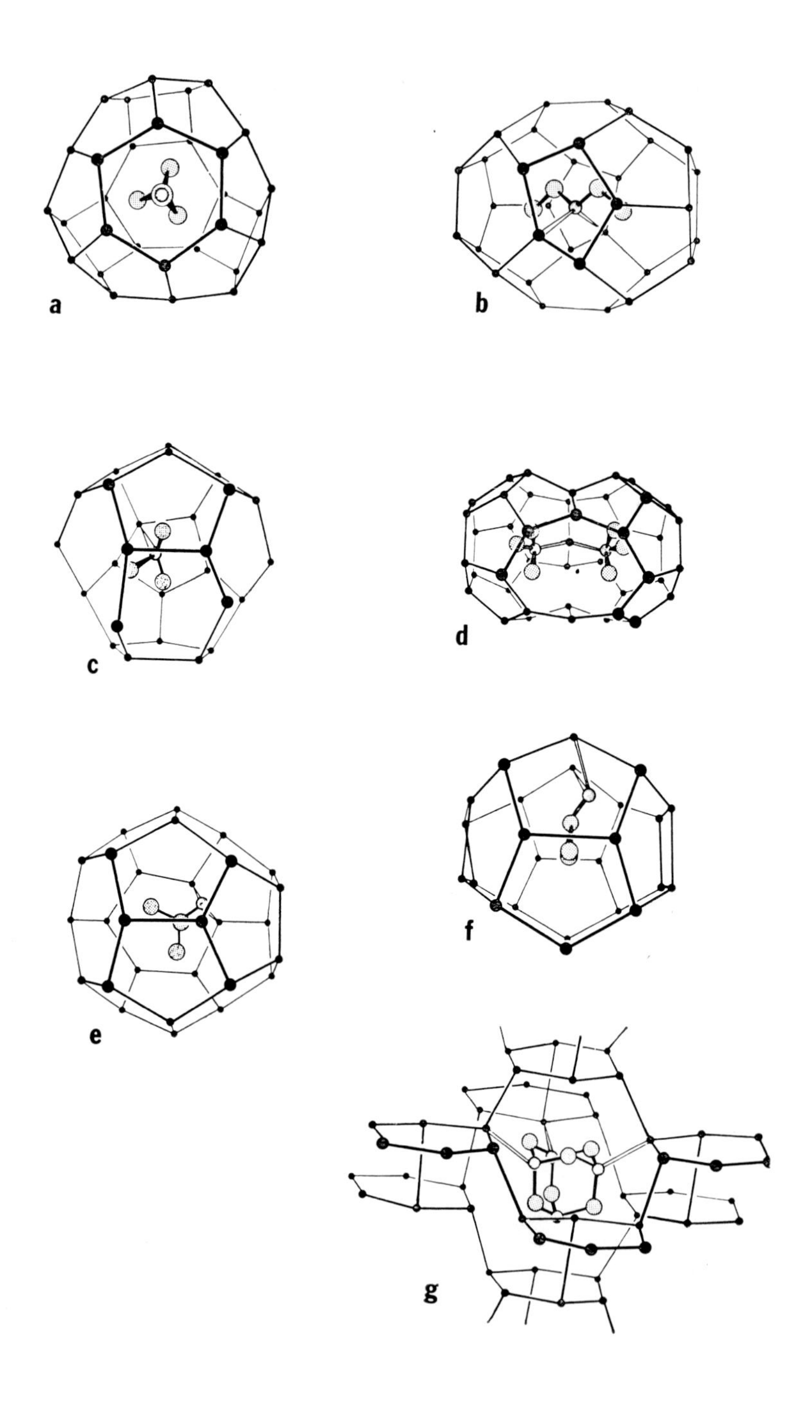
a
b
c
d
e
f
g

More recently newer types of host lattices quite different from those shown above have also been discovered (Jeffrey, 1969; Jordan and Mak, 1967; Mak, 1965; McLean and Jeffrey, 1967; Panke, 1968; Schwarzenbach, 1968). One of the most interesting of these is found in the hydrate of hexamethylenetetra-amine (Mak, 1965; Mak and Jeffrey, 1965) whose structure is shown (together with those of other amine hydrates) in Fig. 8. In this case the host lattice cannot be described in terms of a regular or even a semi-regular polyhedron. Furthermore, in contrast to the other polyhedral hydrates, in hexamethylenetetra-amine hydrate there is definite hydrogen bond formation between three of the four nitrogen atoms in the guest molecule and the water molecules in the host lattice. The interaction, therefore, between the water and the guest molecule is thus stereochemically specific.

Another interesting clathrate structure is seen in tetramethylammonium hydroxide pentahydrate (Fig. 9). The host molecules in this case are arranged at the vertices of truncated octahedra (McMullan, Mak and Jeffrey, 1966). A truncated octahedron is constituted of eight hexagonal faces and six square faces and contains 24 vertices which act as the sites of oxygen atoms of 24 water molecules. In $(CH_3)_4N^+OH^- \cdot 5H_2O$ there is a deficiency of one proton needed to form a full regular truncated octahedron. In consequence the shell is open at the bottom, as shown in Fig. 9. As a result the polyhedron can also expand, as is actually necessary in order to provide enough room for the large cationic guest molecule.

Thus it is apparent that water is a remarkably versatile substance in regard to hydrate formation, being capable of forming a large variety of cage-like structures to accommodate itself to a whole gamut of solute molecules.

Fig. 8. Structures of some amine clathrate hydrates. (See Jeffrey, 1969.) Carbon and nitrogen atoms are shaded, water oxygen atoms are black circles. Black lines are O···H···O bonds; double lines are N···H···O bonds.

(*a*) $(CH_3)_3CNH_2 \cdot 9 \cdot 75H_2O$; amine non-bonded within 17-hedron.
(*b*) $(C_2H_5)_2NH \cdot 8 \cdot 66H_2O$; amine nitrogen bonded to host water molecules within 18-hedron.
(*c*) $(CH_3)_3N \cdot 4 \cdot 25H_2O$; amine nitrogen bonded to host water molecule within 15-hedron.
(*d*) $(CH_3)_3N \cdot 4 \cdot 25H_2O$; two amines bonded to additional water molecule within 26-hedron.
(*e*) $(CH_3)_2CHNH_2 \cdot 10H_2O$; amine nitrogen bonded to host water molecule within 16-hedron.
(*f*) $(CH_3)_2CHNH_2 \cdot 10H_2O$; amine nitrogen bonded to host water molecule within 14-hedron.
(*g*) Structure of hexamethylenetetramine hydrate, $(CH_2)_6N_4 \cdot 6H_2O$.

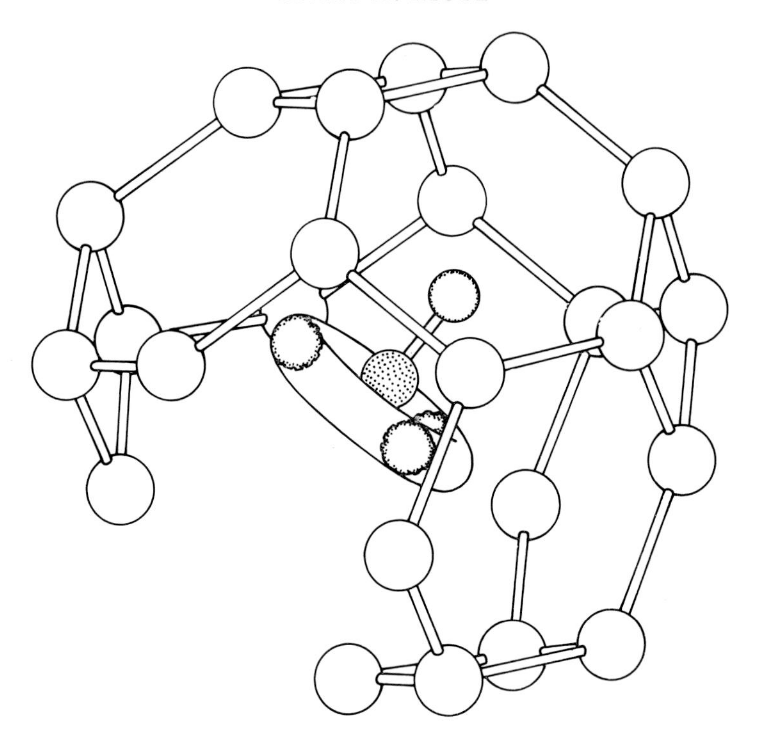

FIG. 9. Orientation of tetramethylammonium cation within cage of hydrate in $(CH_3)_4N^+OH^- \cdot 5H_2O$. Open circles are oxygen atoms of water molecules in lattice. The torus represents the rotation of the molecule about one N–C axis. (See McMullan, Mak and Jeffrey, 1966.)

SMALL MOLECULES IN SOLUTION

Given this background to the properties and structure of crystalline hydrates we might ask, what is the situation when the same solutes or corresponding ones are dissolved in liquid water? In fact, dilute aqueous solutions of non-polar solutes exhibit a number of properties which indicate stabilization of the water structure of the solvent. Table II, for example, compares the solubilization heat (ΔH°) in liquid water with the heat of formation of the clathrate hydrate for a number of non-polar solutes (Glew, 1962). For methane, for example, ΔH° when one mole of gas is dissolved in liquid water is -4621 calories ($-19\,334$ J); correspondingly the ΔH° when one mole of gas is incorporated into solid ice to form the hydrate is -4553 calories ($-19\,050$ J). It is apparent, therefore, that the energy of interaction of methane molecules in aqueous solution with the water in the hydration shell is essentially the same as

TABLE II

SOLUBILITY AND HYDRATES AT 0 °C★

Gas	$-\Delta H°$ *cal †/mole* *Gas dissolving in liquid* $H_2O \rightarrow$ *solution*	$-\Delta H°$ *cal †/mole* *Gas dissolving in ice* $\rightarrow$ *hydrate*
CH_4	4621	4553
H_2S	5140	5550
C_2H_6	5560	5850
Cl_2	6180	6500
SO_2	7420	7700
CH_3Br	7390	8100
Br_2	8750	8300
C_3H_8	6860	6300
CH_3I	8500	7300
C_2H_5Cl	8400	8700

★ Taken from Glew (1962). † 1 cal/mole = 4·184 joules/mole.

that of methane with the water molecules in the solid polyhedra of the hydrate clathrate. This equivalence implies that the environment in both cases, in solution and in the solid hydrate, is similar with respect to co-ordination and spatial orientation of the water molecules in the hydration shells. Likewise the correspondence in the values of the two columns of $\Delta H°$ listed in Table II for the other molecules implies that for these more polarizable solutes also the structure of the hydration shell in the liquid is similar to that in the solid hydrate. Thus these solutes stabilize and orient water molecules in their hydration shells in aqueous solution in the same way as they do water molecules in the hydrogen-bonded solid hydrate lattices.

More recently other physical-chemical properties of aqueous solutions of one clathrate-former, ethylene oxide, have been studied in great detail and with high precision (Glew, Mak and Rath, 1967). This study has revealed a variety of anomalous properties of water in these solutions, maximum deviations occurring at about 4 mole per cent concentration of solute. One particularly striking illustration is given in Fig. 10, which presents the excess partial molar volumes of water in these solutions of non-polar molecules. The positive values of these molar volumes indicate that effectively water occupies a larger volume in these solutions than in the pure liquid. The expansion of the water arises from the orientation of the molecules in the hydration shell in a manner similar to that found in solid hydrates. Solute molecules occupy interstitial spaces in this hydration shell.

Interestingly enough the concentration at which maxima or minima in many physical-chemical properties occur, about 4 mole per cent, cor-

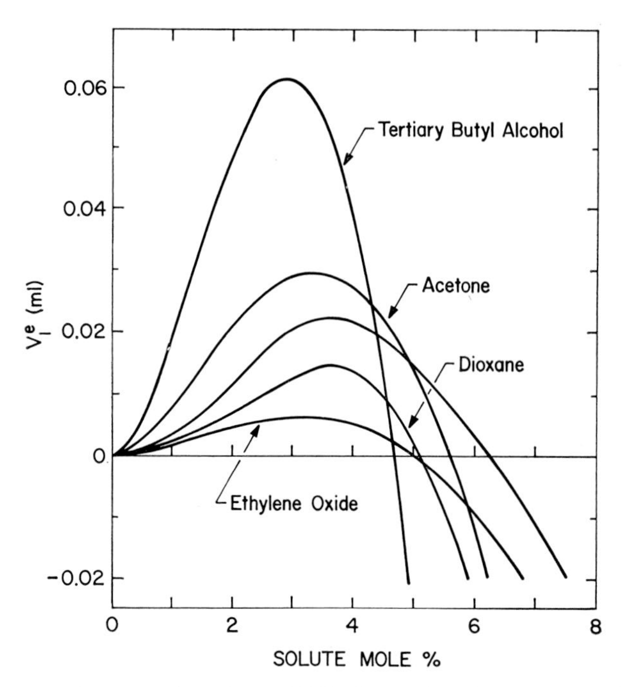

FIG. 10. Excess molar volumes (V_1^e) of water in dilute aqueous solutions of ethylene oxide at 10 °C, dioxane at 25 °C, acetone at 0 °C, tetrahydrofuran at 10 °C and tertiary butyl alcohol at 15 °C. (See Glew, Mak and Rath, 1968.)

responds to approximately 24 to 28 water molecules per molecule of solute, a coordination number which is very near that found in the solid clathrate hydrate. Thus it seems that these dilute aqueous solutions of a non-polar molecule consist of solute molecules surrounded by a solvation shell consisting of water in ordered structures similar to those found in the solid hydrates.

MACROMOLECULES IN SOLUTION

Evidence of ordered water around macromolecules is being provided by several physical and chemical probes. Particularly interesting are recent measurements of ultrasonic attenuation in aqueous solutions of polymers (Hammes and Schimmel, 1967). From the frequency dependence of the sound absorption one can calculate a relaxation time (of the order of 10^{-9} sec) which reflects the structure of the solvent and responds to perturbations thereof. In the presence of a polymer, the (single) relaxation time, τ, is increased, the increment depending on molecular weight of the polymer until a plateau value is reached above 5000 molecular weight. This increase in τ probably reflects the cooperative formation and breakdown of the structured hydration layer around the polymer. Interestingly enough, upon addition of increasing quantities of urea, τ shows no change until a concentration near 2 M is reached, whereupon τ drops steeply as the molarity of urea rises and then levels off once again at concentrations above 3 M. Such behaviour clearly suggests that urea is perturbing the solvent structure.

There is a long list of investigations of bound water in biological macromolecules by nuclear magnetic resonance. In principle this is a very attractive method since the proton resonance should definitely reflect any change in the state of the water. The expected broadenings in signal have been observed repeatedly. (For references to some of these studies see Klotz, 1965.)

Chemical probes also indicate that water at the surface of a polymer is different from bulk water. For example, the exchange of labile H atoms by a D atom from D_2O in CONH groups is very much slowed down when these groups are attached to a polymeric matrix as compared to corresponding small molecules (Klotz, 1968). The transfer of H (or D) to, and from, D_2O is modified in the neighbourhood of the polymer, and it seems very likely that the self-dissociation constant, K_w, of the polymer hydration water is lowered substantially.

Finally, of special interest in a symposium on "The Frozen Cell", should be some recent experiments (Kuntz *et al.*, 1969) in which aqueous solutions of proteins and of nucleic acids were frozen at $-35°C$, and the nature of the H_2O magnetic resonance signals was compared with that of pure ice. Frozen solutions of these biomacromolecules show relatively narrow proton resonance spectra in contrast to those of ice or frozen salt solutions, whose resonance range is so wide that it was not visible with the techniques used. These signals must reflect the presence of bound water on the macromolecule. The very presence of a signal shows that the hydration water cannot be truly ice-like. On the other hand, the broad line-width indicates that this bound water is definitely less mobile than liquid water. No freezing point for this bound water was detectable down to temperatures of $-50°C$.

Summing up, we may say that non-ionic solutes in aqueous solution interact with water to form a diversity of hydrate structures. These may stabilize water lattices in stereochemical conformations markedly different from that of normal ice. It may be relevant, therefore, to consider these structures in any examination of the properties of cells as they are cooled to freezing temperatures.

SUMMARY

Water is a remarkably versatile solvent able to form hydrates with a variety of ionic, polar and apolar solutes. Particularly interesting from a cellular viewpoint are the crystalline polyhedral hydrates of non-ionic solutes in which the "host" water molecules form cages which enclose "guest" solute molecules (M). On the basis of structure studies as well as chemical stoicheiometry one can classify these polyhedral hydrates into three major groups. These may be labelled as: I, $M_2 \cdot M_6 \cdot 46H_2O$; II, $M_8 \cdot M'_{16} \cdot 136H_2O$; III, $M \cdot (5\text{–}40)H_2O$. Guest molecules in Class I include small inorganic and organic molecules such as argon, ethane, ethylene oxide, in Class II larger molecules such as isobutane, benzene, dioxane, and in Class III molecules such as tetralkylammonium derivatives and some amines.

Molecular structures are known for many examples of each class. The pentagonal dodecahedron is the most pervasive polyhedron found in these crystal structures. Other simple polyhedra observed are 14-, 15-, and 16-sided. Face-sharing combinations of these provide larger enclosures with 44, 45, 46 or 60 faces, and even larger irregular polyhedra may also appear. Volumes enclosed by crystallographically known hydrate polyhedra span a range of 160–1600 Å^3.

Some properties of non-polar molecules in aqueous solution strongly suggest that similar hydrate structures exist in the liquid state.

It is thus apparent that water is a very adaptable substance in regard to hydrate formation, being capable of forming a large variety of cage-like structures to accommodate itself to a wide gamut of non-ionic solute molecules.

REFERENCES

Bernal, J. D., and Fowler, R. H. (1933). *J. chem. Phys.*, **1**, 515.
Beurskens, P. T., and Jeffrey, G. A. (1964). *J. chem. Phys.*, **40**, 2800.
Claussen, W. F. (1951). *J. chem. Phys.*, **19**, 1425.
Davy, H. (1811*a*). *Annls Chim.*, **78**, 298.
Davy, H. (1811*b*). *Annls Chim.*, **79**, 5.
Faraday, M. (1823). *Q. Jl Sci.*, **15**, 71.
Forcrand, R. de (1925). *C. r. hebd. Séanc. Acad. Sci., Paris*, **181**, 15.
Glew, D. N. (1962). *J. phys. Chem., Ithaca*, **66**, 605.
Glew, D. N., Mak, H. D., and Rath, N. S. (1967). *Can. J. Chem.*, **45**, 3059.
Glew, D. N., Mak, H. D., and Rath, N. S. (1968). In *Hydrogen-Bonded Solvent Systems*, pp. 195–210, ed. Covington, A. K., and Jones, P. London: Taylor & Francis.
Hammes, G. G., and Schimmel, P. R. (1967). *J. Am. chem. Soc.*, **89**, 442.
Jeffrey, G. A. (1969). *Accounts chem. Res.*, **2**, 344.
Jeffrey, G. A., and McMullan, R. K. (1967). *Prog. inorg. Chem.*, **8**, 43.
Jordan, T. H., and Mak, T. C. W. (1967). *J. chem. Phys.*, **47**, 1222.

KLOTZ, I. M. (1965). *Fedn Proc. Fedn Am. Socs exp. Biol.*, **24**, suppl. 15, S-24.
KLOTZ, I. M. (1968). *J. Colloid Interface Sci.*, **27**, 804.
KUNTZ, I., Jr., BRASSFIELD, T. S., LAW, G. D., and PURCELL, G. V. (1969). *Science*, **163**, 1329.
McLEAN, W. J., and JEFFREY, G. A. (1967). *J. chem. Phys.*, **47**, 414.
McMULLAN, R., and JEFFREY, G. A. (1959). *J. chem. Phys.*, **31**, 1231.
McMULLAN, R. K., and JEFFREY, G. A. (1965). *J. chem. Phys.*, **42**, 2725.
McMULLAN, R. K., JORDAN, T. H., and JEFFREY, G. A. (1967). *J. chem. Phys.*, **47**, 1218.
McMULLAN, R. K., MAK, T. C. W., and JEFFREY, G. A. (1966). *J. chem. Phys.*, **44**, 2338.
MAK, T. C. W. (1965). *J. chem. Phys.*, **43**, 2799.
MAK, T. C. W., and JEFFREY, G. A. (1965). *Science*, **149**, 178.
MAK, T. C. W., and McMULLAN, R. K. (1965). *J. chem. Phys.*, **42**, 2732.
PANKE, D. (1968). *J. chem. Phys.*, **48**, 2990.
PAULING, L., and MARSH, R. E. (1952). *Proc. natn. Acad. Sci. U.S.A.*, **38**, 112.
SCHWARZENBACH, D. (1968). *J. chem. Phys.*, **48**, 4134.
STACKELBERG, M. VON, and MEINHOLD, W. (1954). *Z. Elektrochem.*, **58**, 40.
STACKELBERG, M. VON, and MÜLLER, H. R. (1954). *Z. Elektrochem.*, **58**, 25.
VILLARD, P. (1896). *C. r. hebd. Séanc. Acad. Sci., Paris*, **123**, 377.

DISCUSSION

Mazur: The energetic relation between clathrate ice and ordinary ice has always interested me. Is the difference in energy very small?

Klotz: A clathrate hydrate without a guest molecule is an unstable structure. At temperatures below 0 °C it would go over to ordinary hexagonal ice, and at temperatures above 0 °C it would melt directly to water. However when such a hydrate has a guest molecule in the enclosure this inert molecule compensates for the internal pressure which would otherwise make the structure collapse. So clathrate hydrates may be stable to temperatures as high as +40 °C, depending on the guest molecule. From the practical point of view clathrate hydrates are more stable than ordinary ice above 0 °C, but the melting energies are not very different from those of ordinary ice, since the bond energies of the two are similar.

Mazur: Will clathrate ice nucleate supercooled water?

Klotz: It would depend on which molecules are in the supercooled water. Clathrate ice can't exist unless a solute is present. For example, methane in supercooled water would produce methane hydrate if the stoicheiometry of guest molecule in relation to host molecule is right. I assume this could not nucleate hexagonal water.

The way to form these hydrates is to put a compound which is known to form a hydrate into water at a concentration at which the ratio of water molecules to guest molecules is correct for forming the hydrate, and cool the solution to the appropriate temperature. We then get a slush, which doesn't look like ordinary ice. Jeffrey presumably needs larger crystals for X-ray work (Jeffrey and McMullan, 1967).

Huggins: Would you comment on the structure of glycerol or DMSO-ice?

Klotz: I think the glycerol of glycerol hydrate interacts directly with the water molecules around it (Jeffrey, 1969); the glycerol molecule is definitely not within an enclosure of the polyhedral clathrate type. My guess would be that DMSO too would form a hydrate of the glycerol class, with a direct interaction, rather than form a polyhedral clathrate hydrate, because DMSO is a very strong hydrogen bonder, as is glycerol, and I suspect that the direct hydrogen bond interactions are dominant.

Luyet: I know of only a few incidental references in the literature to the biological effects of clathrate formation, such as the suggestion that anaesthesia may be caused by the presence of some clathrate. The cryobiologist interested in the preservation of biological material in the frozen state would welcome information of a more general nature on properties of clathrates which he could use in cryopreservation. Since the clathrates can be considered as a form of crystallization, do you think that the damage to living matter may be decreased by freezing its water into clathrates instead of into ordinary ice?

Klotz: My guess would be no, because a structure would be imposed on the system which is satisfactory for some of the solute and therefore for parts of the cell, but not satisfactory for other parts. Let me repeat that, in principle, there might be two ways of avoiding freezing injury. One is to find a thermodynamically stable structure, hexagonal or pentagonal, or some combination thereof, or an amorphous type of structure which might, by being more adaptable, fit a whole set of solute molecules. It might also be just as effective to try to prevent injury by a kinetically acting method. Slowing down the rate of crystallization into an injurious structure might satisfactorily prevent low temperature injuries.

Luyet: Can one melt clathrates in the same way as one melts ice, simply by raising the temperature?

Klotz: Pure clathrates of the types I have listed have precisely defined melting points, just as ice does, at a specific pressure of the solute.

Luyet: Does the formation of clathrates involve a change in volume comparable to that which occurs in the formation of hexagonal ice?

Klotz: The volume of the clathrate as such, not including the guest molecule, is larger than that of hexagonal ice. If one subtracts the volume of the guest molecule the volume of clathrated water is greater than that of hexagonal water.

Farrant: What happens to these clathrates if you add ions to the system?

Klotz: I am pretty certain that ions other than the guest molecules, which themselves are occasionally ions, will at high enough concentration melt or dissolve the clathrate.

Farrant: So in a biological solution one would be unlikely to get any clathrates?

Klotz: I think so, except perhaps in the neighbourhood of a macromolecule where the ions may not be able to reach the concentration which is locally necessary to melt the hydrate.

Mazur: Do clathrates show normal phase relationships in a mixture of methane and water with, say, sodium chloride?

Klotz: Sodium chloride hasn't been investigated, but pure clathrates with common guest molecules show normal phase relationships.

Elford: Has the presence of clathrates been demonstrated in biological systems during anaesthesia?

Klotz: The anaesthesia idea is a speculation of Pauling's (1961) and of Stanley Miller's (1961). There are experiments which indirectly can be argued to support the presence of clathrates in anaesthesia, and we have done other types of experiments with macromolecules which we rationalize in similar terms of hydrate formation, but other investigators would rationalize the same observations in a different way.

Elford: But is it likely that clathrates could be formed in the ionic conditions found in nerve fibres, for example?

Klotz: They could probably be formed in the neighbourhood of particular macromolecules in the presence of substances such as xenon, which is the point of the anaesthesia theory. Xenon is known to be an anaesthetic and at the time Pauling and Miller had this idea they could make the very dramatic statement that "xenon has no chemistry" because at that time the xenon compounds hadn't been discovered. Essentially it still has no aqueous chemistry of any biological significance, so how could it have a biological effect if it has no chemistry? The only interactions of xenon known at that time were hydrate formation; and so it was attractive to Pauling and Miller to say that xenon would not form hydrates in normal aqueous media but that in the neighbourhood of some crucial impulse-conducting protein or biological polymer an extension and stabilization of clathrate formation would occur. This would certainly change conduction properties in the system and could in some vague way alter the transmission of the electrical impulse. It would be extremely difficult to demonstrate clathrate formation in these systems. It is even extremely difficult to demonstrate clathrate formation in a simple solution of, let us say, methane in liquid water, in place of the solid, isolated crystal. Certain of the properties of methane in liquid water closely parallel the properties of methane in the crystal hydrate. These argue in favour of similar clathrate formation in the liquid solution. However, except in some very concentrated salt solutions I know no way of determining the hydration structure by X-ray diffraction of something dissolved in liquid water.

Luyet: Do you see any relationship between substances which serve as guests in clathrate formation and cryoprotective substances, in particular those of high molecular weight, such as polyvinylpyrrolidone (PVP)?

Klotz: I wasn't aware that PVP is a cryoprotective agent. It may work by a kinetic mechanism, but on the other hand PVP may form pentagonal hydrates around the various side-chain groups. At least groups attached to PVP have chemical properties which are quite different from the same groups as small molecules in bulk water, so in the neighbourhood of PVP the characteristics of the water must be different from those in bulk water.

Huggins: Is direct hydrogen bond interaction between glycerol or DMSO and water molecules a unique property of cryophylactic agents?

Klotz: I don't think there is any unifying feature among compounds protecting against freezing and thawing, because almost every hydrate is unique in its structure and interaction with water. Admittedly the class I and class II hydrates all form similar structures, but class III hydrates and those that don't form polyhedral clathrate compounds tend to be unique in structure and interactions, each having a different stoicheiometric proportion of solute to solvent molecules.

Another aspect of "antifreeze" capabilities is shown in an antifreeze protein recently discovered in certain fish in the Antarctic (DeVries and Wohlschlag, 1969). This unusual polypeptide has only two amino acids, alanine and threonine, and equimolar quantities of galactose and galactosamine, which make it a peculiar glycopeptide. Nevertheless it is extremely effective in keeping the blood of these particular fish (*Trematomus borchgrevinki*, *T. bernacchi* and *T. hansoni*) from freezing. My first thought was that this polypeptide acted by forming polyhedral clathrate hydrates around various side-chains of the protein, but recent data (R. Feeney, 1969, personal communication) now make me think the activity is more of a kinetic phenomenon; the polypeptide may actually cover the surface of certain nucleation particles in the plasma and in effect remove the nucleating sites so that freezing doesn't occur until a much lower temperature is reached. Many polymeric molecules, for example, ought to be able to inhibit nucleation in this way. This may be a totally different mechanism for slowing down the rate of freezing, but one which in effect may be just as good as a real equilibrium phenomenon, which I assume is how glycerol acts.

For reasons like this I am reluctant to believe that there is a single principle which explains all the antifreeze phenomena.

Mazur: Most antifreeze compounds are of course highly soluble, which implies that they interact strongly with water.

Davies: Could you expand on this idea of how peptides could inhibit nucleation of ice crystals?

Klotz: A particular nucleation site may have the necessary atomic arrangements so that water molecules in a normal hexagonal array fit well and consequently tend to extend out and form ice crystals. If the site is then covered by an antifreeze protein or peptide, the arrangement favourable for hexagonal water may be removed. One has removed the three-dimensional geometry which would serve as a locus for the expansion of ice in three dimensions and as a result the ice won't form since it doesn't have a nucleus to start it.

Davies: In fact you are suggesting that a molecule on the surface or within a cell may be seeding the ice formation?

Klotz: I think this is what the antifreeze protein in fish blood is interfering with. From its solute content the expected freezing point of blood in these species is about $-0{\cdot}7$°C. The observed freezing point of blood from these particular fish from the Antarctic is approximately $-2{\cdot}01$°C. So about $1{\cdot}2$°C or $1{\cdot}3$°C of the lowering of the freezing point is unaccounted for by normal solutes. The antifreeze protein has a very high molecular weight, probably of the order of 20 000. This should show little colligative effect on the lowering of the freezing point. So the protein can only work, I think, in two ways. It could form hydrates of the polyhedral type described here and in effect remove a lot of water from the solution, so that the water volume available to the normal salt is now so much reduced that the effective salt concentration becomes much higher. Or it may work in a totally different way by covering nucleation sites.

Levitt: Could the protein break down into smaller subunits at the lower temperature, as some proteins do, and so be more effective osmotically?

Klotz: Subunit systems in proteins generally are systems in which a molecule of a molecular weight of several hundred thousand is composed of subunits of weight near perhaps 15 000. To account for all the extra lowering of freezing point the antifreeze protein must have a molecular weight of the order of only 13. There is effectively nothing in the blood with a molecular weight of 13—even sodium ion is 23—so a subunit explanation is completely beyond the range of possibility.

Meryman: What is the melting point?

Klotz: That question, what is the melting temperature of the frozen fish blood, is what led me to the idea that a nucleation phenomenon is involved. If this were truly an equilibrium phenomenon, then the melting point ought to be the same as the freezing point—one ought to get the same answer in both directions. But the melting point is quite a bit higher, a result which indicates that the exceptional freezing point is not really an equilibrium phenomenon but a kinetic one.

Mazur: Was the freezing temperature the supercooling point or the actual equilibrium freezing point?

Klotz: It is the temperature at which during cooling one first sees crystals.

Mazur: This could be the supercooling point.

REFERENCES

DeVries, A. L., and Wohlschlag, D. E. (1969). *Science*, **163**, 1073.
Jeffrey, G. A. (1969). *Accounts chem. Res.*, **2**, 344.
Jeffrey, G. A., and McMullan, R. K. (1967). *Prog. inorg. Chem.*, **8**, 43.
Miller, S. L. (1961). *Proc. natn. Acad. Sci. U.S.A.*, **47**, 1515.
Pauling, L. (1961). *Science*, **134**, 15.

PHYSICAL CHANGES OCCURRING IN FROZEN SOLUTIONS DURING REWARMING AND MELTING*

B. J. LUYET

Cryobiology Research Institute, American Foundation for Biological Research, Madison, Wisconsin

ONE of the directive ideas of this conference on "The Frozen Cell" is to correlate the physical and the biological changes caused by the freezing process. In order to bring more sharply into focus the points that I intend to discuss, I shall classify the changes on the basis of their occurrence (*a*) in physical or in biological systems (in solutions or in cells or tissues), and (*b*) during freezing or during rewarming or thawing. Of the four resulting categories of changes:

(1) Physical changes produced by freezing
(2) Physical changes produced by rewarming
(3) Biological changes produced by freezing
(4) Biological changes produced by rewarming

I select the second, physical changes during rewarming, as my main topic. But, since what happens on rewarming depends on what had happened on cooling, I shall devote an introductory part of this paper to a survey of the changes of the first category. A few representative cases in which physical changes similar to those reported in solutions were observed in cells or tissues will be presented in an Appendix.

PHYSICAL CHANGES PRODUCED BY FREEZING IN SOLUTIONS

Three questions will be examined: (1) the types of crystallization units developed, (2) the amount of ice formed, (3) the configuration and relative distribution of the amorphous and the crystalline phases.

* This paper being primarily a survey of the topics to be discussed at the symposium, I shall present in a condensed form the data reported *in extenso* in original papers. The reader is referred to them for further information and for a more adequate illustration of the changes involved in freezing and thawing.

The research on which this survey is based was supported mostly by the Office of Naval Research of the U.S. Navy.

Types of crystallization units developed

The combined action of four main factors—the lowest temperature reached, the cooling rates, the nature and the concentration of the solute—determines the type of ice formation which will develop. The various

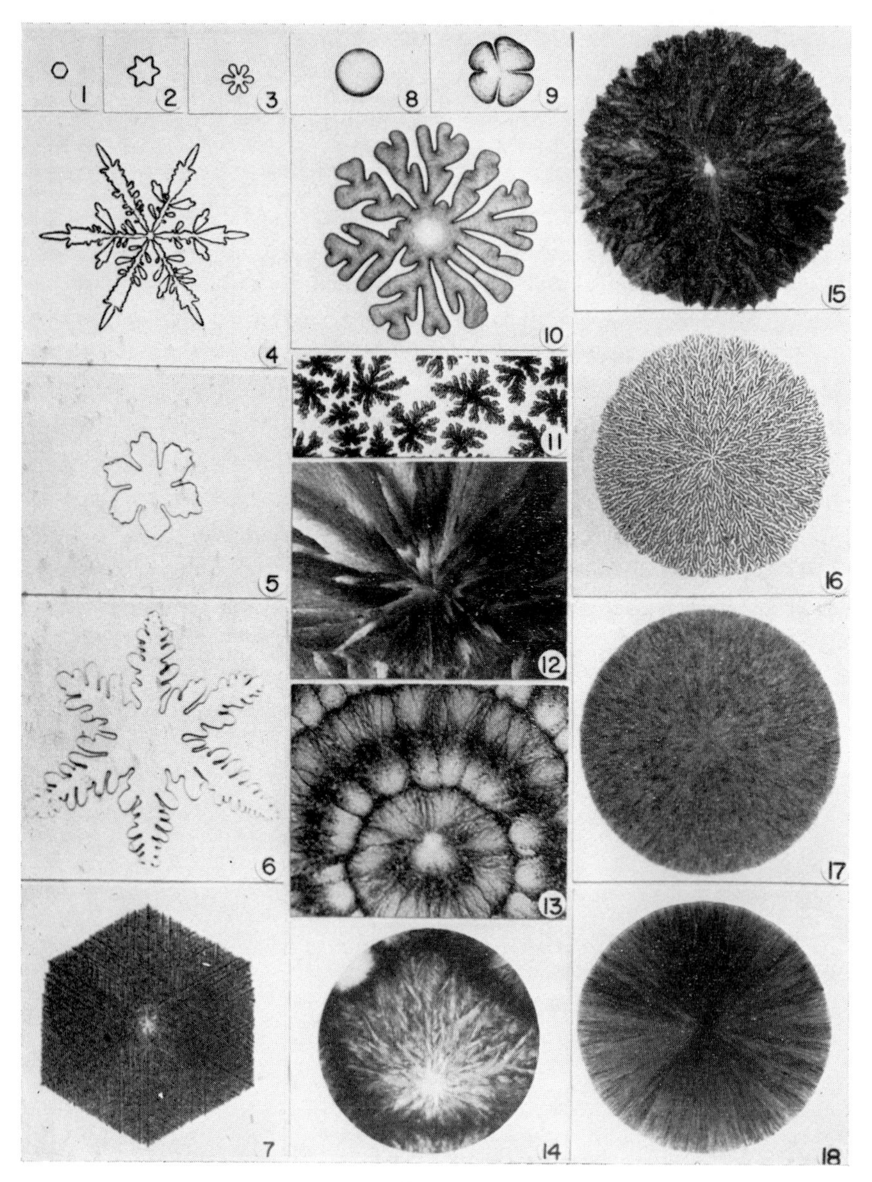

FIG. 1

types can be classified into two categories: the ice phase is either in the form of compact solid bodies without "entering angles" (Fig. 1: 1), or in the form of arborescences. The latter, in turn, can be classified into three groups: (*a*) under certain conditions, mostly when the cooling rate is low, one obtains crystalline formations with well-preserved hexagonal symmetry (Fig. 1: 2 to 7); (*b*) more rapid cooling results in irregularly constructed arborescent structures (Fig. 1: 9 to 14); still more rapid cooling gives rise to spherulitic crystalline formations (Fig. 1: 17 and 18; the patterns represented in 15 and 16 may be considered transitional forms from irregular dendrites to spherulites).

Between the three fundamental groups there is a great variety of intermediate forms. The groups themselves are not essentially different; they all have an arborescent structure, the differences between them consisting of the mode of arrangement and the number and dimensions of the branches and sub-branches. (The arborescent nature of the spherulites will be discussed later on.)

Amounts of ice formed and of material remaining unfrozen

Three cases will be distinguished on the basis of the solute concentration and of the attainment or non-attainment of the phase-transition equilibrium.

(*a*) *Very high solute concentrations.* The simplest case is apparently that in which the solute concentration is so high that no water freezes out at any temperature, no matter how low the cooling rate. The entire

---

FIG. 1. Types of ice formations developed at various temperatures, cooling rates and solute concentrations. All the photographs reproduced here were taken on thin layers of solutions through the AFBR cryomicroscope ($\times 60 \pm 5$). Nos. 1–7: regular hexagonal forms; Nos. 8–14: irregular forms. Nos. 15–18: transitional forms and spherulites. Nos. 7, 14 and 18 show characteristic representatives of these three types. The formation shown in No. 8, in which the ice is laid in concentric layers, is not a spherulite. It is comparable in its mode of development to the forms represented in Nos. 9 and 10, although the central body was unable to grow lobes. The pattern shown in No. 13 has been selected as a typical case of the occurrence of periodicity in the crystallization process. (For further details consult the original paper cited below.)

Freezing temperatures (°C) *and solute concentrations*: Nos. 1-4: −1·5°, 35% bovine albumin; Nos. 5 and 6: −2·5°, 11·2% bovine albumin; No. 7: −35°, 6 M-glycerol; Nos. 8 and 9: −30°, 50% gelatin; No. 10: −25°, 40% gelatin; No. 11: −40°, 50% polyvinylpyrrolidone; No. 12: −100°, 11·2% myosin; No. 13: −40°, 10% albumin; No. 14: −65°, 10% sucrose; No. 15: −25°, 50% sucrose; No. 16: −25°, 40% gelatin; No. 17: −30°, 50% sucrose; No. 18: −60°, 6 M-glycerol.

(From Luyet and Rapatz, 1958; by permission of *Biodynamica*.)

solution will solidify in the amorphous state. With glycerol as solute, one has the conditions indicated in Fig. 2, where the rectangular area above 73 per cent concentration represents the solution from which no water freezes out. (It is assumed that neither the solute nor the eutectic mixture crystallizes.)

(*b*) *Low and intermediate solute concentrations, under conditions of equilibrium.* When enough time is allowed for a freezing solution to reach the phase-transition equilibrium, the amounts of ice formed are determined by the original concentration and the lowest temperature attained. They can be calculated from the melting curve of the phase diagram. The corresponding amounts of non-frozen solution can also be calculated for each original concentration and each final temperature. Graphs of the type represented in Fig. 2 will give such values. Thus, a 30 per cent glycerol solution frozen at $-30°$C will have 47·4 per cent of its total weight transformed into ice (the amount represented by the ordinate at abscissa 30 up to its intersection with the line marked $-30°$), with 52·6 per cent

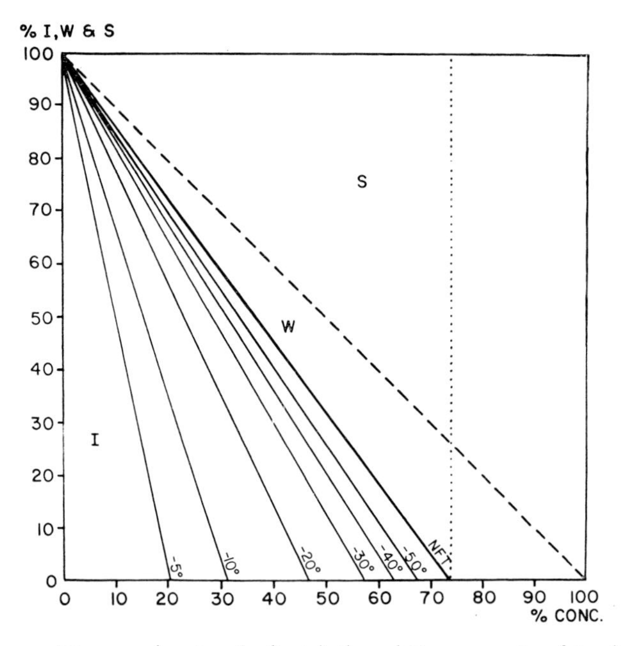

FIG. 2. Diagram showing (ordinate) the relative amounts of ice (I), water (W) and solute (S) present (1) when glycerol solutions of various initial concentrations (abscissa) are frozen at various temperatures (oblique lines), (2) at concentrations above which no ice can develop (line NFT, non-freezing temperature). (For details see original papers: Luyet and Gehenio, 1952; Luyet, 1969; by permission of *Biodynamica*.)

of the total weight remaining unfrozen (amount represented by the same ordinate above that intersection).

If the temperature is lowered to or below the point corresponding to the concentration at which no more ice is formed (−58° corresponding to about 73 per cent, as represented in Fig. 2 by the line NFT), the amounts of ice present and of material unfrozen for any particular original concentration can be calculated in the same manner as at higher temperatures. A 45 per cent glycerol solution, for example, cooled slowly to any temperature below about −58° will contain approximately 39 per cent ice and 61 per cent vitreous solution.

(*c*) *Low and intermediate solute concentrations, under non-equilibrium conditions.* In cases in which the phase-transition equilibrium is not attained, the proportion of ice and of non-frozen material will depend on the extent to which the cooling rate can prevent or limit the formation of ice; it cannot be calculated on the basis of an established rule and has to be determined in each particular case. (An example of such a determination is given in Fig. 3; see explanation in the legend.)

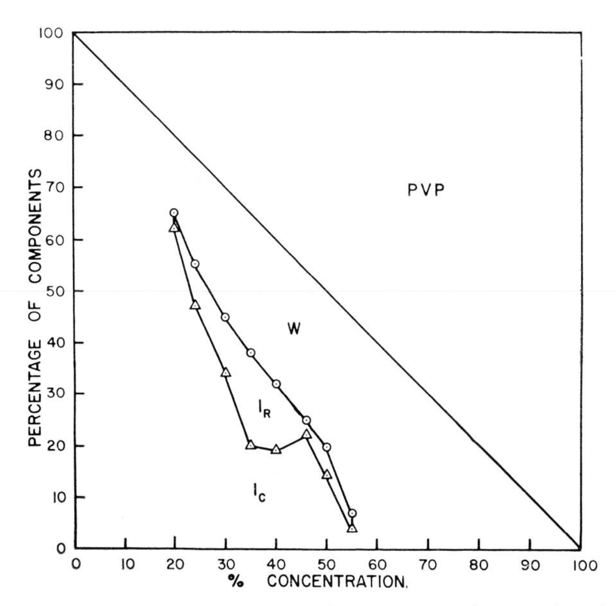

FIG. 3. Diagram giving (ordinate) the percentage by weight of ice formed during cooling (I_C), ice formed during rewarming (I_R), non-frozen water (W) and solute, polyvinylpyrrolidone (PVP), at initial solute concentrations varying from 20 to 55 per cent. The triangles represent the values obtained after cooling and the circles the values obtained after rewarming. (From Luyet, Vehse and Gehenio, 1969; by permission of *Biodynamica*.)

Configuration and relative distribution of the amorphous and crystalline phases

Electron microscopy (see Fig. 5) has shown that the spherulites are arborescent structures comparable in their general architecture to the other two types of ice formation, the regular and irregular dendrites (MacKenzie and Luyet, 1962). All these structures develop according to the same fundamental principle: from a centre of crystallization there grow either six or a variable number of spears which branch out according to a more or less regular pattern. While growing, they reject the concentrated solution in which they remain embedded. Morphologically, the

FIG. 4. Photograph of an ice skeleton, illustrating the interpenetration of the ice phase and the non-frozen concentrated medium. The structure, developed in 35 per cent albumin solution, started at $-2°$ and continued growing at $-3\,°C$ ($\times$ 85). (From Luyet and Rapatz, 1958; by permission of *Biodynamica*.)

non-frozen medium is then a negative replica of the crystalline framework which grows in it. The two phases interpenetrate each other and, since they generally consist of thin branches or narrow channels, their surfaces of contact may be enormously large.

The mode of interpenetration of the two phases is well illustrated in Figs. 4 and 5, which represent the configuration of the crystalline framework in, respectively, a regular dendrite of hexagonal symmetry and a spherulite. An examination of the details of the mode of invasion of the medium by the ice phase shows that every branch or sub-branch proceeds as far as it can go in occupying new territory until it is stopped by the presence of a "first occupant". The regularity in the symmetrical arrangement could be preserved only to the extent that enough territory was left unoccupied for a spear or a branch to grow unhindered.

Another point which deserves attention in this connexion is the gradient of concentration through the thin layers or channels separating the ice spears or branches. I shall only cite one case, which is illustrated in Fig. 6. In a frozen gelatin gel, the solute concentration through the unfrozen phase varies from 40 to 67 per cent over a distance of 250 μm

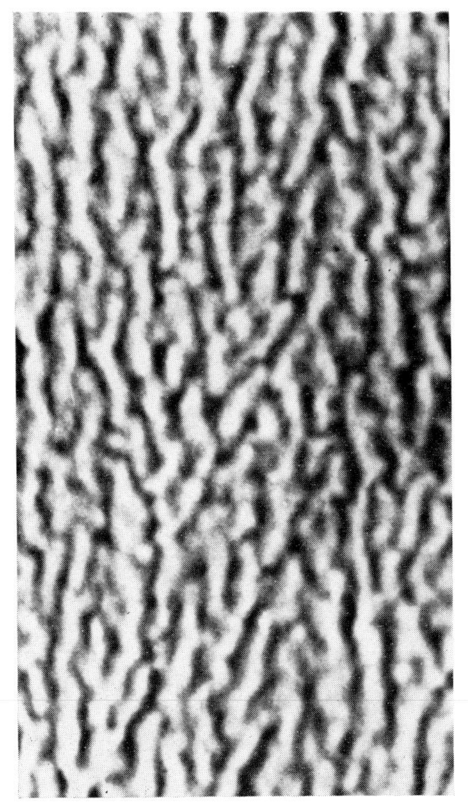

FIG. 5. Electron micrograph illustrating the mode of branching and the relative size of the ramifications and of the interstitial spaces in a spherulitic ice formation obtained when a layer of gelatin gel of about 15 per cent concentration and approximately 10 μm thick was frozen at −150 °C and vacuum-sublimed (× 7200). (From MacKenzie and Luyet, 1962; by permission of *Biodynamica*.)

(Persidsky and Luyet, 1959). On account of the slow diffusion at low temperatures, the concentration gradients may be extremely slow in reaching equilibrium.

When the ice phase is in the form of compact crystals, the frozen solution consists of a suspension of such crystals in a non-crystalline medium.

To summarize: the frozen solutions, in which we are to study the physical changes produced by rewarming and thawing, consist of a framework of ice, often of a very fine structural design, interpenetrating a non-frozen phase of considerable total volume, often finely divided and therefore separated from the ice by enormously large contact areas; steep concentration gradients may be maintained for long periods of time in the narrow channels of non-frozen material.

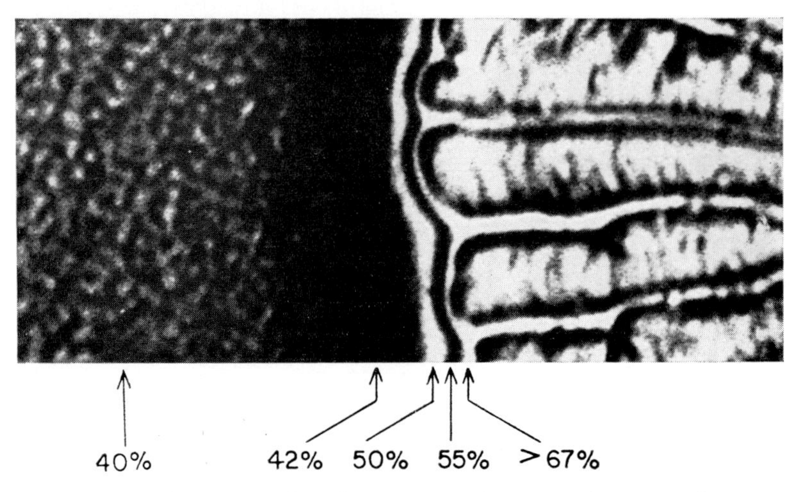

FIG. 6. Enlargement of a photomicrograph representing parts of a rosette of ice (the 3 to 4 lobes on the right-hand side) growing into a 40 per cent gelatin gel (grey background on the left-hand side). The gel was cooled to $-30°$, then to $-100°$, and rewarmed to $-30°$C. (This procedure permits the formation of the rosette and leads to the determination of the concentrations of the intermediate points; for details see the original paper.) The figures underneath the photographs indicate the values of the concentrations resulting from the growth of the rosette at the points marked by the arrows ($\times$ 980). (From Persidsky and Luyet, 1959; by permission of *Biodynamica*.)

PHYSICAL CHANGES OCCURRING DURING THE REWARMING OF FROZEN SOLUTIONS

The changes produced by rewarming in solutions solidified at low temperatures may be classified into two categories: some are the result merely of the rise in temperature through ranges in which no particular event of the phase-transition type takes place; others result from the occurrence of a phase transition of some sort. The former are usually uniform and gradual, the latter are often abrupt. After a glance at the changes of the first category, I shall here examine in some detail those of the second. Five series of changes involving a transition will be considered.

Three seem well identified and are recognized as such in the literature on solid-state physics, namely glass transformation, devitrification, and melting. The fourth, recrystallization, which has been extensively studied in our laboratories, is hardly mentioned in the physical literature in a sense

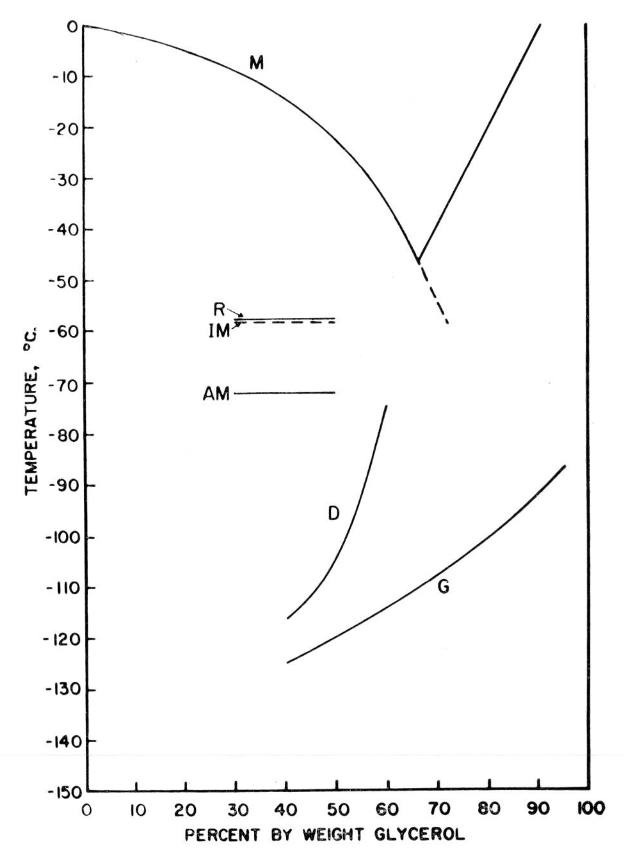

FIG. 7. "Phase diagram" giving the temperatures at which the glass transformation (G), devitrification (D), a transition of unknown nature designated as "ante-melting" (AM), incipient melting (IM), recrystallization (R), and melting (M) take place in solutions of glycerol of the concentrations indicated on the abscissa. (Adapted from Luyet and Rasmussen, 1968, and from a paper in press; by permission of *Biodynamica*.)

which would fit our findings. The existence of changes of the fifth series, recently brought out in our differential thermal analysis (DTA) studies (Luyet and Rasmussen, 1968), seems well established experimentally, but little is known about their nature. The essential facts about the changes of this last series are described below under the heading *An ante-melting transition.* For glycerol solutions, the five series of changes are represented in Fig. 7 by the curves marked G, D, AM, R and M.

I. Gradual Changes Caused Merely by Warming in Ranges of Temperatures at which No Particular Transition Occurs

Within such ranges, a rise in temperature means primarily a gradual increase in the mobility of the molecules, either in their vibratory motion around positions of equilibrium or in rotatory and translatory motions. To that increase in mobility will correspond changes in physical properties, such as viscosity, cohesion, surface forces and diffusibility. Thus, when the temperature of a solution rises from, say, -150° to 0°C, we would expect continuous and gradual changes in properties, interrupted at certain points or ranges by discontinuous, sometimes abrupt changes.

II. Changes Resulting from Particular Transitions Occurring During Rewarming

The glass transformation

The general tendency in discussions between cryobiologists is to disregard the glass transformation as irrelevant, because of the insignificant quantity of glass formed and because the structural changes involved appear, at first glance, to be of minor importance. Let us consider briefly these two points.

(*a*) *Proportion of vitreous material in solutions solidified at low temperatures.* Our differential thermal analysis curves (Luyet and Rasmussen, 1967, 1968) showed the endothermic depression characteristic of the glass transformation in most cases in which the solution was frozen very slowly for the purpose of permitting as complete a crystallization as possible. Calculations from the graphs presented and from similar graphs for other solutes indicate that one often obtains large proportions of amorphous material. For glycerol (Fig. 2) above a concentration of 73 per cent the proportion of the amorphous phase is 100 per cent; for an initial concentration decreasing from 73 to 30 per cent, the proportion decreases from 100 to 41 per cent.

(*b*) *Nature of the changes occurring at the glass transformation.* When a substance in the vitreous state passes through the glass transformation stage, it undergoes a sudden change in several physical properties, such as specific heat and index of refraction. These changes, which are well observed experimentally, are interpreted by the physicists as being due to a passage of the molecules from a state where they were capable only of vibratory motion to a state in which they acquire translational and rotational motion. Thus, the glass transformation would suddenly give rise to a significant increase in molecular mobility which would account for the great changes in physical properties reported.

To emphasize the difference in properties below and above the temperature of the glass transformation, the students of substances liable to undergo that transformation (such as some polymers) have adopted a nomenclature in which they reserve the term vitreous for material held below the temperature of the transformation, while they call supercooled liquids the materials which had been warmed above that temperature.

Crystallization upon rewarming (devitrification)

When the temperature is high enough, that is, when the molecular mobility has increased sufficiently, the molecules of water become able to move from their positions in the disorderly liquid state to new positions in the orderly arranged crystals. This passage is often characterized by a marked exothermic surge in the differential thermogram (see Fig. 8A). The temperature at which it occurs is usually several tens of degrees centigrade above that of the glass transition.

In solutions of polyvinylpyrrolidone, in which the curve for the temperature of devitrification in terms of initial concentration is about parallel to the curve for glass transformation, an approximate evaluation of the decrease in viscosity required to permit a devitrification led to the conclusion that, if the glass transformation occurs when the viscosity is of the order of 10^{12} N s m^{-2} (10^{13} poises), as generally admitted, the crystallization of water would take place when, upon warming, the viscosity has decreased about 10 000 times, that is, has reached some 10^{8} N s m^{-2} (10^{9} poises). A reasoning of this sort may apply to other solutions, although one should note that the curves of the glass transformation and of devitrification are parallel only within limited ranges of concentrations, and that, with some solutes, they depart greatly from parallelism at increasing concentrations (see, for example, Fig. 7 for glycerol solutions).

One should distinguish two cases of crystallization upon warming: (1) the system may have been nucleated during the previous cooling, so that the process consists of a resumption of the growth of pre-existing crystallites when the molecules have acquired enough mobility, or (2) the system may undergo, upon being rewarmed, the two stages of nucleation and crystal growth (and these two stages may take place at different

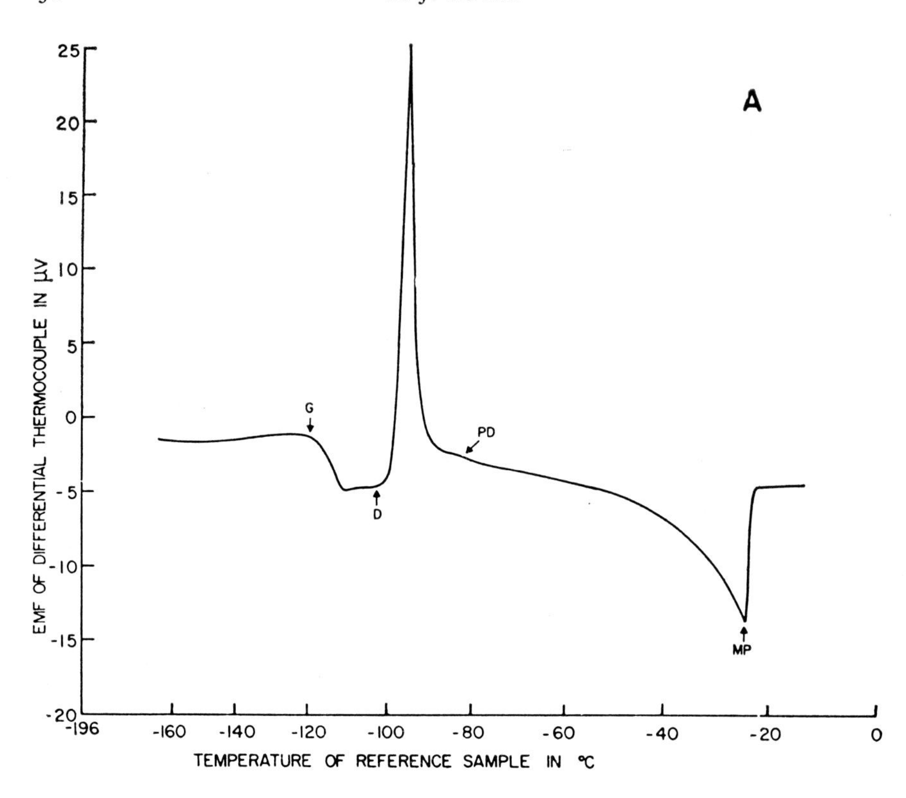
A
EMF OF DIFFERENTIAL THERMOCOUPLE IN μV
25
20
15
10
5
0
-5
-10
-15
-20
G
D
PD
MP
-196
-160
-140
-120
-100
-80
-60
-40
-20
0
TEMPERATURE OF REFERENCE SAMPLE IN °C

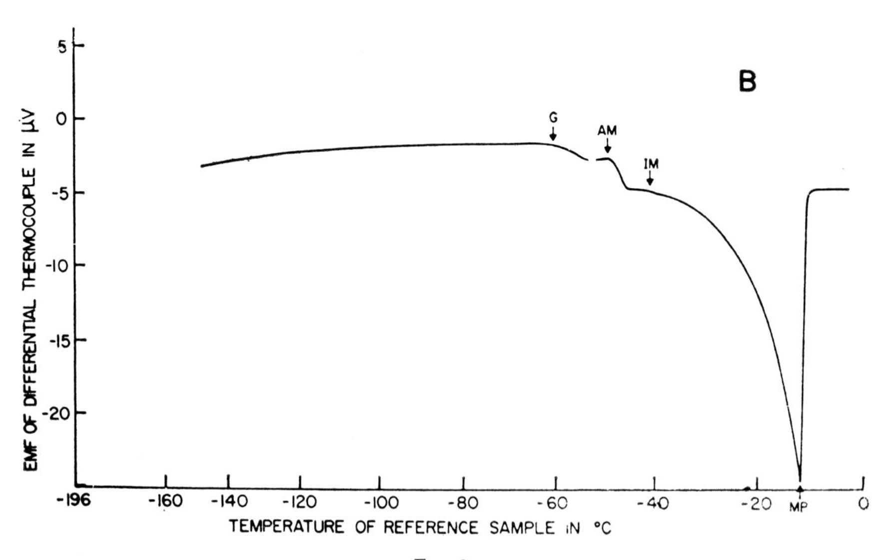
B
EMF OF DIFFERENTIAL THERMOCOUPLE IN μV
5
0
-5
-10
-15
-20
G
AM
IM
MP
-196
-160
-140
-120
-100
-80
-60
-40
-20
0
TEMPERATURE OF REFERENCE SAMPLE IN °C

FIG. 8

temperatures). In both cases (of growth alone and of nucleation and growth) there is a passage from the vitreous to the crystalline state.

An "ante-melting" transition

In a great number of differential thermal analysis determinations of the thermal events taking place during the rewarming of frozen solutions of various kinds, the pattern represented in Fig. 8B, in which an endothermic change (marked AM) occurs some 10 to 15 °C below incipient melting (IM), has consistently been observed. The phenomenon has provisionally been named "ante-melting" transition. Its existence seems quite definite; but we know too little about it to say anything more now on its nature and its role. My associate D. Rasmussen is studying the possible connexions between the appearance of that pre-melting transition and the formation of "liquid-like" layers of molecules at the surface of small ice crystals, a phenomenon which is itself related to the sintering process. A preparation which has been cooled rapidly has a devitrification peak in its differential thermogram (Fig. 8A) and no AM endotherm.

The ante-melting transition is apparently independent of the solute concentration; the average values obtained at concentrations extending from 30 to 50 per cent glycerol are represented in Fig. 7 by a line parallel to the X axis.

Recrystallization

The phenomenon of recrystallization appears in various forms, two of which, designated respectively as "irruptive" and "slow-pace" recrystallizations, are particularly spectacular. "Irruptive recrystallization" occurs when a thin layer of an aqueous solution (of a great variety of

---

FIG. 8. A: Copy of the differential thermogram obtained during the slow rewarming of a 50 per cent solution of glycerol which had been rapidly cooled. Ordinate: electromotive force (EMF) in μV; abscissa: temperature of reference sample. The points marked by vertical arrows, at G and D, represent the beginning of, respectively, the glass transformation and devitrification; the oblique arrow, at PD, indicates a change of unknown nature designated as post-devitrification change; MP denotes the melting point.

B: Copy of the differential thermogram obtained during the slow rewarming of a 50 per cent solution of glucose which had been cooled in such a way that maximum crystallization was attained. Ordinate: electromotive force in μV; abscissa: temperature of reference sample. The arrows at G and IM point to the beginning of, respectively, the glass transformation and melting; the arrow at AM points to a change of unknown nature designated as ante-melting. MP indicates the melting point. (From Luyet and Rasmusssen, 1968; by permission of *Biodynamica*.)

solutes), at concentrations extending from about 25 to 60 per cent, is rewarmed slowly after having been cooled rapidly. The preparation, which was transparent at low temperatures, becomes suddenly opaque when its temperature reaches a certain value—the "recrystallization temperature". "Slow-pace recrystallization" consists of the gradual coarsening of the grain and in the subsequent increase in size of the large particles in the crystalline population, at the expense of the small ones which vanish out of existence, when the specimen is warmed gradually from the temperature of irruptive recrystallization to the melting point of the solution.

The two forms of recrystallization are apparently two manifestations of the same fundamental process and consist of the transfer of molecules of water from the surface of the small particles to that of the large ones. The principal factor responsible for this migration would be a too large surface-to-volume ratio in the small particles which renders them unstable. When the particles of ice grow from dimensions at which they are not visible into the range of visibility (the borderline being at about 0·5 μm, the wavelength of visible light), they become observable under ordinary light as a cloud, and the preparation is seen to turn opaque. This is irruptive recrystallization. At gradually higher temperatures, where the particles are easily observable under the microscope, one actually sees the large pieces grow larger and the small ones decrease in size and disappear. This process is illustrated in Fig. 9. (For more details about recrystallization, as observed in physical systems, see my review: Luyet, 1967.)

Melting

The physical changes occurring when a frozen solution is being melted under equilibrium conditions have been extensively studied and are fairly well known. As was said above, the relationship between temperature and concentration is then expressed in the melting curves of the classical phase diagrams. But some of our observations on recrystallization and on the transitions detectable by differential thermal analysis indicate that often a system becomes unstable during rewarming at temperatures far below the lowest point in the melting curve of the phase diagram (the eutectic point). Thus the temperature at which the melting endotherm begins in glycerol solutions (line IM in Fig. 7) is more than 10°C below the eutectic, and so is the recrystallization temperature (line R, Fig. 7). As was pointed out above (p. 39), "ante-melting" changes (line AM) take place some 15°C below the incipient melting point IM.

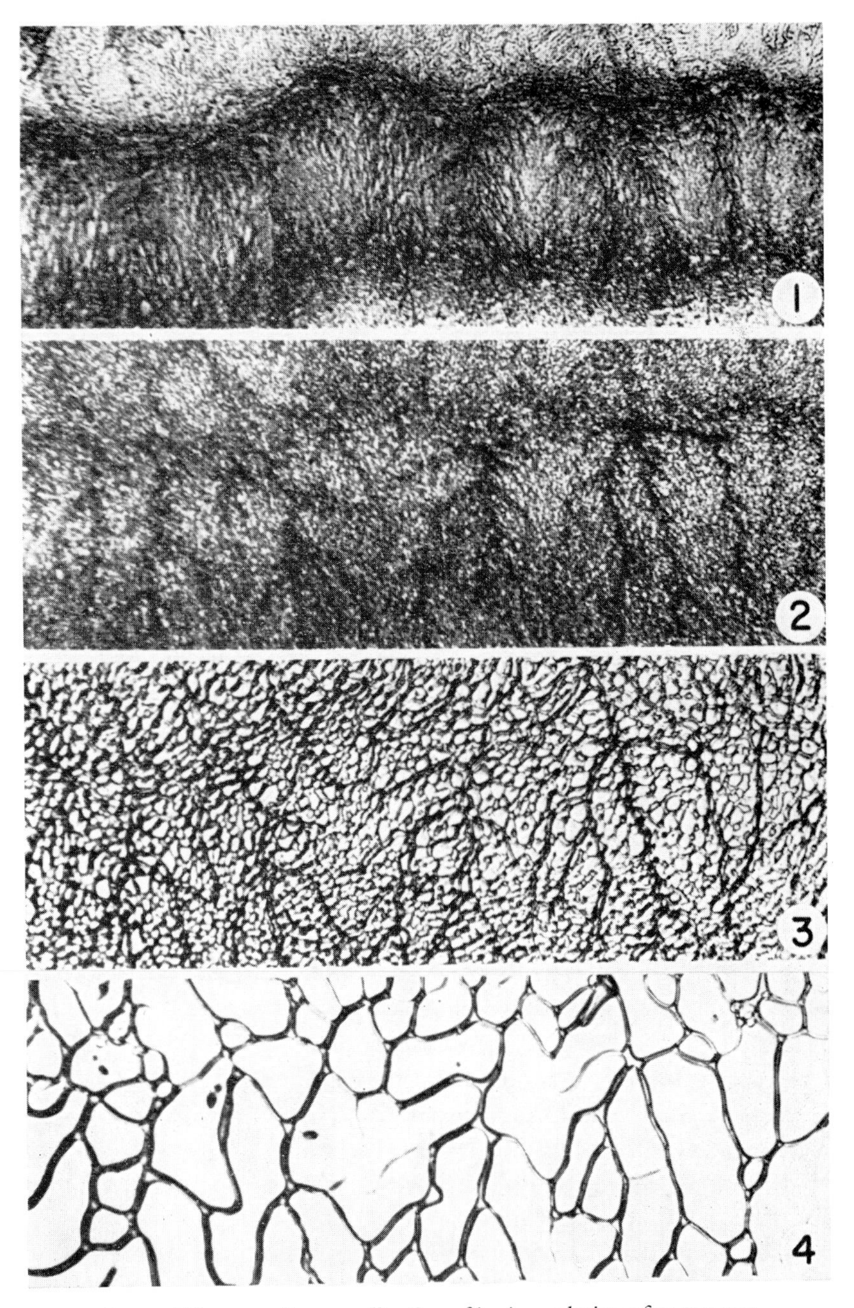

FIG. 9. "Slow-pace" recrystallization of ice in a solution of 10 per cent albumin. (1): specimen frozen at $-40°$ and warmed to -7 °C; (2): same field as in (1) after the temperature was raised to -3 °C; (3): same field after the temperature was raised to -2 °C; (4): same field after the temperature was raised to -1 °C (× 100). (From Luyet, 1966; by permission of *Biodynamica*.)

The fact that, at glycerol concentrations of from 20 to 50 per cent, the temperatures at which these three changes take place are independent of the concentration led Luyet and Rasmussen (1968) to the idea that it is the highly concentrated solutions in contact with the ice phase, not the 20 to 50 per cent solutions, which are responsible for the AM, IM and R changes. On that basis, the temperatures of incipient melting IM were plotted at the highest concentration at which ice is still formed and the points so determined fall on the prolongation of the melting curve (segment drawn in dashed line, Fig. 7).

The instability of the system far below the eutectic is probably due to the few layers of molecules which separate the ice near its melting point from the highly concentrated solution.

SUMMARY

I. Physical changes occurring during freezing: The ice separating from a solution assumes four principal forms: compact solid masses, regularly arranged dendrites, irregular dendrites, or spherulites. The latter three have essentially the same structure; they are arborescent systems which differ in the size and number of branches. A frozen solution consists thus of a network of ice, often with extremely fine ramifications, bathed in a medium in which there are steep concentration gradients. The proportion of solution remaining unfrozen is often considerably more than is generally thought, so that, even upon slow freezing, a relatively large fraction of a solution may be in the vitreous state. The vitreous and crystalline phases are closely interpenetrated.

II. Physical changes occurring during rewarming: Five transitions may occur in the course of rewarming: the glass transformation, devitrification, a particular transition designated as "ante-melting", of which the nature is unknown, recrystallization and melting. (The first three and the last were detected by differential thermal analysis.) The glass transformation involves a marked change in molecular mobility and in several physical properties. The quantity of material to undergo glass transformation and devitrification is not negligible, as often assumed. Melting of the ice phase is frequently observed far below the eutectic temperature. There is generally no eutectic freezing in the presence of cryoprotective agents. Recrystallization is characterized by noticeable changes in the size of the ice particles and of the channels of medium between the particles. The enormously large surfaces of contact between the finely branched arborescent network of ice and the surrounding non-frozen medium, the steep concentration gradients through the layers of that medium, and slow diffusion deserve more attention in the evaluation of the factors producing instability in frozen solutions.

Acknowledgements

The author wishes to thank his colleagues, Dr G. Rapatz, Dr A. MacKenzie and Mr D. Rasmussen, and his former colleagues, Dr L. Menz, Dr P. M. Gehenio, and Mr M. Persidsky, who made this study possible by contributing to the establishment of the facts on which it is based. Dr Rapatz, whose publications in collaboration with the author (see references) supplied most of the illustrations reproduced in the paper, deserves a particular mention.

REFERENCES

LUYET, B. (1966). *Biodynamica*, **10**, 1–32.
LUYET, B. (1967). In *Physics of Snow and Ice*, vol. 1, part 1, pp. 51–70, ed. Oura, H. Sapporo: Institute of Low Temperature Science, Hokkaido University.
LUYET, B. (1969). *Biodynamica*, in press.
LUYET, B., and GEHENIO, P. M. (1952). *Biodynamica*, **7**, 107–118.
LUYET, B., and RAPATZ, G. (1958). *Biodynamica*, **8**, 1–68.
LUYET, B., and RASMUSSEN, D. (1967). *Biodynamica*, **10**, 137–147.
LUYET, B., and RASMUSSEN, D. (1968). *Biodynamica*, **10**, 167–192.
LUYET, B., VEHSE, R. C., and GEHENIO, P. M. (1969). *Biodynamica*, in press.
MACKENZIE, A. P., and LUYET, B. J. (1962). *Biodynamica*, **9**, 47–69.
MENZ, L. J., and LUYET, B. J. (1961). *Biodynamica*, **8**, 261–294.
PERSIDSKY, M. D., and LUYET, B. J. (1959). *Biodynamica*, **8**, 107–120.
RAPATZ, G., and LUYET, B. (1959). *Biodynamica*, **8**, 121–144.
RAPATZ, G., and LUYET, B. (1960). *Biodynamica*, **8**, 195–239.
RAPATZ, G., and LUYET, B. (1961). *Biodynamica*, **8**, 295–315.
RAPATZ, G., NATH, J., and LUYET, B. (1963). *Biodynamica*, **9**, 83–94.

APPENDIX

CASES IN WHICH PHYSICAL CHANGES COMPARABLE TO THOSE REPORTED IN SOLUTIONS WERE OBSERVED IN CELLS AND TISSUES

One of the essential differences between solutions and cells or tissues is the presence, in the latter, of membranes or of structural networks which hinder the free development of the ice. A few cases representative of two tissues, blood and muscle, which we studied more extensively in our laboratories will serve to illustrate the main similarities and dissimilarities in the behaviour of the physical and the biological systems.

Among the factors that seem to be of prime importance in the freezing of both systems, the cooling rates and the freezing temperatures, which determine the number of nuclei formed, deserve particular attention.

In biological systems, slow freezing at high subzero temperatures, where very few crystalline nuclei are generated, results in the formation of large irregular masses of ice in the intercellular spaces; rapid freezing at low temperatures, which permits the generation of a large number of nuclei, results generally in the formation of intracellular ice.

For blood-cell suspensions these alternatives are illustrated in Fig. 10A, 1, which shows the cells caught in channels between large masses of ice, and in Fig. 10A, 2, which shows intracellular ice. In this latter instance,

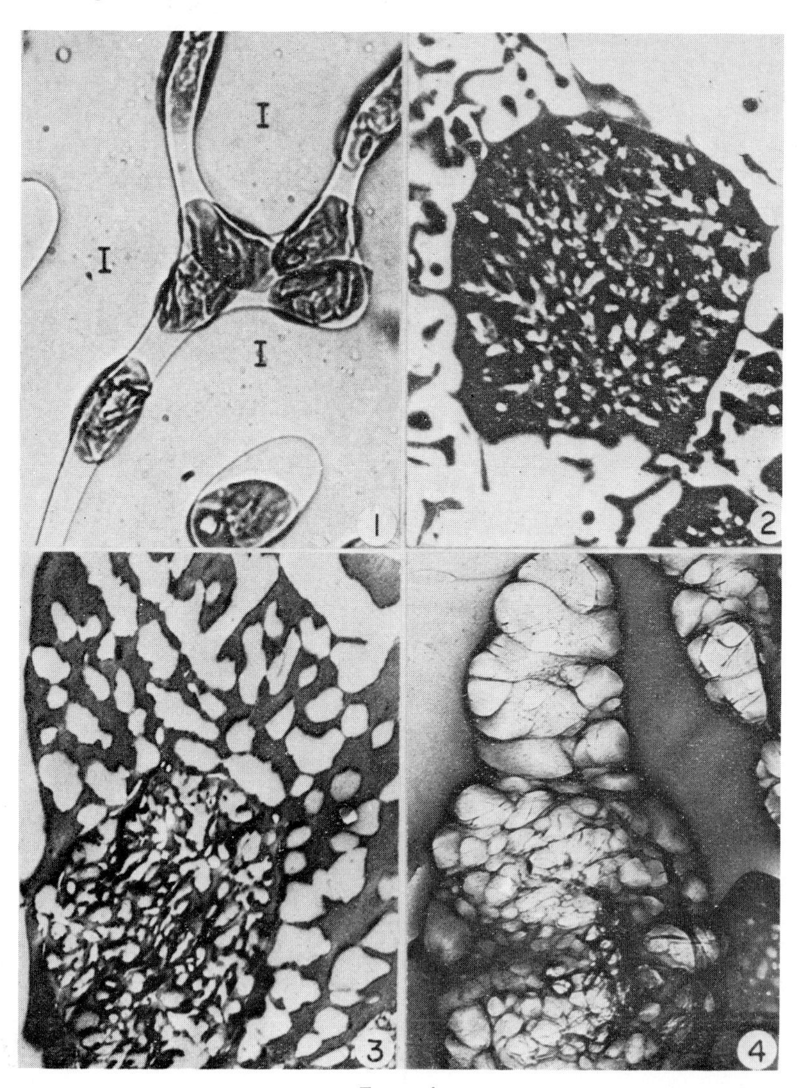

FIG. 10A

one may note, ice developed into arborescent structures. Cells which possess a nucleus, like frog erythrocytes, may show differences in the size or the configuration of the ice particles in the nucleus and the cytoplasm (Fig. 10A, 3). Fig. 10A, 4, illustrates the formation of relatively large particles of ice in frog erythrocytes frozen and rewarmed to the recrystallization temperature.

The modification of the freezing pattern by the cellular structure is particularly marked in muscle tissue, where the ice spears are led to grow preferentially in a direction parallel to the axes of the fibres (Fig. 10B, 1). In rapid freezing at low temperatures the distribution of the ice particles is conditioned also by the cross-striations, as shown in Fig. 10B, 2. However, well-ramified ice formations are sometimes encountered (Fig. 10B, 3). Fig. 10B, 4 shows, in a cross-section of a very rapidly frozen fibre, the difference in the size of ice spears or particles (blank spaces) between the interior of the fibre, at A, and its periphery, at E, where cooling was faster. Freezing at a sufficiently high rate leads to the formation of ice particles of such small size that the fibres remain semi-transparent (Fig. 10B, 5), but turn opaque (Fig. 10B, 6) when warmed to the recrystallization temperature. Fig. 10B, 7 and 8 illustrate in a cross-section and a longitudinal section, respectively, the growth of some ice particles (blank areas) when rapidly frozen muscle fibres are rewarmed to recrystallization temperatures. (Note that all the photographs of Fig. 10B, except 1, 5 and 6, are electron micrographs representing, at high magnifications, the patterns of ice formation and of growth in single muscle fibres frozen rapidly and either vacuum-sublimed or freeze-substituted.)

FIG. 10A. Mode of action of freezing on red blood cells. (1): Effects of slow freezing at -5 °C on blood; the photograph shows the cells (frog erythrocytes) confined in narrow channels between large masses of ice, labelled I. (2): Electron micrograph of a section through a red cell in bovine blood frozen rapidly at -150 °C; the micrograph illustrates the arborescent arrangement of the ice particles within the cell (the white specks represent cavities from which ice has been removed by freeze-substitution.) (3): Electron micrograph of a section through a frog erythrocyte frozen rapidly at -150 °C, showing the difference in the size and the distribution of the ice particles (white specks) in the cytoplasm and the nucleus (the latter is at the lower left-hand side of the photograph. (4): Electron micrograph of a replica of a frog erythrocyte frozen at $-80°$ and recrystallized at -10°C. The spheroidal portion on the lower half of the photograph represents the main body of the erythrocyte which contains the nucleus; the upper portion represents one of the two narrow ends of the oval cell. Note that the two parts are filled with ice particles. (Magnifications: (1) × 655; (2) × 14 500; (3) × 11 900; (4) × 6900.) (From Rapatz and Luyet, 1960, 1961; and Rapatz, Nath and Luyet, 1963; by permission of *Biodynamica*.)

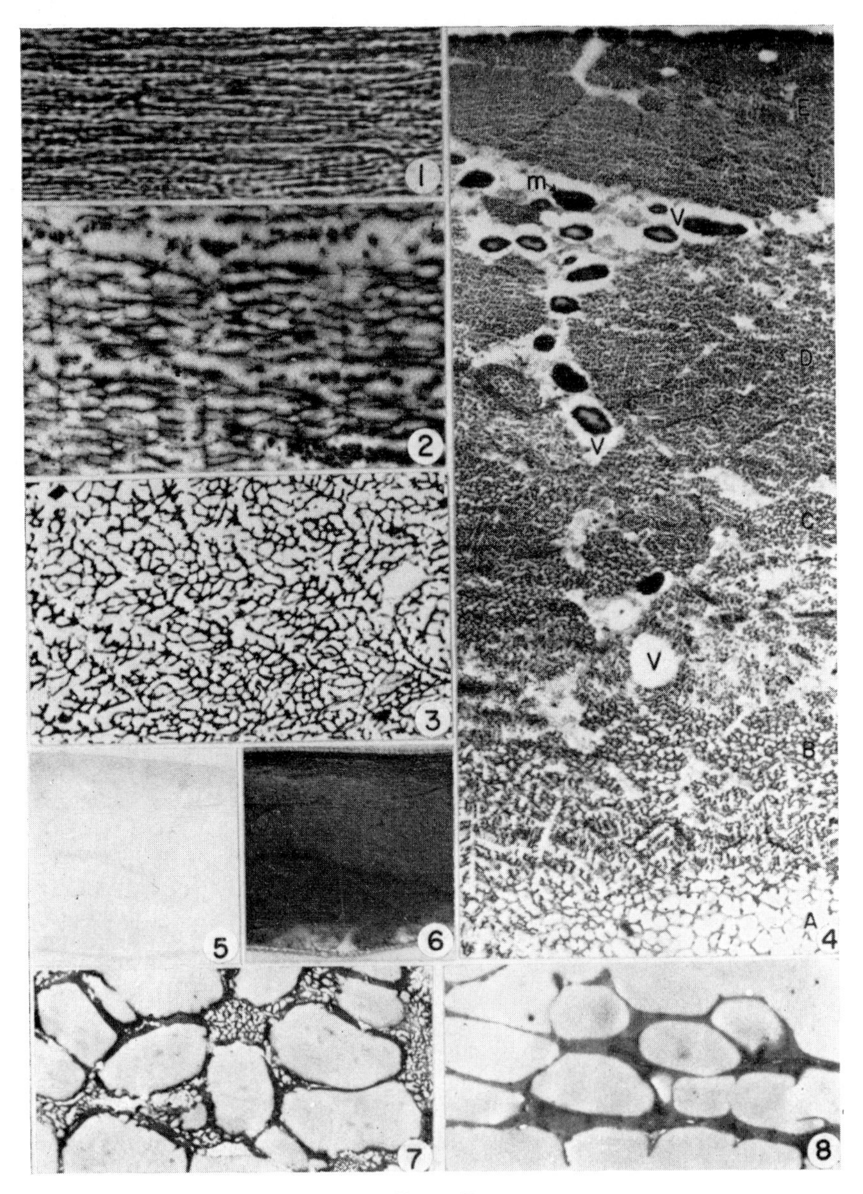

FIG. 10B

DISCUSSION

Klotz: What do you mean by a supercooled liquid? It seems to me thermodynamically impossible to obtain it on raising the temperature of a crystalline phase.

Luyet: The answer to that question calls for a clarification of two points: (1) what does in fact happen when a substance in the vitreous state is warmed to the "glass transformation" temperature (T_g)? (2) what do we call supercooled liquid? (a problem of semantics).

(1) As I said in my paper, important changes in physical properties take place at T_g, which are interpreted by physicists as due to the acquisition by the molecules of translational and rotational motions. But the material, which was amorphous below T_g, remains amorphous above T_g. (2) The physicists, to emphasize the fact that a rapidly cooled liquid remains amorphous far below the melting point (T_m), say that it is a supercooled liquid. Thus Pugh (1962), talking of substances in the vitreous state, writes: "The structure is that of a supercooled liquid of very high viscosity." According to this definition, ordinary window glass and vitreous water at, say, -150°C, are supercooled liquids. But, when it became desirable to distinguish the conditions of a substance below and above the glass transformation temperature, the term supercooled liquid

---

FIG. 10B. Mode of action of freezing on muscle. (1): Photomicrograph of a longitudinal section of a freeze-dried muscle fibre, showing that the channels previously occupied by ice were oriented in the direction of the fibres. (2): Electron micrograph of a longitudinal section of a muscle fibre frozen at -150°C, showing the effect of the myofilaments and of the cross-striations in determining the shape, the size and the orientation of the ice particles (the white areas from which ice has been removed). (3): Electron micrograph of a cross-section of a fibre frozen at -150°C and freeze-substituted, exhibiting the arborescent type of ice formation. Apparently the presence of myofilaments did not disturb the freezing process sufficiently to obliterate the arborescent pattern. (4): Electron micrograph of a cross-section through a single muscle fibre frozen at -150°C, illustrating the progressive decrease in size of the ice cavities (white specks) from the interior of the fibre, at A, to its periphery, at E. (Vacuolar areas, V, often surround the mitochondria, m.) (5) and (6): Photographs of a piece of a muscle fibre cooled rapidly to -150°C (5), then warmed to -10°C (6). The turning opaque of the specimen upon rewarming is characteristic of recrystallization. (7) and (8): Electron micrographs of, respectively, a cross-section and a longitudinal section through a recrystallized muscle fibre. The tissue, previously frozen at -150°C, was transferred to a bath at -15°C in which it remained for one minute. (Magnifications: (1) × 725; (2) × 12 800; (3) × 5450; (4) × 12 350; (5, 6) × 320; (7, 8) × 5000). (From Menz and Luyet, 1961; and Rapatz and Luyet, 1959; by permission of *Biodynamica*.)

was limited to the latter. Kennedy (1962) summarizes this situation as follows: "Below T_g the substance is described as a glass; between T_g and T_m it is a supercooled liquid." In that sense it is correct to say that one obtains a supercooled liquid by raising the temperature.

Brandts: You showed liquid channels of about a third of a micron wide in fast-cooled ice. Do any smaller channels penetrate into the ice?

Luyet: Dr MacKenzie took that picture and other pictures at higher magnifications.

MacKenzie: When 20 per cent aqueous gelatin, in the form of a very thin film, was frozen by rapid immersion in isopentane precooled to $-140°C$, freeze-dried, and examined in the electron microscope, the smallest discrete cavities resolved measured about 5 nm in diameter. From what was known of the nature of the freezing and the recrystallization of ice in gelatin during cooling we concluded that freezing must first have resulted in the formation of dendrites composed of branches much smaller in diameter than 5 nm. So far we have been unable to resolve these supposedly finer structures.

Mazur: In the pictures taken at one-minute intervals just below the melting point, how do you know that crystal growth resulted from recrystallization rather than crystallization? The formation of the equilibrium amount of ice at those temperatures could require several minutes.

Luyet: I used the pictures taken at one-minute intervals near the freezing point to emphasize the predominant effect of recrystallization. I agree with you that, if we were to study the rate of ice formation, we would have, in the first instants after nucleation, a confusing superimposition of the effects of crystallization and recrystallization.

In connexion with the notion of the availability of freezable water, I would like to raise a question which often comes up for consideration in the study of the freezing of solutions, namely, that of the effects of gradually increasing concentrations of solutes, in particular of high-molecular-weight solutes, on the freezing point depression and on the osmotic pressure of the solutions. Let me take a specific case. In dilatometric determinations of the amounts of ice formed in 10 per cent gelatin solutions, we found that a certain amount of ice (81 per cent by weight) would form at $-5°C$; then, when we lowered the temperature to $-10°C$, more ice (83·5 per cent) formed; after a further cooling to $-20°C$, still more ice appeared (85 per cent); but cooling to lower temperatures did not cause the separation of more ice. Since solutes like gelatin have a very limited osmotic activity, I would like to raise before this group the question of the factors responsible for the lowering of the freezing point and for the osmotic pressure in cases like the one mentioned.

Klotz: I assume that it is a kinetic phenomenon and not an equilibrium phenomenon, and that gelatin interferes somehow with crystallization by a surface action of some kind. It is hard to believe this is an equilibrium phenomenon, and that is what the lowering of freezing point amounts to. Gelatin surely could not lower the freezing point of water to −20°C by colligative means.

MacKenzie: We are, after all, discussing systems in which the solute content has increased to values in the range 60 to 80 per cent weight/weight. We cannot expect these systems to act in accordance with Raoult's Law. We can, on the other hand, employ the concept (well established by numerous measurements at higher-than-freezing temperatures) of the *desorption isotherm.* (This, for our purposes, is a curve describing the dependence of the water content on the water activity—or the relative humidity—where the latter was slowly reduced at constant temperature.)

Essentially, we can describe the very slow cooling of an ice-containing polymer system by the passage of a point across a series of desorption isotherms. Clearly, the point must cross these isotherms at water activities equal always to those of ice at the temperatures in question. Riedel (1961) discussed this question in some detail and was, I believe, the first person to define the extent of "unbinding" of bound water in terms of an equilibrium phenomenon. Difficulties of an obviously kinetic nature may, of course, arise during an experimental determination of desorption equilibrium data.

Meryman: In our X-ray powder patterns of frozen gelatin solutions, if we freeze reasonably rapidly we get no hexagonal ice pattern (Meryman, 1958). When we warm the solution up, the transition point that Dr Luyet is talking about we think coincides with the temperature at which the gelatin matrix itself becomes flexible. It is possible, at temperatures below −20°C, to sublime the water out of the gelatin without ever getting the appearance, by X-ray powder patterns, of hexagonal ice. We interpreted this as indicating that rapid freezing produced either incomplete crystallization or crystals too small to be detected. When we warmed the specimen, even though the vapour pressure of the water was high enough for it to sublime, the rigidity of the gelatin matrix was such that there simply wasn't the physical room for the water to re-arrange into the ice structure.

Klotz: That is a kinetic phenomenon!

Meryman: Yes, in a sense, except that the rigidity of the gelatin is to all intents and purposes permanent at about −25 °C. It depends on what you mean by kinetic.

Klotz: I mean a system with a thermodynamic free energy which without restraints will allow the water to go from one state to the other. But the restraints introduce the kinetic phenomenon.

Meryman: Yes. It is possible to distil the water off at temperatures as low as −60 °C, and obviously the water is not under any thermodynamic restraint.

REFERENCES

KENNEDY, S. W. (1962). In *Encyclopaedic Dictionary of Physics*, vol. 3, pp. 474 *et seq.*, ed. Thewlis, J. Oxford: Pergamon Press.

MERYMAN, H. T. (1958). *Biodynamica*, **8**, 69–72.

PUGH, S. F. (1962). In *Encyclopaedic Dictionary of Physics*, vol. 7, p. 661, ed. Thewlis, J. Oxford: Pergamon Press.

RIEDEL, L. (1961). *Kältetechnik*, **13**, 122–128.

THE EXCEEDING OF A MINIMUM TOLERABLE CELL VOLUME IN HYPERTONIC SUSPENSION AS A CAUSE OF FREEZING INJURY *

HAROLD T. MERYMAN

American National Red Cross, Blood Research Laboratory, Bethesda, Maryland

SEVERAL hypotheses have been advanced for the mechanism by which freezing injures living cells. Probably the most self-evident was the proposal that the growth of ice crystals caused mechanical injury by crushing or spearing cells. The massive dislocation of tissue architecture produced by the growth of extracellular ice would appear to provide ample evidence for this. However, subsequent experience has shown that tissues can tolerate the presence of extracellular ice which, despite histological appearances, probably exerts no mechanical pressure on the cells and certainly does not puncture them (Luyet and Gehenio, 1940). Intracellular ice is a special situation which probably does not occur in animal tissues save in extreme laboratory conditions producing very high rates of cooling. In plant tissues intracellular freezing may occur even at low rates of cooling and is generally considered to be inevitably destructive (Levitt, 1966). Mazur (1966) has conducted many elegant studies on the incidence and effect of intracellular ice formation in yeast and other organisms. Although there have been occasional reports of animal cells surviving after intracellular freezing (Lozina-Lozinsky, 1967; Sherman, 1964), we must presume for the moment that they are rare exceptions and restrict the scope of this paper to the consideration of injury from slow freezing with extracellular ice formation.

Other hypotheses for the mechanism of injury are based, not on the ice itself, but on the dehydration and concentration produced by it. Some investigators have concerned themselves with that portion of cell water which is involved in functions other than solvation. It is well known that proteins may have associated with them an adsorbed monolayer of water

* Contribution No. 137 from the Blood Research Laboratory, American National Red Cross.

equal to between 0·2 and 0·5 the dry weight of the protein. It has also been postulated that additional water is confined in some structured order at the protein surface, possibly contributing in some way to the stability of the protein itself (Sinanoglu and Abdulnur, 1965). It has been suggested that the formation of ice strips this water from the protein surface, leading to denaturation (Karow and Webb, 1965).

Levitt (1962) has proposed that the removal of cell water can lead to the distance between proteins or portions of a protein being reduced and that, as they get nearer to each other, abnormal disulphide bonds can be formed either through disulphide interchange or the oxidation of thiol (sulphydryl) bonds.

Probably the simplest and most valuable hypothesis has been that of Lovelock (1953*a*) who, with particular reference to red cells, proposed that it was the concentration of the electrolytes produced by the freezing out of water which led to cell injury. His conclusion was based on the observation that freezing haemolysis in red cells began to be evident whenever the extracellular salt concentration reached 0·8 M-sodium chloride. The introduction of glycerol into the suspending medium was shown to reduce the amount of ice formed at any temperature and therefore to reduce the concentration of electrolytes produced. In these conditions, red cell haemolysis was still seen to occur only when the salt concentration exceeded 0·8 M, regardless of the temperature (Lovelock, 1953*b*). The only legitimate quarrel with this hypothesis is that a biochemical injury such as the "lyotrophic" effect proposed by Lovelock has not been demonstrated. Purified proteins commonly withstand this ionic concentration and many higher animals survive normally in saline media approaching 0·8 M. Sea water, for example, has an osmolality equivalent to 0·56 M-NaCl (1015 m-osmolal) and molluscs have been acclimatized readily to 150 per cent sea water (0·74 M-NaCl) and on several occasions in our laboratory to 190 per cent sea water (1·0 M-NaCl, 1810 m-osmolal).

EFFECT OF HYPERTONIC SOLUTIONS ON RED CELLS

To a first approximation Lovelock's experiments (1953*a*,*b*) are easily duplicated and there appears to be little question but that the onset of red cell haemolysis is related to a critical concentration of sodium chloride when it constitutes the sole osmotic support for the cells. Substituting other osmotically active solutes for sodium chloride results in qualitatively similar effects, although the critical concentration producing haemolysis varies somewhat from one compound to another. It has been suggested that, since haemolysis is seen at roughly the same osmolar concentration of sugar as of salt, this argues against the concept that electrolyte concentration is a specific cause of injury. However, it can be observed in rebuttal that intracellular electrolytes, principally potassium chloride, are being identically concentrated regardless of the non-ionic nature of the extracellular solute. The experiments summarized in this paper were conducted in an effort to determine whether cell injury from freezing is, in fact, related to an absolute salt concentration or whether some other factor is responsible.

Our initial studies were conducted primarily at room temperature with human red cells suspended in concentrated solutes to simulate the solute concentrations produced by freezing. This approach enabled us to assay the effects of solute concentration alone without having to superimpose the return to isotonic suspension which would occur with thawing. We were also able to discriminate between any possible contribution to injury by either the ice crystals or the reduced temperature.

Studies (Meryman, 1968) of the mean corpuscular volume of erythrocytes suspended in steadily increasing concentrations of sodium chloride or plasma with added NaCl showed the expected decrease in volume as the suspending solution was increased to 1100 m-osmolal, or about four times isotonic. The volume then remained nearly constant to about 1300 m-osmolal, after which a volume increase without haemolysis was seen. This response in the face of increasing extracellular osmolality implies the passage of extracellular solute into the cell. When cell sodium and potassium were studied as a function of increasing osmolality in both salt and sugar suspensions it was found that at between 1100 and 1500 m-osmolal (four to five times isotonic) there was a sudden rise in intracellular sodium to approximately the concentration of the suspending medium and a steady loss of potassium at a rate sufficient to deplete the cell within 60 to 90 minutes. Subsequent experiments indicated that this was probably an all-or-none response for each individual cell. The proportion of cells that succumbed to the concentrated solute and became leaky was related both to the osmolality of the solution and to the duration of the exposure.

Although cells which have become irreversibly leaky do not haemolyse while in hypertonic suspension, on return to an isotonic medium, as would occur with thawing, the greatly increased intracellular solute content produces grossly hypertonic cells, leading to immediate osmotic lysis. This response to hypertonic suspension appears to be an adequate explanation for the freeze-thaw haemolysis reported by Lovelock, although these experiments do not, of course, shed any light on the cause

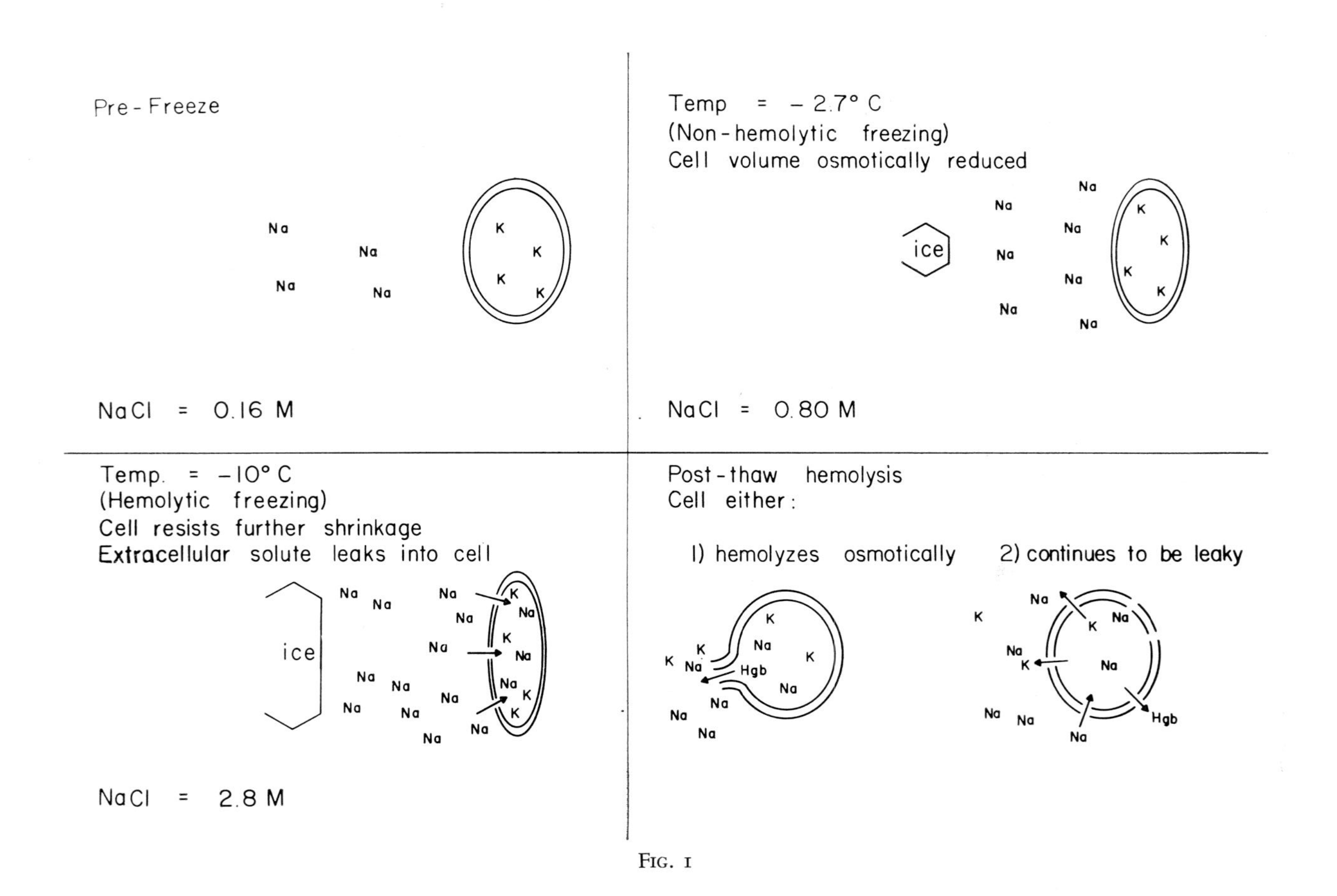
Pre-Freeze
NaCl = 0.16 M
Temp = − 2.7° C
(Non-hemolytic freezing)
Cell volume osmotically reduced
ice
NaCl = 0.80 M
Temp. = −10° C
(Hemolytic freezing)
Cell resists further shrinkage
Extracellular solute leaks into cell
ice
NaCl = 2.8 M
Post-thaw hemolysis
Cell either:
1) hemolyzes osmotically
2) continues to be leaky
Hgb

FIG. 1

of the apparent membrane injury. Lovelock has reported an increasing concentration of cholesterol and phospholipids in the suspending medium when cells are exposed to hypertonic salt solution (Lovelock, 1955) and it is probable that these are lost from the membranes at the time of breakdown. Our reluctance to accept the suggestion that this membrane alteration was due to some denaturing effect of the salts led us to consider other alternatives.

"MINIMUM CELL VOLUME" HYPOTHESIS

The fact that the nature of the extracellular solute can be altered at will, yet still create membrane injury when sufficiently concentrated, suggested to us that injury did not result from a biochemical effect of the solutes themselves but rather was a result of exceeding a maximum tolerable osmotic pressure (Meryman, 1968). We proposed that, as the cell is reduced in size by increasing osmolality, there is an increasing resistance to shrinkage due to the compression of cell contents. Any inability of the cell volume to change in response to a concentration gradient must inevitably result in an osmotic pressure difference across the membrane. The development of a hydrostatic pressure across the membrane would appear intuitively to be incompatible with continued membrane integrity. In other words, we proposed what might be termed a "minimum cell volume" hypothesis in which membrane damage results from osmotic stresses imposed on the cell membrane when the cell volume is decreased beyond a minimum tolerable size (Fig. 1).

FIG. 1. Schematic representation of the response of a red cell to extracellular freezing.

Upper left: The cell is suspended in isotonic (0·16 M) NaCl. For illustrative purposes it is presumed that potassium is the intracellular osmotic solute.

Upper right: Red cells will tolerate freezing to about −2·7°C. At this temperature sufficient water has been frozen out to concentrate the extracellular solute to 0·8 M. Intracellular solute is presumably concentrated to a similar osmolality. The cell is reduced in volume but membrane integrity is maintained.

Lower left: Freezing to −10°C produces complete haemolysis after thawing. It is proposed that resistance to further shrinkage of the cells has led to the development of an osmotic pressure gradient. This has resulted in a breakdown of the membrane and the influx of large quantities of sodium.

Lower right: On thawing and return to isotonic suspension, the excessive sodium content can lead to hypotonic lysis. Alternatively, the cell membrane may be so permeable as to permit haemolysis without membrane rupture.

If it is the compression of cell contents which leads to the development of stress in hypertonic suspension, then a reduction in the volume of cell contents should also accordingly modify the circumstances under which injury occurs. One experimental approach to this is through the red cell ghost. Teorell (1952) has reported that it is possible to haemolyse red cells and yet recover an intact ghost which displays normal osmometric characteristics and is capable of selective ion transport. These red cell ghosts behaved as osmometers over a 100-fold range of concentration, with a reduction in cell size to 10 per cent of that in isotonic saline but without any departure from the calculated osmometric response or any loss of initial membrane permeability characteristics. This suggests that the absence of cell contents permits a passive and reversible response to hypertonic solutions not tolerated by normal cells.

If osmotic pressure rather than electrolyte concentration is the agent of injury, then a penetrating salt should be tolerated by the red cell at concentrations well above 0·8 M. Although cations in general will not penetrate the red cell membrane, the ammonium ion can enter as the neutral molecule, ammonia, acquiring a proton inside the cell to regain its ionized form (Jacobs, 1926). When coupled with a penetrating anion such as chloride or acetate, this becomes in effect a penetrating salt. Red cells were shown to tolerate concentrations of NH_4Cl as high as 4 M.

The fact that penetrating ammonium salts are tolerated in high concentration tends to confirm the conclusion that absolute electrolyte concentration is not responsible for injury. However, it could still be argued that it is the concentration of intracellular potassium which is specifically responsible for membrane injury and that the concentration of ammonium cannot be considered equivalent to that of intracellular potassium. In order to test this argument, experiments were designed whereby cell size and intracellular potassium concentration could be independently controlled (Meryman, 1969*a*).

It is well known that red cells suspended in isotonic sucrose lose potassium and that this alteration in membrane permeability is reversible (Davson, 1939). Potassium-depleted cells reinfused into the donor will re-establish their original sodium and potassium content (Crawford and Mollison, 1955). When cells are suspended in hypotonic sucrose there is a net loss of potassium before osmotic haemolysis and it is possible almost totally to deplete the cells of sodium and potassium without other evidence of injury. Such cells, when returned to isotonic suspension, are smaller than normal cells because of their reduced content of osmotically active solute. Using such a preparation, suspension in increasing osmotic strength should bring such cells to their "minimum volume" at a lesser osmolality than would be required for a normal cell. When cells were depleted in potassium to 35 per cent of normal, it was found that the departure from predicted osmometric behaviour, the loss of intracellular potassium and the haemolysis on resuspension in isotonic solution developed at about 700 m-osmolal as compared to about 1300 m-osmolal for the normal cell and that, furthermore, the development of membrane changes occurred in both series at approximately the same mean corpuscular volume. When potassium-depleted cells were frozen and thawed it was found that haemolysis was seen after freezing to about $-2°C$ as compared to $-3°C$ for normal cells. Calculations of the intracellular potassium concentration at the concentrations injurious for potassium-depleted and for normal cells show that there is an even greater discrepancy in potassium concentration than there is in the extracellular osmolality, because of the contribution of haemoglobin to the intracellular osmotic pressure. This further supports the proposal that, in the red cell, the membrane injury associated with suspension in hypertonic solutions (as by freezing) is related to a degree of dehydration and reduction in cell volume beyond a tolerable minimum rather than to an absolute concentration of any solute.

HYPERTONIC STRESS IN OTHER ANIMAL CELLS

Studies of human blood platelets have also revealed a behaviour in hypertonic solution essentially identical to that of erythrocytes (Lundberg, 1969). The platelets are reduced in volume in predictable fashion until the concentration of the suspending solution approaches about five times isotonic. Above this concentration, there is apparently an inward leak of extracellular solute as evidenced by a volume increase. There is a loss of intracellular potassium and of intracellular serotonin, a failure of serotonin uptake and a reduction in clot retraction capacity. Observations similar to those made on potassium-depleted red cells have been carried out with potassium-depleted platelets. As with red cells, membrane failure occurs at a lower concentration than for normal platelets.

Experiments with non-mammalian tissues also appear to confirm a relationship between freezing injury and the loss of a fixed proportion of total cell water. Williams (1968) studied two species of molluscs: *Venus mercenaria*, which can survive to $-6°C$ but no lower, and *Mytilus edulis*, which will tolerate freezing to $-10°C$. Calorimetric studies of these two organisms show that at $-6°C$ *Venus* has lost 64 per cent of total tissue water to ice, while at −10°C *Mytilus* has also lost 64 per cent of its total water. Fig. 2 compares the calorimetric data for these two organisms with that from human red cells, which can survive freezing only to −3°C and where we found the same proportion of water, 64 per cent, to be frozen out. The freezing curve for *Venus* is similar to that of sea water. The

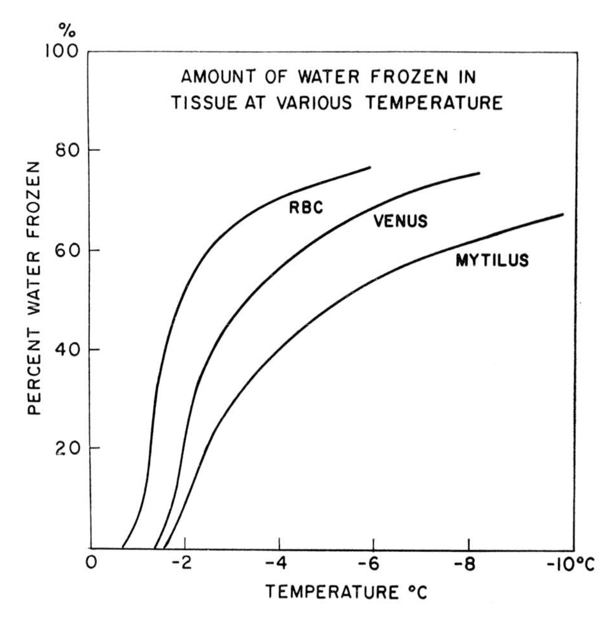

Fig. 2. The proportion of water frozen out as a function of temperature is illustrated for the human red cell and for the molluscs, *Venus mercenaria* and *Mytilus edulis*. The red cell is isotonic at 0·16 M-NaCl and therefore freezes at −0·56°C. *Venus* is isotonic at 0·57 M-NaCl and therefore begins to freeze at −1·8°C. Its freezing curve is nearly identical to that of sea water. *Mytilus* is also isotonic with sea water but has an elevated osmotic coefficient due to the presence of bound water (roughly 18 per cent of its total water). This means that intracellular solutes must be at a higher concentration to equal the proportion of total water loss seen in *Venus*. The minimum temperatures tolerated by the three specimens are, respectively, −3, −6, and −10°C. Approximately 64 per cent of cell water has been frozen out at the minimum tolerable temperature for each system.

shift in the freezing curve of *Mytilus* and its ability to survive to −10°C was shown to be due to the binding of about 18 per cent of its total water. Through this device the same proportion of water loss and of reduction in cell size develops at −10°C in *Mytilus* as in *Venus* at −6°C, despite an almost two-to-one difference in the absolute concentration of extracellular solute. This is additional evidence that freezing injury is related, not to the absolute concentration of either intra- or extracellular solute, but to the removal of a critical proportion of total water and/or a corresponding reduction in cell size.

CRYOPROTECTION BY PENETRATING AGENTS

On the minimum cell volume hypothesis, the cryoprotective properties of glycerol, dimethylsulphoxide and other penetrating cryoprotective agents can be explained on the basis proposed originally by Lovelock (1953*b*). Since these compounds penetrate the cell membrane, they are osmotically indifferent and do not contribute to the osmotic support of the cell. Therefore they can be present in quite large concentrations without altering cell volume. However, when the total solute concentration is increased by the addition of the cryoprotective agent, less ice will be formed at any subfreezing temperature. This will in turn result in less of an increase in concentration not only of total solute but of the osmotically active constituents originally present at isotonic concentration. This is illustrated by Fig. 3, in which it can be seen that even at the temperature of −40°C, where sufficient water will be frozen out to produce a solution of 21 500 m-osmolal concentration, or roughly 70 times isotonic, the osmotically active sodium chloride has been concentrated only by a factor of 2·6. On the basis of this model, the only criteria for a cryoprotective agent are that it must penetrate the membrane with sufficient freedom to have no lasting osmotic effect and that it must be non-toxic to the cell in the multimolar concentrations necessary.

Using these simple criteria we investigated a number of solutes. Of particular interest were two electrolytes which fulfilled the above requirements. The first of these, ammonium acetate, was found to afford cryoprotection comparable to that of glycerol (Meryman, 1968). Since ammonium chloride has a eutectic point at about −12°C, it is unsuitable as a cryoprotective agent. The other cryoprotecting electrolyte, trimethylamine acetate, is presumed to have a mechanism of penetration similar to that of ammonium salts, with the cation entering the cell as a neutral molecule. Although possessing some undesirable properties, trimethylamine acetate was found to be an excellent cryoprotective agent for red cells, affording complete protection at concentrations slightly in excess of 1 M.

CRYOPROTECTION BY NON-PENETRATING AGENTS

Heber (1967) has reported that sucrose will protect grana of spinach chloroplast against freezing injury at any temperature. This has been

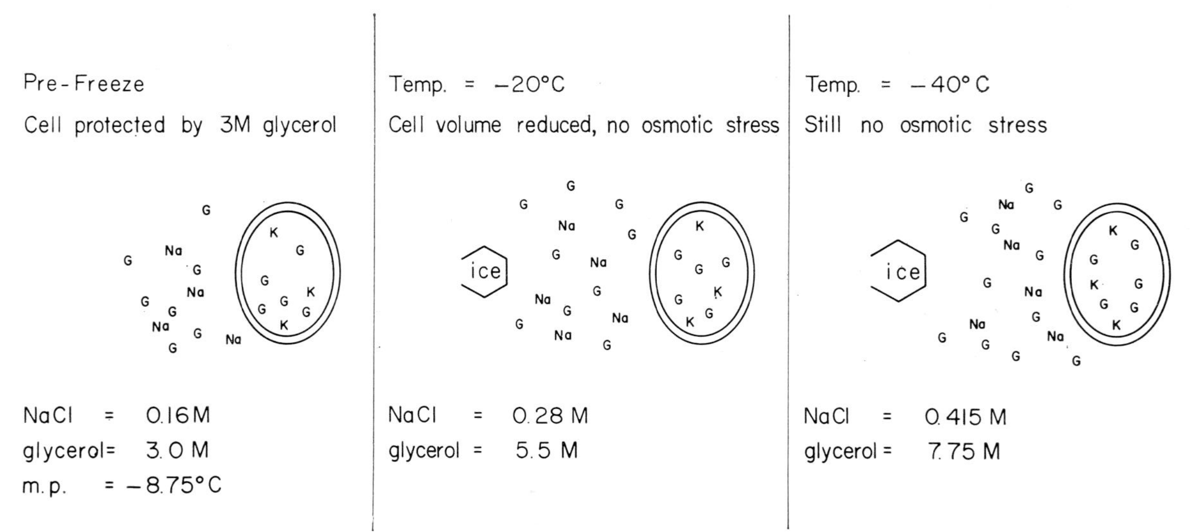

FIG. 3. Cryoprotection by glycerol.
Left: The cell is suspended in isotonic saline plus 3 M-glycerol which has come to equilibrium across the membrane.
Centre: Because of the high solute concentration, only a small amount of ice has formed at −20° C. The glycerol and sodium chloride are only concentrated 1·75 times and a modest cell volume reduction results.
Right: The amount of water which must be frozen out to lower the freezing point of the remaining solution to −40 °C is insufficient to concentrate the solution by a factor of more than 2·6. The sodium chloride has thus been concentrated only to 0·36 M, producing insufficient cell shrinkage to result in hypertonic stress.

confirmed by us for mannitol as well. However, the concentrations of sugar, 100 to 300 mM, which protect are far lower than those necessary to afford protection on a colligative basis, and the mechanism by which these low concentrations of nominally non-penetrating solutes could provide cryoprotection was therefore studied (Williams and Meryman, 1969). When grana are suspended in sodium chloride solutions of increasing osmolality, a progressive loss of osmometric properties and of the ability to photophosphorylate and to pump protons is seen as the suspending solution exceeds 2000 m-osmolal (or when the grana are frozen to $-5\,^{\circ}C$). When grana are suspended in hypertonic solutions of mannitol (equivalent to freezing in mannitol) they are initially reduced in volume, as expected. However, beyond 800 m-osmolal further shrinkage ceases, nor is there loss of either the ability to photophosphorylate or osmometric integrity. On return to lesser concentrations, the grana are found to be larger in volume. It is apparent that, at 800 m-osmolal, these cells have permitted an influx of solute which forestalls further volume reduction with further concentration and loss of membrane function has been prevented.

Heber and Ernst (1967) have isolated from hardened spinach leaves a protein which in micromolar concentration confers freezing protection on washed non-hardy grana. When we suspended grana in hypertonic salt in the presence of this protein they were found, at 800 m-osmolal or higher, to leak solute inward without loss of function. We postulate that this naturally occurring protective agent, like sugar, reduces the threshold of membrane permeability to small molecules and permits the influx of extracellular solute as minimum critical size is approached, thus preventing the excess dehydration responsible for injury.

These observations take on particular significance when one attempts to discuss the mechanisms by which dehydration may be responsible for the membrane changes observed. Clearly the experimental evidence with red cells, platelets and molluscs merely indicates that membrane injury is related to the loss of more than a critical proportion of total cell water. Whether the crucial element is the reduction in volume of the cell or the dehydration *per se* is not revealed by these experiments. On the other hand, the fact that grana suspended in sugar do not leak until the cell approaches its maximum tolerable osmolality would indicate that some additional stress has been introduced by the increased osmotic pressure. These data are compatible with the proposition that an osmotic gradient across the membrane has been developed by excessive cell shrinkage. Furthermore, they also imply that sugar, or the Heber protein, can reduce the threshold for small molecule leak so that solute influx can relieve the osmotic gradient before it reaches a magnitude sufficient to overstress the membrane.

THE INFLUENCE OF SUGARS ON MEMBRANE PERMEABILITY IN RED CELLS

In view of the alterations in membrane permeability induced by sugars in grana, the possibility of similar phenomena in red cells was considered. Red cells suspended in isotonic sugar solutions leak potassium. The presence of 0·1 M-NH_4Cl (which freely penetrates the membrane and therefore exerts no osmotic effect) prevents spontaneous potassium leak in sugar concentrations ranging from 150 to 1200 m-osmolal. However, when cells were suspended in hypotonic sugar solutions as low as 60 m-osmolal, which should have been haemolytic, haemolysis was not seen. Instead sufficient intracellular potassium was lost to obviate further volume increase. The amount of potassium lost at any given osmolality was just sufficient to maintain maximum cell volume with no further loss seen over a period of hours (Meryman, 1969*b*). From this we conclude that the sugar has served to reduce the threshold of membrane permeability so that an osmotic stress which, in salt solution, would lead to irreversible injury and haemolysis is relieved by solute leak before the stress has reached a destructive magnitude.

CONCLUSIONS

We have therefore come to the conclusion that freezing injury in red cells and probably in most cells is unrelated to any absolute concentration of any particular solute, but is instead related to the removal of a critical proportion of total cell water and the associated reduction in cell size beyond a critical volume. Although the evidence is still indirect, we feel that it is more likely that the injury results from the reduction in cell volume than from the absolute reduction in cell water. Hypotheses of injury which are related to solute activity such as the salt concentration theory or theories related to the removal of structurally significant water are incompatible with our findings. Levitt's hypothesis (1962) of abnormal chemical bonds which result when proteins are brought into closer approximation with dehydration is compatible with our data, although we are inclined to prefer the simpler model of mechanical stress to one which depends on the formation of disulphide bonds which may be present with widely varying frequency in membranes of various species.

The effect of sugars and certain other compounds on the threshold for solute leak across the membrane provides support, in our opinion, for the osmotic stress model and, in addition, provides a possible explanation for the effect, in plant cells, of non-penetrating compounds which are cryoprotective in less than colligatively effective concentrations. The similarity between red cells and spinach grana in this regard encourages us to speculate that this lowering of the threshold for small molecule leak by compounds present in low concentration may be a very general phenomenon of cell membranes.

SUMMARY

Evidence has been presented in support of a "minimum cell volume" model for the mechanism of injury from extracellular freezing. It is proposed that the development of extracellular ice leads to a concentration of those extracellular solutes which do not normally penetrate the cell. Water leaves the cell and osmotic equilibrium is maintained across the membrane. With continuing cell volume reduction and the compression of cell contents, a resistance to further cell shrinkage develops. If the cell cannot shrink freely in response to the concentration gradient, then an osmotic pressure difference must develop across the membrane. When this pressure gradient exceeds the tolerance of the membrane, irreversible changes in membrane permeability result.

Penetrating cryoprotective agents act on a purely colligative basis by reducing the amount of ice formed, so that solute concentration and the resulting reduction in cell volume are insufficient to exceed the tolerance of the cell.

Under certain circumstances non-penetrating compounds may have cryoprotective capability at less than colligatively significant concentration. A possible mechanism for this is found in the ability of sugars and other compounds to alter the permeability characteristics of the membrane in such a way that an osmotic pressure gradient leads to a reversible solute leak. This relieves the osmotic stress and prevents it from increasing sufficiently to cause irreversible membrane damage.

REFERENCES

CRAWFORD, H., and MOLLISON, P. L. (1955). *J. Physiol., Lond.*, **129**, 639–647.

DAVSON, H. (1939). *Biochem. J.*, **33**, 389–401.

HEBER, U. (1967). *Pl. Physiol., Lancaster*, **42**, 1343–1350.

HEBER, U., and ERNST, R. (1967). In *Cellular Injury and Resistance in Freezing Organisms*, pp. 63–77, ed. Asahina, E. Sapporo: Institute of Low Temperature Science, Hokkaido University.

JACOBS, M. (1926). *Harvey Lect.*, **22**, 146–164.

KAROW, A. M., Jr., and WEBB, W. R. (1965). *Cryobiology*, **2**, 99–108.
LEVITT, J. (1962). *J. theoret. Biol.*, **3**, 355–391.
LEVITT, J. (1966). *The Hardiness of Plants*. New York: Academic Press.
LOVELOCK, J. E. (1953*a*). *Biochim. biophys. Acta*, **10**, 414–426.
LOVELOCK, J. E. (1953*b*). *Biochim. biophys. Acta*, **11**, 28–36.
LOVELOCK, J. E. (1955). *Br. J. Haemat.*, **1**, 117–129.
LOZINA-LOZINSKY, L. K. (1967). In *The Cell and Environmental Temperature*, pp. 90–97, ed. Troshin, A. S. Oxford: Pergamon.
LUNDBERG, A. (1969). *Fedn Proc. Fedn Am. Socs exp. Biol.*, **28**, 792 (abstract).
LUYET, B. J., and GEHENIO, P. M. (1940). *Life and Death at Low Temperatures*, 261 pp. Normandy, Mo.: Biodynamica.
MAZUR, P. (1966). In *Cryobiology*, pp. 213–315, ed. Meryman, H. T. New York and London: Academic Press.
MERYMAN, H. T. (1968). *Nature, Lond.*, **218**, 333–336.
MERYMAN, H. T. (1969*a*). Manuscript in preparation.
MERYMAN, H. T. (1969*b*). II Ann. Sci. Symp. Washington, D.C.: American National Red Cross. To be published.
SHERMAN, J. K. (1964). *Cryobiology*, **1**, 103–129.
SINANOGLU, O., and ABDULNUR, S. (1965). *Fedn Proc. Fedn Am. Socs exp. Biol.*, **24**, suppl. 15, S12–23.
TEORELL, T. (1952). *J. gen. Physiol.*, **35**, 669–701.
WILLIAMS, R. J. (1968). *Cryobiology*, **4**, 250 (abstract).
WILLIAMS, R. J., and MERYMAN, H. T. (1969). In preparation.

DISCUSSION

Mazur: Cryoprotection is seen with many different additives. Certain sugars will protect the red cell, but only at higher cooling rates. Why don't they protect at low cooling rates?

Meryman: They do, but they don't protect enough to be useful.

Mazur: But they do protect the granum, as Dr Heber shows (see pp. 175–186).

Meryman: That is where being a plant is so nice. Grana suspended in hypertonic sugar shrink just so far, then begin to leak instead. When they are returned to isotonic conditions they are bigger than they were initially because they contain a great deal more solute. For a red cell this would be disastrous, and it would undergo osmotic hypotonic haemolysis. When the plant cell fills the volume enclosed by the cell wall, it can't expand any further. There will then be an osmotic gradient in the opposite direction and the solute will leak out.

Mazur: But the granum still functions.

Meryman: Yes, the granum in sugar appears to be able to leak solute in or out without loss of membrane characteristics. It will not do this in salt alone since osmotic stress will build up to the point where the membrane ruptures. In the presence of sugar the membrane threshold to solute leakage has been reduced sufficiently so that when there begins to be a stress, it leaks solute before the stress reaches a destructive level.

Levitt: Did you say that leakage prevented further size reduction? If a plant cell leaks solutes surely it should contract more than if it didn't leak solutes.

Meryman: The extracellular solute concentration is being increased by freezing. Water is leaving the cell and the cell is getting smaller in response to the increasing extracellular concentration. As the extracellular concentration continues to increase, the cell must continue to get smaller, but if extracellular solute can leak into the cell then the intra- and extracellular concentrations can remain equal without the necessity for cell shrinkage.

Levitt: Of course with the plant cell this wouldn't happen, because there is no solution outside the cell, only air and ice.

Meryman: But a little bit of solute (0·001 M or so) is present, and if you freeze enough water out you can concentrate this as far as you want.

Levitt: No, you are not freezing the water of the intercellular spaces since this is insignificant in quantity.

The loss of electrolyte has been used for 35 years as a method of measuring injury to plant cells, in other words the greater the loss of electrolyte the greater the injury. This seems to be opposed to your hypothesis.

Meryman: No, because the injury is a loss of membrane integrity, which will result in a loss of intracellular electrolyte on thawing.

Levitt: That isn't what happens. There are different degrees of resistance. One plant may be killed at −15°C, and another at −20°C. If both are frozen at −8°C, which does no damage whatsoever, a greater conductivity is still measured from the one with the lower resistance to freezing damage. This means that there is a certain amount of leakage, but the cell isn't injured. This is just the opposite of what you describe.

Meryman: No, this is still what we see in the presence of a cryoprotective compound. With Heber's protein (see pp. 175–186), depending on the concentration we use, we can get varying degrees of protection. With an amount not sufficient to protect we will get some leakage and some offsetting of the reduction in volume. I am probably making a serious mistake in trying to project our data to plants as a whole, since our observations are only with grana.

Levitt: Data that we had some 30 years ago, if you will accept them, are I think opposed to this concept (Levitt, 1939). The extent of shrinkage can be measured from the amount of ice formed in the same kind of plant cell both when it is not resistant and when it is resistant to freezing. The more resistant cell was able to shrink to a greater degree without injury. So it is not just a matter of both shrinking to the same degree.

Meryman: That would certainly be contrary to this model.

Heber: The granum as a simplified test system is more closely comparable to a red cell than to an ordinary vacuolized plant cell. It is part of an intracellular organelle of leaf cells.

Levitt: But the theory must apply to the whole cell.

Heber: Santarius's observation of reversible inactivation of photophosphorylation by high concentrations of sucrose supports your conclusion, Dr Meryman, that the grana become leaky under the influence of high concentrations of sugars. It is known that leakiness results in the loss of phosphorylation. At a sucrose concentration of 1 to 2 M phosphorylation breaks down, but returns to full activity if the system is brought back to iso- or hypotonic conditions (Santarius and Ernst, 1967). However, in contrast to red cells, grana cannot be protected by salts such as ammonium acetate.

Meryman: Ammonium acetate protects red cells on a colligative basis, not by changing membrane permeability as is the case with sugars.

Heber: Grana frozen in the presence of sucrose swell, after the removal of sucrose by washing, to about the same size as unfrozen controls. This may speak against significant entry of sucrose into the granum compartment during freezing.

Mazur: Is the only evidence of sucrose permeation the fact that the grana swell to a larger size after thawing?

Meryman: The only evidence we have at the moment is that if we plot their size at increasing osmolalities in salt there is a continual reduction and eventually a loss of membrane properties, whereas in sucrose at 800 m-osmolal the size remains constant even though the extracellular osmolality is increasing. There really is no other explanation, other than extraordinary changes in the osmotic coefficient, than that there has been penetration of solutes.

Davies: When you return the grana from hypertonic sucrose solutions back into less hypertonic or isotonic solutions, do the grana swell up and in fact pass their original size?

Meryman: On return to very low osmolalities, some of them leak solute out again. This is a little indistinct. We have to deal with a sort of a "granacrit" measurement, so we are looking at the average size of a very large number of cells.

Luyet: If you use a non-penetrating solute, such as PVP, at gradually increasing concentrations, do you obtain sufficient osmotic pressure to reduce the volume of the red cells to the same extent as you would with a 4·5 times isotonic solution of a non-penetrating solute like sucrose?

Meryman: I have never made these measurements with PVP and red cells because of the viscosity of the solution.

Farrant: We shall provide evidence later (pp. 97–114) that solutions of PVP become progressively less ideal at high concentrations, giving them appreciable osmotic pressure.

Levitt: From our freezing point measurements of high concentrations of PVP we can calculate values of $30 \cdot 4 \times 10^5$ Newtons per square metre (30 atmospheres).

Luyet: We are back to the question already discussed, of high molecular weight substances permitting the development of high osmotic pressures.

Farrant: It depends on the concentration.

Meryman: It depends on the mole fraction. If one removes enough water, the mole fraction of solute can be elevated sufficiently to give as low a vapour pressure as desired.

Luyet: Dr Rinfret, you had much experience with the effects of PVP on red blood cells. What observations did you make on the reduction in cell volume?

Rinfret: At the concentrations used, about 7 per cent w/v PVP, we did not observe significant reduction in cell volume.

Bricka and Bessis (1955) showed that they could irreversibly damage red cells in the presence of a 35 per cent concentration of PVP without freezing, simply because the concentration was great enough. I don't know the haematocrit in this context.

REFERENCES

Bricka, M., and Bessis, M. (1955). *C.r. Séanc. Soc. Biol.*, **149,** 875.
Levitt, J. (1939). *Pl. Physiol., Lancaster*, **14,** 93–112.
Santarius, K. A., and Ernst, R. (1967). *Planta*, **73,** 91–108.

INTERACTIONS OF COOLING RATE, WARMING RATE AND PROTECTIVE ADDITIVE ON THE SURVIVAL OF FROZEN MAMMALIAN CELLS*

PETER MAZUR, STANLEY P. LEIBO, JOHN FARRANT,† E. H. Y. CHU, M. G. HANNA, JR, and L. H. SMITH

Biology Division, Oak Ridge National Laboratory, Oak Ridge, Tennessee

WHEN an investigator subjects a cell to freezing, he must select the final temperature to be attained, the cooling rate to that temperature, the storage time, and the warming rate. Variations in all these factors have been shown to affect the survival of one or more types of cells, and the effects are sometimes profound. In spite of this, most work in cryobiology, especially with mammalian cells, has been restricted to final temperatures of −78 or −196°C, cooling rates of about 1°C/min, and warming velocities of several hundred degrees per minute. We thought it desirable to determine the consequences of varying these factors over much wider ranges, not only to define the optimal conditions for maximal survival, but also to gain an understanding of the causes of injury and the means of preventing it.

Detailed studies of survival versus cooling rate have been published for a few cells, including mammalian erythrocytes and yeast, and in general the curves are of the form shown in Fig. 1. That is, survival is low after either rapid or slow cooling and becomes maximal at some intermediate optimal velocity. Warming velocity, as shown by the dotted lines, affects the survival of rapidly cooled cells more than it affects the survival of slowly cooled cells.

The shapes of the curves and the numerical values of the cooling rates corresponding to the various portions of the curve depend on the particular cell and on the type and concentration of protective compound added. Thus, the optimal cooling rates for unprotected yeast and red

* Research jointly sponsored by the National Cancer Institute and by the U.S. Atomic Energy Commission under contract with the Union Carbide Corporation.

† Permanent address: Division of Low Temperature Biology, Clinical Research Centre Laboratories, National Institute for Medical Research, Mill Hill, London, England.

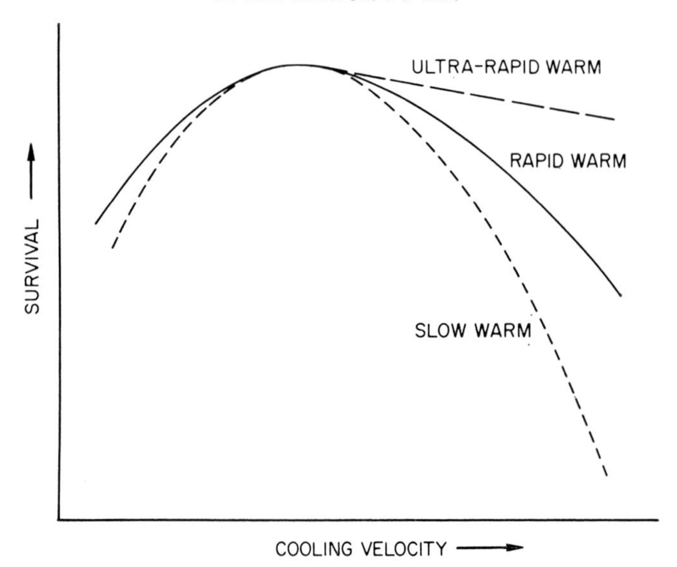

FIG. 1. Schematic relation between the survival of frozen cells and cooling velocity.

cells differ by a factor of about 300, the values being 10°C/min and 2500–3000°C/min, respectively (Mazur and Schmidt, 1968; Rapatz, Sullivan and Luyet, 1968; Mazur, 1968). However, the addition of glycerol to the red cell shifts the whole survival curve to lower velocities; for example, the presence of 20 per cent glycerol lowers the optimum cooling rate from 3000 to below 1500°C/min (Rapatz and Luyet, 1965; Mazur, 1968).

The existence of an optimum cooling velocity must mean that survival is affected by at least two factors that depend oppositely on cooling rate, and the suggestion has been made (Mazur, 1965) that the two factors are "solution effects" and intracellular freezing. According to this hypothesis, "solution effects" are responsible for the low survivals of cells cooled at rates lower than optimal, and intracellular freezing is responsible for the low survivals of cells cooled faster than optimal. By "solution effects" we mean the changes in a solution or in protoplasm that result from the progressive removal of water during freezing, i.e. any one or a combination of effects such as dehydration, increased solute concentration, changes in pH, and precipitation of solutes.

Intracellular freezing is suggested as the factor responsible for the killing of cells cooled more rapidly than optimal, because thermodynamic and kinetic arguments indicate that although it is an improbable event at low cooling velocities, it becomes a near certainty at high velocities (Mazur, 1963). Furthermore, there are close correlations in yeast and red cells between loss in survival from over-rapid cooling and the appearance of intracellular ice (Mazur and Schmidt, 1968; Rapatz, Nath and Luyet, 1963). In addition, the occurrence of intracellular ice provides an explanation of why rapidly cooled cells are sensitive to slow warming; namely, slow warming permits small innocuous ice crystals to grow to damaging size or shape by the process of recrystallization (Mazur, 1966).

Regardless of the cooling and warming rate, most nucleated mammalian cells fail to survive freezing in the absence of a protective additive (Meryman, 1966). Apparently, cooling rates low enough to prevent intracellular freezing produce an exposure to solution effects that is long enough to be lethal. An additive could protect in four ways: (1) It could decrease the magnitude of the solution effects. (2) It could decrease the susceptibility of a cell or cellular organelle to solution effects. (3) It could decrease the likelihood of intracellular freezing. (4) It could decrease the sensitivity of the cell to damage from intracellular freezing. Lovelock's explanation of protection by glycerol and dimethylsulphoxide (DMSO) is an example of the first category. He concluded that the red cell becomes leaky when the concentration of extra- or intracellular electrolyte exceeds mole fraction 0·014 and haemolyses when subsequently transferred to isotonic media. He further suggested that additives protect on a molar basis by reducing the electrolyte concentration in the residual unfrozen solution in and around a cell at any temperature. To protect the cell interior, the additive must permeate (Lovelock, 1953). Meryman (1968) has put forward a modified hypothesis, in which the cause of injury is not electrolyte concentration *per se* but is rather an osmotic pressure differential brought about by the concentration of extracellular solutes. He too explains the protection by additives on a colligative basis, and his explanation, like Lovelock's, requires an additive to permeate the cell.

Interestingly, although glycerol and DMSO protect the red cell against haemolysis by solution effects during slow cooling, they actually increase haemolysis at high cooling rates (Rapatz and Luyet, 1965; Mazur, 1968). In terms of Fig. 1, this observation would suggest that glycerol and DMSO enhance the probability and extent of intracellular freezing at high cooling rates. Farrant and Woolgar (1968) have suggested a theoretical basis for such a supposition.

FREEZING OF NUCLEATED MAMMALIAN CELLS

As mentioned, most studies on nucleated mammalian cells have only covered a limited portion of Fig. 1, i.e. cooling rates have usually been around 1°C/min. Furthermore, most workers have used glycerol and DMSO as additives, but have used them without knowing whether these compounds can in fact permeate the cell under study. Our present efforts have been made to determine the survival of nucleated mammalian cells as functions of wide ranges of cooling velocity, warming velocity, and type and concentration of additive, and as functions of the ability of the additive to permeate. More specifically, we wished to obtain answers to the following questions:

(1) Do survival curves exhibit an optimal cooling velocity which would indicate the involvement of two or more factors? Are the results consistent with the belief that the two factors are solution effects and intracellular freezing?

(2) Is the optimal cooling velocity always around 1°C/min, or is the optimal rate dependent on the type of cell and the type and concentration of additive?

(3) Is the protection conferred by an additive solely a function of its molarity and its ability to permeate the cell?

The cells selected for study were mouse marrow stem cells and Chinese hamster tissue culture cells. The chief reason for their selection was that there exist precise viability assays for both.

The general plan of the experiments was as follows. Cells in balanced salt solutions were mixed with salt solutions containing twice the desired final concentration of additive. After 45–90 minutes samples were seeded at $-2°$ to $-3°$C and then cooled to $-196°$C at rates ranging from 0·3° to 600°C/min. They were usually held at $-196°$C for 15 minutes to three hours and then warmed at high, moderate, or low rates. The concentration of additive in the thawed samples was reduced fivefold by sequential dilution with Tyrode's or minimal Eagle's medium to avoid possible osmotic shock, and the percentage of surviving cells was then determined (Leibo *et al.*, 1969; Mazur *et al.*, 1969). The viability of hamster cells was determined from their plating efficiency (Chu *et al.*, 1969).

The viability of stem cells in marrow was measured by the spleen colony assay of Till and McCulloch (1961). In this assay, the treated cells are injected into lethally irradiated mice. Eight days later the spleens are removed from the mice and the number of surface colonies determined. A colony has been shown by Till and McCulloch to represent the descendants of a single viable stem cell. Viability of the entire marrow population was also assessed by eosin exclusion. The resulting viability levels were generally similar to stem cell survival but showed some differences (Leibo *et al.*, 1969).

Four additives were studied: glycerol and DMSO were chosen as examples of low molecular weight, presumably permeating solutes; sucrose was chosen as a low molecular weight, presumably non-permeating additive; and polyvinylpyrrolidone (PVP) (mol.wt. ~ 40 000) was chosen as a high molecular weight, non-permeating additive.

EFFECTS OF FREEZING ON MOUSE MARROW STEM CELLS

Fewer than 2 per cent of the marrow cells survived freezing when suspended in Tyrode's solution alone, regardless of the cooling rate, but a considerable percentage survived freezing in glycerol. As shown in Fig. 2, the percentage of surviving cells depended on both cooling rate

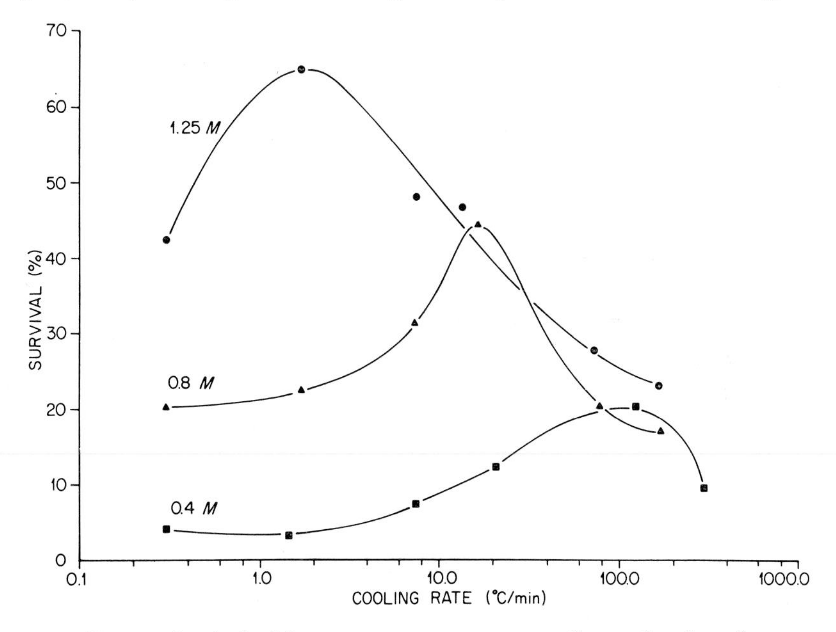

FIG. 2. Survival of frozen mouse marrow stem cells as a function of cooling velocity. The cells were suspended in Tyrode's solution containing the indicated concentrations of glycerol and cooled to −196°C. Subsequent thawing was rapid (~ 1000°C/min).

and glycerol concentration. For a given concentration, survival was maximal at some given cooling velocity, but the optimal velocity decreased with increasing concentration. It was approximately 100°, 20°, and 2°C/min with 0·4, 0·8, and 1·25 M-glycerol, respectively. Thus, the results are consistent with the current belief that cooling at about

1°C/min is optimal for marrow cells in 10–15 per cent glycerol, but they indicate that the optimal rate for nucleated cells may depend on the concentration of additive, as is the case with mammalian red cells (Rapatz and Luyet, 1965; Mazur, 1968).

Changes in glycerol concentration appear to have little effect on cells cooled faster than optimal, as indicated by the coincidence of the right-hand limbs of the curves. The chief effect appears to be on cells cooled slower than optimal. As illustrated in Fig. 3, survival curves of this sort can be accounted for in terms of a "two-factor hypothesis" if it is assumed

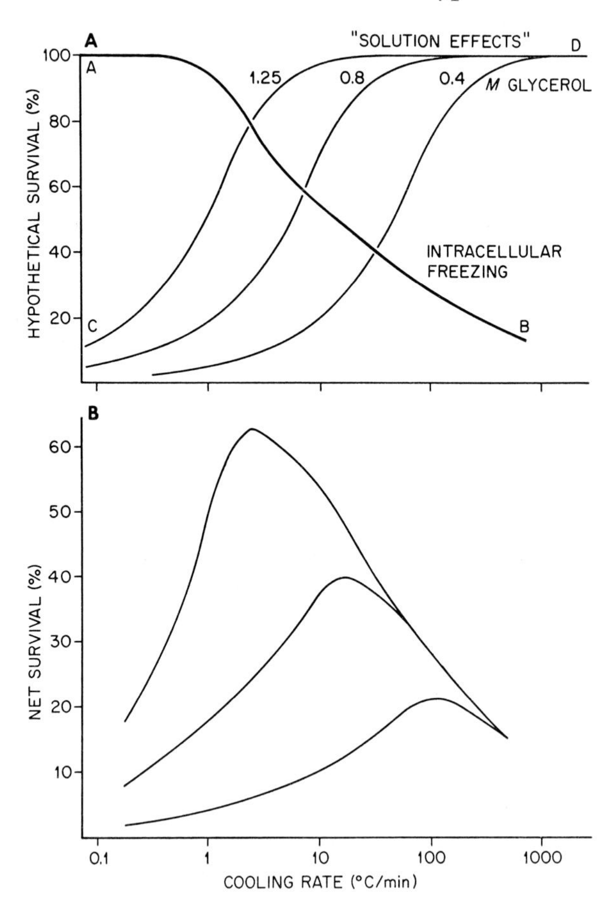

FIG. 3. Hypothetical relation between survival, cooling velocity, and additive concentration on the basis of two injurious factors. (A) Separate contributions of intracellular ice formation (curve AB) and solution effects (curves CD) to injury. The three CD curves show the hypothetical effects of three different concentrations of a protective additive. (B) Calculated survival as a result of the contribution of both intracellular freezing and solution effects.

that glycerol influences only solution effects and not intracellular freezing. Curve AB in Fig. 3A represents a hypothetical survival curve if intracellular freezing were the only cause of injury. Curves CD represent hypothetical survival curves if solution effects were the only cause of injury. We are assuming that increases in glycerol concentration do not affect curve AB but cause curve CD to move to the left. That is to say, an increase in glycerol concentration makes the cells less susceptible to slow cooling. The resulting survivals from the combination of both intracellular freezing and solution effects are shown in Fig. 3B. They were calculated from 3A by simply multiplying a survival on curve AB for a given cooling rate by the corresponding survival on curves CD for the same rate. The shapes of the resulting synthesized curves are similar to those observed in Fig. 2, so two factors, one of which is sensitive to glycerol concentration and the other not, seem sufficient to account for the results observed in Fig. 2. However, no special significance should be placed on the degree of similarity between Figs. 3B and 2, since that depends on the exact shapes and positions chosen for the curves in Fig. 3A.

If intracellular freezing is occurring in cells cooled faster than optimal, then, as suggested in connexion with Fig. 1, rapidly cooled cells ought to be more damaged by slow warming than by rapid warming. Furthermore, rapidly cooled cells ought to be more susceptible to slow warming than are slowly cooled cells. Both of these suppositions held true in stem cells (Table I).

TABLE I

EFFECT OF WARMING RATE ON SURVIVAL OF STEM CELLS COOLED SLOWLY OR RAPIDLY IN 1·25 M-GLYCEROL

Cooling rate (°C/min)	*Warming rate (°C/min)*	*Survival (%)*
1·7	910	65·1 ± 6·6
	1·8	62·3 ± 6·1
295	910	23·0 ± 2·5
	1·8	4·6 ± 1·4

If solution effects are responsible for the reduced survival of stem cells cooled slower than optimal and if glycerol protects, as Lovelock suggests, by reducing the concentration of electrolyte at a given temperature, then the presence of glycerol ought to lower the temperature at which inactivation occurs. As shown in Fig. 4, it does so. In the absence of glycerol, inactivation occurred between −2° and −7°C. In the presence of 0·4 M-glycerol, inactivation occurred chiefly between −10° and −20°C.

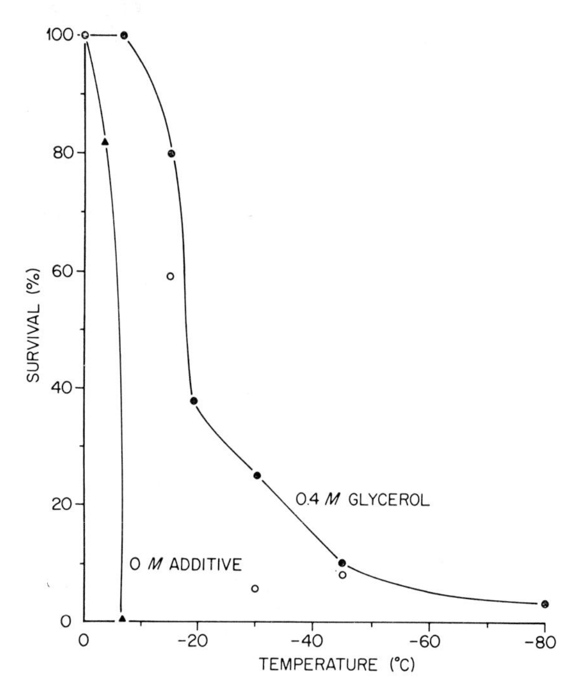

FIG. 4. Relation between the survival of marrow stem cells and the minimum temperature to which they were cooled. The cells were suspended in Tyrode's solution or in Tyrode's containing 0·4 M-glycerol, cooled at 1·6°C/min to the indicated temperatures, and immediately thawed rapidly (▲, ●), or held 30 minutes and then thawed rapidly (○).

The available evidence is thus consistent with survival being dependent on two factors and with the two factors being solution effects and intracellular ice formation. Of course, the present evidence does not constitute proof.

The assumption that survival requires a permeating additive was tested by freezing the cells in Tyrode's solution that contained sucrose or PVP. The molecular weight of the PVP was probably too high for it to permeate passively, and the temperature (<2°C) was probably too low for it to permeate by phagocytosis, suppositions that are supported by the observations of Persidsky and Richards (1962) on the whole marrow population. Unfortunately, there is at present no direct way to determine the permeability of the stem cells to sucrose and PVP, since they constitute only a small fraction of the marrow population and have not been identified morphologically. The best we could do with sucrose was to minimize the possibility of its permeation by minimizing the time the cells were in contact with it before freezing. The time was kept under five minutes.

The survivals of stem cells after freezing in PVP and sucrose are compared in Fig. 5 with the survival after freezing in glycerol. Several features of the results are interesting: (1) As with glycerol, there was an optimal cooling rate for maximum survival, a finding that is again consistent with the belief that survival is affected by two factors depending oppositely on cooling velocity. (2) The optimum velocity was not 1°C/min. It was about 10°C/min for cells in 15 per cent PVP and 20–70°C/min

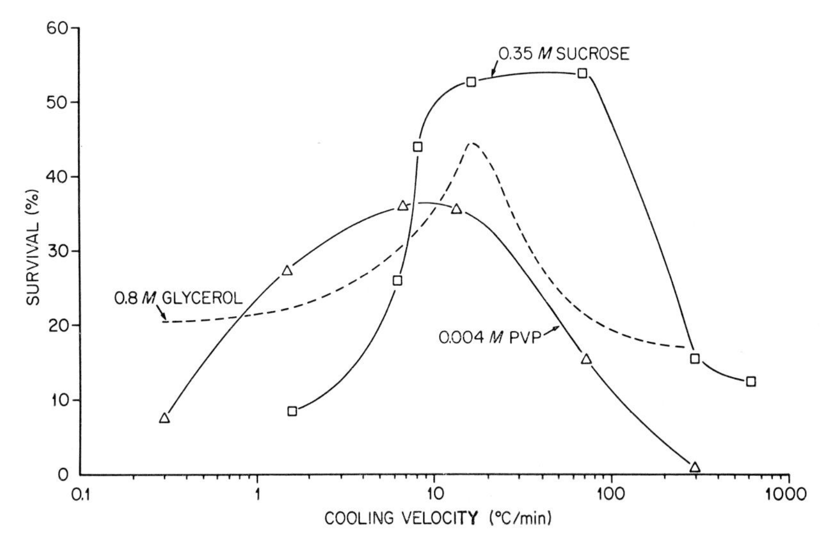

Fig. 5. Relation between the survival of marrow stem cells frozen at various rates and the protective additive. The concentration of PVP was 15 per cent (w/v). Thawing was rapid.

for cells in 0·35 M-sucrose. (3) Survivals after cooling at optimal rates, although not as high as with 1·25 M-glycerol, were nevertheless appreciable. In fact, if one compares the efficacy of the three additives on a molar basis, the protection decreases in the order PVP > sucrose > glycerol. For example, 15 per cent (w/v) PVP (0·004 M) was as protective as a 200-fold higher molar concentration of glycerol (0·8 M), and 0·35 M-sucrose was somewhat more protective than 0·8 M-glycerol. (4) Sucrose was more protective at cooling rates of around 100°C/min than were either PVP or glycerol, but was less protective for cells cooled at 2°C/min. In terms of our "two-factor" hypothesis, this would suggest that sucrose was less able to eliminate solution effects but was better able to reduce the probability or extent of intracellular freezing. A reduction in the extent of intracellular freezing could have come about because of the osmotic dehydration produced by 0·35 M-sucrose in Tyrode's solution. Calculations (P. Mazur, unpublished) indicate that, when a cell is partially dehydrated before freezing, there can be an increase in the cooling rate required to produce intracellular ice.

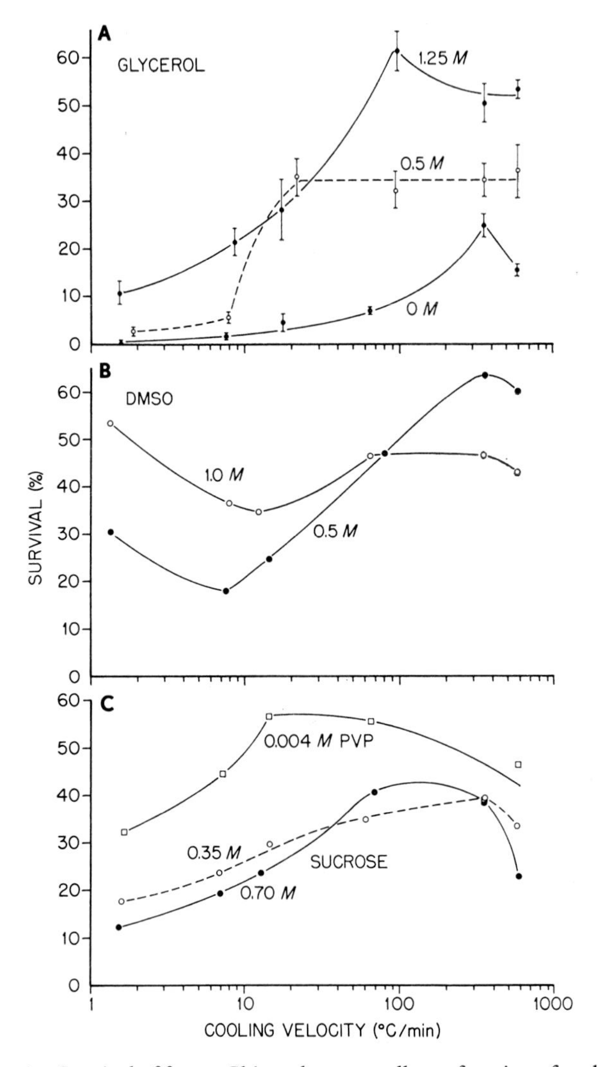

FIG. 6. Survival of frozen Chinese hamster cells as a function of cooling velocity and the type and concentration of additive. Cells were suspended in Hank's balanced salt solution (HBSS) containing the indicated concentrations of additive, cooled to −196°C at the indicated rates, and thawed rapidly (~ 1000°C/min).

EFFECTS OF FREEZING ON CHINESE HAMSTER CELLS

The response of hamster cells to freezing and rapid warming differed in several respects from that of marrow stem cells. First of all (Fig. 6A), some 25 per cent of the cells survived freezing at 300°C/min in the absence of any additive. Second, even with additives, maximal survival occurred with cooling rates above 10°C/min in every case but one (1 M-DMSO). In fact, cooling at 1·6°C/min was the most deleterious of all the rates tested for cells in glycerol, PVP and sucrose. Third, unlike the marrow cells, the maximum survival did not occur at a sharply defined cooling

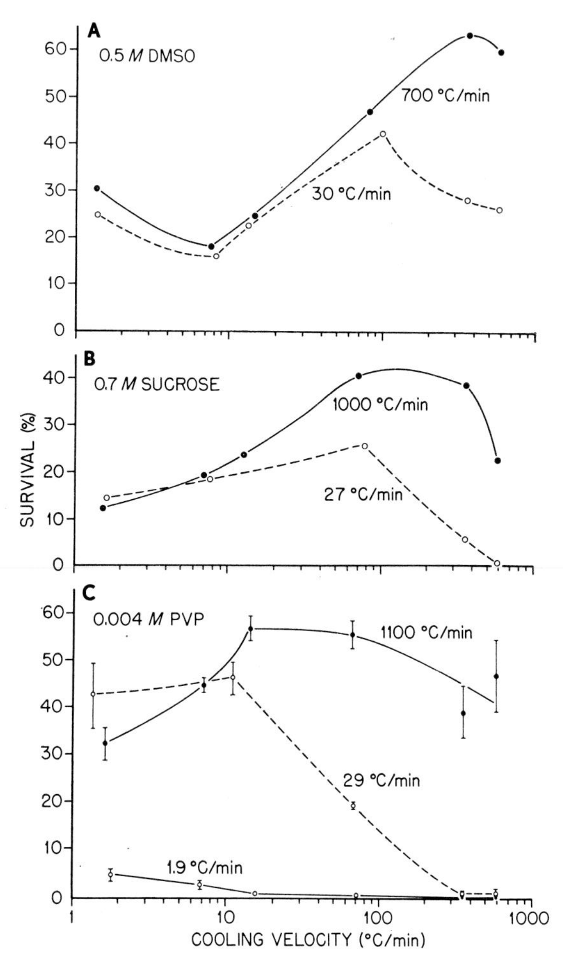

FIG. 7. Effect of warming velocity on the survival of hamster cells suspended in HBSS containing DMSO, sucrose, or PVP, and cooled at various rates to −196°C.

rate but rather over a broader range of rates. But the results were similar to those for stem cells in one respect: survival increased to a maximum with increasing cooling rate, and then decreased, if only slightly, with still more rapid cooling.

If these results are to be interpreted in terms of solution effects and intracellular freezing, one would have to argue that the relatively high survival of rapidly cooled cells means either that cooling was not fast enough to cause intracellular freezing, or that rapid warming prevented or reduced the recrystallization of intracellular ice. The latter supposition is supported by a comparison of the effects of slow and rapid warming. We see in Fig. 7 that a reduction in warming velocity was invariably more deleterious to rapidly cooled cells than to slowly cooled cells. This sort of differential effect of slow warming is easily explicable in terms of the recrystallization of intracellular ice, but it is not easily explicable on the basis of other suggested hypotheses of injury (Mazur and Schmidt, 1968). Fortunately it is a hypothesis that can be tested by examining freeze-substituted cells with an electron microscope.

Aside from the question of the mechanistic basis of injury, there are several significant aspects about the relative efficacy of the several additives in protecting cells from injury. First, although all the additives protected rapidly cooled–rapidly warmed cells about equally well, they differed in their ability to protect slowly cooled cells. For cells cooled at 1·6°C/min, the extent of protection decreased in the order 1 M-DMSO > 0·004 M-PVP and 0·5 M-DMSO > 0·35 M and 0·7 M-sucrose > 1·25 M-glycerol > 0·5 M-glycerol. Clearly, the ability to protect slowly cooled cells did not correlate solely with the molar concentration of the additive.

Nor did the ability to protect correlate with the ability to permeate the cells. The ability to permeate was determined by measuring the volume of cells after they had been in contact with hyperosmotic concentrations of the additives for various lengths of time. If a cell were completely impermeable to an additive, it should shrink as a result of the osmotic withdrawal of water, and it should remain shrunken. On the other hand, if a cell were completely permeable to an additive, it should shrink initially from loss of water and then expand to its original volume as the additive permeates. The data in Fig. 8 indicate that cells in sucrose responded in the manner expected for a non-permeating additive and that cells in DMSO responded in the way expected for a rapidly permeating additive. The intermediate volumes of cells in glycerol are more difficult to interpret. They suggest that glycerol was partially able to penetrate the cells, and could mean that glycerol permeates by facilitated diffusion (Mazur *et al.*, 1969). Unfortunately, no conclusions about the permeation of PVP can be drawn from the cell volumes. The lack of cell shrinkage in PVP probably reflects the fact that a 17 per cent solution had too low a molar concentration (0·004 M) to produce appreciable osmotic withdrawal of water. However, the fact that the molecular weight of the PVP was about 40 000 makes it unlikely that appreciable quantities could have moved into the cells.

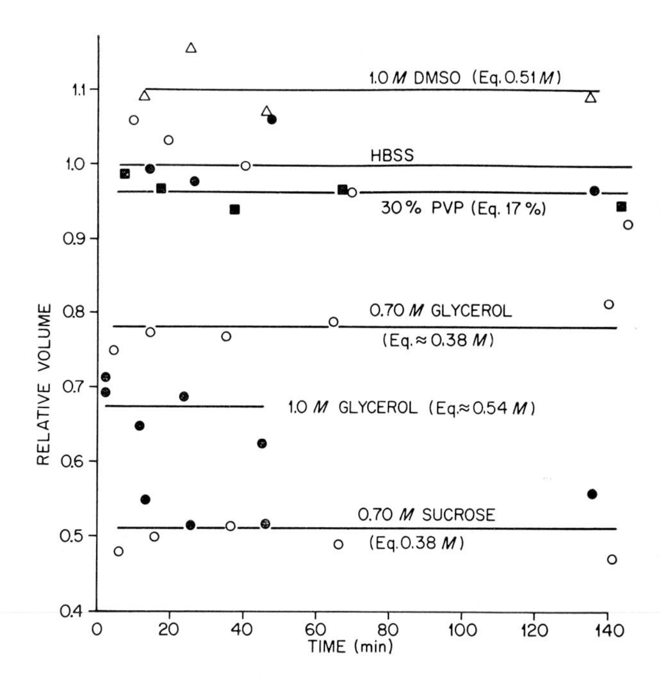

FIG. 8. Relative volumes of Chinese hamster cells at various times after mixing the cells with equal volumes of HBSS containing the indicated initial concentrations of additives. The phrase (Eq ≈) refers to the approximate concentration of additive after it has been mixed with the cells. (○, ●) repeat experiments; (■) data for PVP; (△) data for DMSO.

Before mixing, the volume of the packed cells was 28·4 per cent that of the suspension, and the fractional volume of water in the cells as determined by drying *in vacuo* at 50°C was 0·852. The osmolalities of the HBSS and 0·70 M-sucrose were 0·25 and 0·854, respectively. From these data it can be calculated that after cells in HBSS have been mixed with an equal volume of 0·70 M-sucrose in HBSS, the total osmolality of the suspension becomes 0·652. If the cells are totally impermeable to sucrose, if they behave like perfect osmometers, and if the fraction of the cell pack occupied by trapped medium is constant, the volume of the cells would shrink to 0·48 of that of the cells in HBSS.

In summary, then, sucrose and PVP protected slowly frozen hamster cells better than glycerol and almost as effectively as DMSO, especially when compared on the basis of their molar concentrations. Furthermore, sucrose and PVP both protected the cells in spite of their apparent inability to permeate.

CONCLUSIONS

Although most cryobiologists freeze cells at around 1°C/min, this is not necessarily the cooling rate that yields maximum survival. The survival curves in Fig. 9 show that optimal rates can differ by at least a factor of

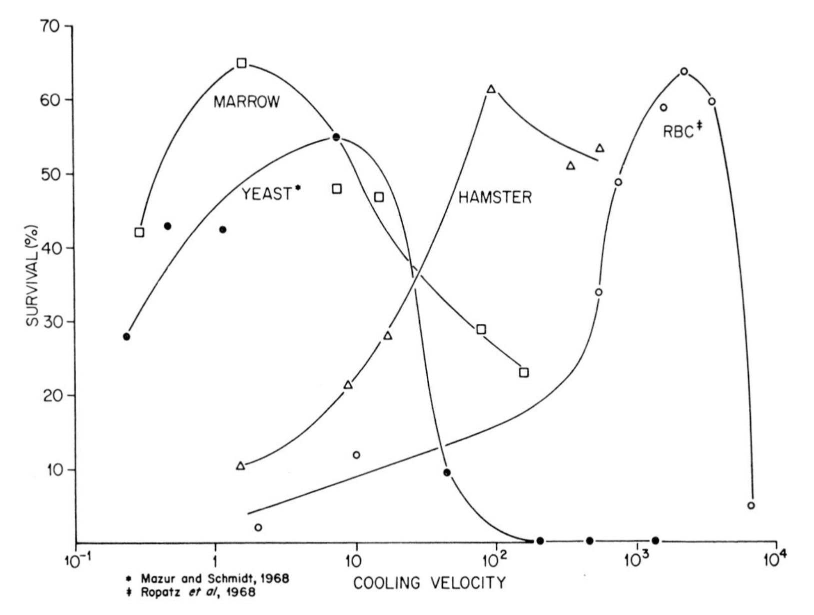

FIG. 9. Comparative effects of cooling velocity on the survival of various cells cooled to −196°C and thawed rapidly. The yeast and human red cells (RBC) were frozen in distilled water and blood, respectively. The marrow and hamster cells were suspended in balanced salt solutions containing 1·25 M-glycerol. The curves for yeast and red cells are redrawn from data of Mazur and Schmidt (1968) and Rapatz, Sullivan and Luyet (1968).

2000 in different cells. Furthermore, even in a given type of cell, the optimum can vary with the conditions of freezing. For example, the curves in Fig. 2 demonstrate a 50-fold shift in the optimum rate for stem cells when the concentration of glycerol is changed threefold, and the curves in Fig. 6 show a sevenfold higher optimal cooling velocity for hamster cells in 1·25 M-glycerol than for cells in 15 per cent PVP.

In spite of the large differences in the cooling rate at which maximum survival occurs, all the survival curves exhibited certain common features. As already stressed, the very existence of an optimum implies that survival is determined by at least two factors oppositely dependent on cooling rate. Furthermore, cells cooled more rapidly than optimal are generally considerably more sensitive to slow warming than are cells cooled more slowly than optimal. The difference in sensitivity is appreciable in stem and hamster cells (Table I; Fig. 7), and it is enormous in yeast (Mazur and Schmidt, 1968). Differential sensitivity has also been observed in the bacterium *Pasteurella tularensis*, in spores of the fungus *Aspergillus flavus* (Mazur, 1966), and in parenchymal cells of mulberry trees (Sakai and Yoshida, 1967).

The evidence that intracellular freezing is responsible for the killing of cells cooled faster than optimal is rather strong, though circumstantial, in red cells and yeast (Mazur, 1968; Mazur and Schmidt, 1968; Rapatz, Sullivan and Luyet, 1968). The evidence for injury from intracellular ice in hamster and stem cells rests at the moment on two facts: (1) The likelihood of intracellular freezing increases with cooling velocity. (2) Cells cooled faster than optimal are much more sensitive to slow warming than are slowly cooled cells, a finding that is consistent with damage from recrystallization of intracellular ice and with observations on yeast and mulberry cells cooled rapidly enough to freeze intracellularly.

The damage observed in marrow and hamster cells cooled slower than optimal can be accounted for in terms of the solution effects described earlier, i.e. it could be caused by exposure to concentrated solutes, by the removal of critical quantities of water, or by osmotic shrinkage. Our present data do not permit a choice among these possibilities. However, our results argue against ascribing damage to the precipitation of sodium chloride below its eutectic point. When stem cells were frozen in the absence of glycerol, all the cells were killed above -7°C, which is about 14°C above the eutectic point of a solution of NaCl (Fig. 4).

One of the more interesting features of our results is the lack of a clear correlation between the ability of additives to protect and either their molar concentration or their ability to permeate the cell. This lack of correlation suggests that protection by additives does not depend solely on their reducing the concentration of electrolyte in partially frozen solutions. The concentration of NaCl at -10°C, for example, will be much higher when the medium contains 0·35 M-sucrose than when it contains 1·25 M-glycerol. In spite of this, 0·35 M-sucrose protected hamster cells as well as did 1·25 M-glycerol.

Furthermore, regardless of its molarity, a non-permeating additive would not be able to prevent the concentration of intracellular electrolytes. The fact that sucrose and PVP were nevertheless able to confer considerable protection suggests that the ability of a cell to survive freezing depends more on maintaining the cell surface than on protecting the cell interior. Perhaps this is because hamster and stem cells normally contain some intracellular compound or compounds in sufficient concentration to protect *intra*cellular membranes and organelles during freezing. Such a possibility has been suggested by Heber's recent finding (1968) that hardy spinach leaves possess a protein which in concentrations as low as 0·05 per cent can prevent freezing injury in isolated chloroplast membranes. In the presence of this protein, frozen and thawed membranes retain their osmotic integrity and retain the ability to carry out photosynthetic phosphorylation. In the absence of the protein, freezing causes the loss of both these properties, and Heber has shown that the damage is to the membranes themselves rather than to membrane-associated enzymes.

There are a number of indications that cell surface membranes are also extremely sensitive to freezing (Mazur, 1966; Levitt, 1966). Therefore, it would not be too surprising to find that additives that maintain cell viability do so chiefly by protecting the surface membrane.

SUMMARY

Current practice assumes that, to obtain high survival, mammalian cells must be suspended in a solution containing a low molecular weight permeating additive and frozen at about 1°C/min. However, recent results with Chinese hamster tissue culture cells and mouse marrow stem cells are not consistent with this assumption. The cells were suspended in balanced salt solutions containing various concentrations of glycerol, dimethylsulphoxide, sucrose, or polyvinylpyrrolidone (mol.wt. ~40 000), frozen to −196°C at rates ranging from 0·3 to 600°C/min, and thawed rapidly or slowly. The optimum cooling rate for rapidly thawed marrow cells ranged from 2 to 100°C/min and that for hamster cells lay between 2 and 350°C/min, depending on the concentration and type of additive. The most deleterious cooling rate for hamster cells in glycerol, polyvinylpyrrolidone, and sucrose was 1·5°C/min. Sucrose protected in spite of the fact that it could not permeate hamster cells. Rapidly cooled cells were more damaged by slow warming than were slowly cooled cells.

These results suggest that two factors underlie injury: (1) Cells cooled more slowly than optimal are injured chiefly by solute concentration or dehydration. (2) Cells cooled more rapidly than optimal are injured chiefly by intracellular freezing and by recrystallization during warming. The fact that extracellular additives can confer protection during slow cooling suggests that the cell surface membrane is especially susceptible to damage from freezing.

REFERENCES

CHU, E. H. Y., BRIMER, P., JACOBSON, K. B., and MERRIAM, E. V. (1969). *Genetics, Princeton*, **62,** 359–377.

FARRANT, J., and WOOLGAR, A. E. (1968). *Cryobiology*, **4,** 248 (abstract).

HEBER, U. (1968). *Cryobiology*, **5,** 188–201.

LEIBO, S. P., FARRANT, J., MAZUR, P., HANNA, M. G., Jr., and SMITH, L. H. (1969). *Cryobiology*, submitted for publication.

LEVITT, J. (1966). In *Cryobiology*, pp. 495–563, ed. Meryman, H. T. London & New York: Academic Press.

LOVELOCK, J. E. (1953). *Biochim. biophys. Acta*, **11,** 28–36.

MAZUR, P. (1963). *J. gen. Physiol.*, **47,** 347–369.

MAZUR, P. (1965). *Fedn Proc. Fedn Am. Socs exp. Biol.*, **24,** suppl. 15, S175–S182.

MAZUR, P. (1966). In *Cryobiology*, pp. 213–315, ed. Meryman, H. T. London & New York: Academic Press.

MAZUR, P. (1968). *Proceedings of the XIth Congress of the International Society of Blood Transfusion*, Sydney, 1966. [Bibliotheca Haematologia, No. 29, pt. 3, pp. 764–777, ed. Hollander, L. P.] Basel & New York: Karger.

MAZUR, P., FARRANT, J., LEIBO, S. P., and CHU, E. H. Y. (1969). *Cryobiology*, **6,** 1–9

MAZUR, P., and SCHMIDT, J. J. (1968). *Cryobiology*, **5,** 1–17.

MERYMAN, H. T. (1966). In *Cryobiology*, pp. 1–114, ed. Meryman, H. T. London & New York: Academic Press.

MERYMAN, H. T. (1968). *Nature, Lond.*, **218,** 333–336.

PERSIDSKY, M., and RICHARDS, V. (1962). *Nature, Lond.*, **196,** 585–586.

RAPATZ, G., and LUYET, B. J. (1965). *Biodynamica*, **9,** 333–350.

RAPATZ, G., NATH, J., and LUYET, B. J. (1963). *Biodynamica*, **9,** 83–94.

RAPATZ, G., SULLIVAN, J. J., and LUYET, B. (1968). *Cryobiology*, **5,** 18–25.

SAKAI, A., and YOSHIDA, S. (1967). *Pl. Physiol., Lancaster*, **42,** 1695–1701.

TILL, J. E., and MCCULLOCH, E. A. (1961). *Radiat. Res.*, **14,** 213–222.

DISCUSSION

Levitt: Perhaps polyvinylpyrrolidone (PVP) penetrates the cell. When PVP solution is supplied to the roots of a plant, the next day the leaves at the top are injured around the margin.

Mazur: In studies with ^{14}C-labelled PVP, Persidsky and Richards (1962) found that PVP protected even when it did not permeate marrow cells. We hope to test this further. Is there any evidence of PVP permeation in red cells?

Rinfret: We never had any evidence of penetration.

Persidsky and Richards showed that some PVP could get into bone marrow cells by pinocytosis (Persidsky and Richards, 1964; Persidsky, Richards and Leef, 1965). Would that explain the plant observation?

Levitt: I doubt it. It is very difficult for anything to get into a plant cell by pinocytosis because the cell has a cellulose wall around the plasma membrane.

Rinfret: But PVP must be permeating that cell wall.

Levitt: Yes, but with pinocytosis there is usually an invagination surrounding the solution before it is absorbed. There is no evidence of that happening to a protoplasm layer right up against the cell wall.

Leibo: Persidsky and Richards (1962) showed that no more than about 0·2 per cent of labelled PVP was associated with the cell. They assumed that this small amount got in by pinocytosis.

Rinfret: That concentration would give no significant protection against freezing to the mammalian cell. The plant cell may have another order of permeability.

Levitt: PVP might be soluble in the lipids of the plasma membrane and get through in this way.

Greaves: Did you use different molecular weights of PVP?

Mazur: No.

Klotz: Although the molecular weight of PVP is given as 40 000, it is a synthetic polymer and there must be wide variations from this weight. Also it is difficult to find out how that number of 40 000 was obtained.

Rinfret: The manufacturers use a K value derived from the relative viscosity of a 1 per cent solution which they relate to an average molecular weight. Our research department has recently devised a method of preparing PVP in which the upper and lower weight limits can be closely controlled, thus providing a very narrow range of molecular weight distribution, unlike anything that is commercially available now (Chiddix and Azorlosa, 1960). Material of this type should lend itself to more exact investigation of the problem under discussion than would the preparations of broader distribution.

Levitt: But it mustn't be too wide a range. Gelatin gel in a PVP solution will shrink to about 10 per cent of its original volume. The water is being removed but the PVP cannot get into the gelatin even though there is no membrane.

Klotz: Polymers in general have a hard time penetrating volumes occupied by other polymers. I have no idea what the distribution in molecular weight is, but in some other synthetic polymers it has been surprisingly broad, far beyond what the manufacturer has indicated.

The molecular weight of PVP is exceedingly difficult to measure. We tried to measure it by osmotic pressure but the polymer is adsorbed on membranes.

Rinfret: Did you use K-30 PVP?

Klotz: Yes.

Levitt: You can get freezing point measurements and calculate the osmotic pressure from them.

Rinfret: A data sheet (L. Light & Co. Ltd, Poyle, Colnbrook, Bucks.) for K-25 PVP, average molecular weight 24 500, shows 5 per cent lying between 32 000 and 38 000, 20 per cent between 13 000 and 20 000 and the remainder between 20 000 and 32 000. Preparations with an average molecular weight of 10 000 show a decline in the cryoprotection afforded to erythrocytes.

Mazur: If our PVP had contained appreciable quantities of low molecular weight material it should have had some osmotic effect, causing the cells to shrink; but it didn't do so.

Rinfret: These very low and very high molecular weights are just the tails of the distribution. Of course they have taken on a clinical significance because of the rate of their clearance from the body. But there is a distribution, lying mostly between 20 000 and 50 000, which is characteristic of the bulk of preparations having an average molecular weight of 40 000. The number K-30 corresponds to this molecular weight average.

Davies: In previous publications, Dr Mazur, you have shown (see Mazur, 1965) that a further increase in survival of yeast cells is obtained when the freezing rate is increased beyond that used in the experiments described today. Have you in fact measured the effect of the various additives on this particular increase in survival with these mammalian cells?

Mazur: No, we haven't.

Leibo: The rise in survival in very rapidly frozen yeast is only apparent when it is plotted on a log scale capable of showing very low survivals. The rise is not visible when survival is plotted on an arithmetic scale as in Fig. 9 (p. 82).

Meryman: Did you use an isotonic concentration of Hank's solution with the known penetrating compounds and with sucrose?

Mazur: Isotonic Hank's solution (0·25 osmolar) was present in every case and the non-electrolytes were dissolved in it to make the complete solution hyperosmotic. Unlike what happens in plant cells, we have observed no examples with mammalian cells or microorganisms in which slow thawing produces less damage than rapid thawing. If the damage were essentially an osmotic effect from an excess of ions within the cell after freezing, very slow thawing ought to be protective.

Levitt: Is it true that even when there is no intracellular freezing the slow thawing has no protective effect?

Mazur: Yes. In fact, slow thawing is usually more damaging.

Farrant: But we never thawed quite as slowly as Dr Meryman (Meryman, 1967).

Meryman: This was an isolated observation in red cells. I have never seen anything like it before and I don't think that it should be thought of as too important.

Levitt: There is a big difference between your method, Dr Meryman, and the method used with plant cells. In plant cells, not only is the thawing slow, but the temperature is raised to only just above the freezing point and maintained there for about 24 hours. If one thaws slowly, but all the way up to room temperature, there is a completely different situation.

REFERENCES

Chiddix, M. E., and Azorlosa, J. L. (1960). In *Encyclopedia of Chemical Technology*, 2nd suppl., pp. 673–674, ed. Kirk, R. E., and Othmer, D. F. New York: Wiley-Interscience.

Mazur, P. (1965). *Fedn Proc. Fedn Am. Socs exp. Biol.*, **24,** suppl. 15, S175–S182.

Meryman, H. T. (1967). In *Cellular Injury and Resistance in Freezing Organisms*, pp. 231–244, ed. Asahina, E. Sapporo: Institute of Low Temperature Science, Hokkaido University.

Persidsky, M., and Richards, V. (1962). *Nature, Lond.*, **196,** 585–586.

Persidsky, M. D., and Richards, V. (1964). *Blood*, **23,** 337.

Persidsky, M. D., Richards, V., and Leef, J. (1965). *Cryobiology*, **2,** 74.

Short communication

DEATH OF FROZEN YEAST IN THE COURSE OF SLOW WARMING*

A. P. MacKenzie

American Foundation for Biological Research, Madison, Wisconsin

The question of injury during rewarming (where cells survive rapid freezing) has a direct bearing on the subject of injury during freezing which we have just considered. I would like (1) to describe the ways in which we observed a 100 per cent survival in yeast by rapid freezing followed by rapid thawing, (2) to describe the results of some experiments based on this observation and to compare them with the results of experiments based on slow freezing, (3) to pursue the comparison in the light of results of differential thermal analyses of rapidly and of slowly frozen samples.

Near-complete recoveries were observed when small numbers of cells, widely dispersed on the surfaces of very thin filters, were subjected to rapid cooling and, afterwards, to rapid warming. More precisely, convenient numbers of washed yeast cells were trapped on presterilized Millipore filter discs (average pore size 0·45 μm) by filtration from suitably diluted suspensions in distilled water.

Yeast-bearing filter discs, clamped at their extreme edges between two thin aluminium rings, were then immersed abruptly in liquid nitrogen maintained either at its freezing point (−210°C) or its boiling point (−196°C). Thawing was effected by rapid transfer to distilled water at about 30°C.

The filters withstood rapid cooling but fragmented upon rapid warming. The thawed cells were in any event detached from the filters or fragments thereof by vigorous agitation, recovered on fresh Millipore filters, and assayed for survival *in situ*.

"Control" experiments were conducted to determine the extent and effects of detachment of the cells from the filter discs.

* These studies were supported by NIH Research Grant No. GM-15143 from the National Institute of General Medical Sciences.

The results are shown in Table I and indicate, in almost every case, a near-100 per cent survival, based on "method controls" (though 10 to 20 per cent losses from "dilution controls" to "method controls" in some cases call for explanation). Actively budding cells and cells in the stationary state survived equally well. Three distinctly different genera of yeast demonstrated similarly high survival*. Further studies will be required, however, to determine whether or not "freezing nitrogen" effects a more nearly complete preservation of *Saccharomyces cerevisiae* than does liquid nitrogen. Yet less rapid cooling rates were not studied.

TABLE I

ULTRA-RAPID FREEZING AND THAWING OF YEASTS

Organism	*Age of culture (hours)*	*Working dilution (counts/ml)*	*"Method control" (counts/filter)*	*Frozen in freezing N_2 (counts/filter)*	*Frozen in liquid N_2 (counts/filter)*
*S. cerevisiae**	24		210	215	
			202	248	
			242	203	
*S. cerevisiae**	24		140	147	
			158	156	
			171	155	
*S. cerevisiae**	72		399	396	
*S. cerevisiae**	24	143	129	132	126
		160	126	129	120
					125
*S. cerevisiae**	72	94	72	80	77
		91	71		90
			82		73
Hansenula anomala† (ATCC 8168)	24		207	211	
			190	206	
Candida utilis† (*Torulopsis utilis*)	24		134	140	
			139	131	

* Courtesy Dr Peter Mazur
† Courtesy University of Wisconsin Bacteriology Department

The cooling rates in liquid nitrogen are probably in the order of 10^5 °C per minute, or possibly higher. Unfortunately, our attempts to measure directly cooling rates on the surface of the filter were not successful. Overlapping metal coatings, vapour-deposited on dry filters, ceased to conduct when the filters were wetted.

Next, we turned our attention to the effects on the very rapidly frozen cells of a slow warming interrupted, at various temperatures, by an ultra-rapid thawing. Thus, eight separate preparations—each consisting of an

* Before this manuscript was completed in final form, 100 $\pm$ 5 per cent survivals were obtained when *Escherichia coli*, *Serratia marcescens*, and *Staphylococcus aureus* were frozen in nitrogen in the same general manner.

aluminium frame, a filter, and about 150 yeast cells—were, one at a time, abruptly immersed in liquid nitrogen and transferred to thin slots in a massive aluminium block precooled to about −100°C. The set-up is shown in Fig. 1. To complete a run, warm air was passed into the copper coil immersed in the bath to produce a steady warming, monitored on

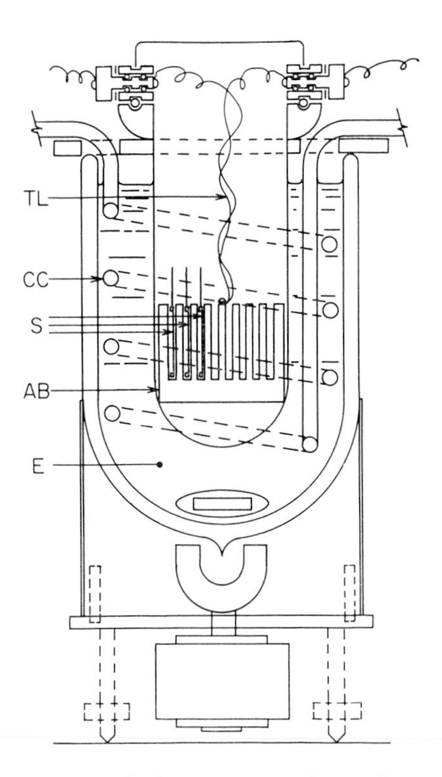

FIG. 1. Apparatus in which *S. cerevisiae*, dispersed on microporous filters, were warmed from about −100°C to temperatures as high as −10°C at rates in the range 1±0·1°C/min. See text for description of initial cooling treatment. AB: aluminium block; CC: copper coil; E: ethanol (anhydrous); S: specimens; TL: thermocouple leads.

the block, of 1·0 ± 0·1°C per minute. Samples were withdrawn at intervals of 10°C and subjected, without delay, to ultra-rapid thawing.

The results of three separate runs are indicated in Table II and, in graphical form, in Fig. 2. A sharp drop in survival, first observed when slow warming was extended from −40 to −30°C, was confirmed, each time in duplicate, in second and third experiments. Evidently the yeast suffers exposure to temperatures as high as −40°C for periods in the order of minutes without adverse effect, but undergoes lethal changes during

TABLE II

ULTRA-RAPID FREEZING, SLOW WARMING (1°C/MIN), AND ULTRA-RAPID THAWING OF *S. cerevisiae*

Experiment	*"Method control" (Counts/filter)*	*Temperature of termination of slow warming (°C)* −196	−80	−70	−60	−50	−40	−30	−20	−10
		Counts/filter								
1	136		170	109	140	130	117	44	7	0
	168									
	139									
2	130	167				171	137	62	4	
	128	158				162	168	69	7	
3	304	344				365	343	48	4	
	342	365				337	350	56	2	

equally brief exposures at temperatures somewhere between −40 and −30°C. The abrupt nature of the onset of damage with increasing temperature is especially evident from Fig. 2.

We then compared these results with those Dr Mazur obtained when he subjected the same yeast to *slow* cooling and slow rewarming interrupted by rapid thawing (Mazur and Schmidt, 1968). His data, expressed and plotted in Fig. 2 in terms of percentages of the number of cells surviving slow cooling (and rapid warming), exhibit a point of inflexion between −55 and −50°C. We observed, in our own data, a corresponding point of inflexion between −35 and −30°C.

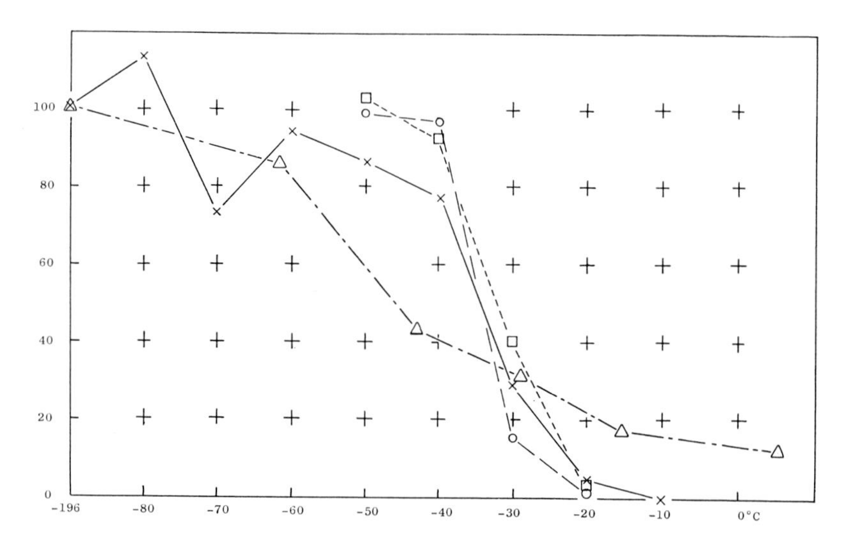

FIG. 2. Survival of *S. cerevisiae*, variously frozen, after slow rewarming, terminated, at various temperatures, by rapid thawing. △: data of Mazur and Schmidt (1968); ×, ○, □: data of MacKenzie and Albrecht.

In an effort to explain these differences, we compared the effects of slow cooling and rapid cooling on yeast by differential thermal analysis. Aliquots of several milligrams of a distilled water/yeast suspension in capillary tubes with an internal diameter of 1 mm, frozen either at 3°C/min or in nitrogen at its freezing point, were placed in one of two small holes in a precooled metal block and rewarmed at about 4°C/min. The temperature difference developed between the sample and the "control" (several milligrams of water in an identical capillary tube placed in the other hole) was obtained as a function of sample temperature with the aid of a high-gain d.c. amplifier and an X–Y recorder. The resulting "thermograms", both readily reproduced, are shown in Figs. 3 and 4.

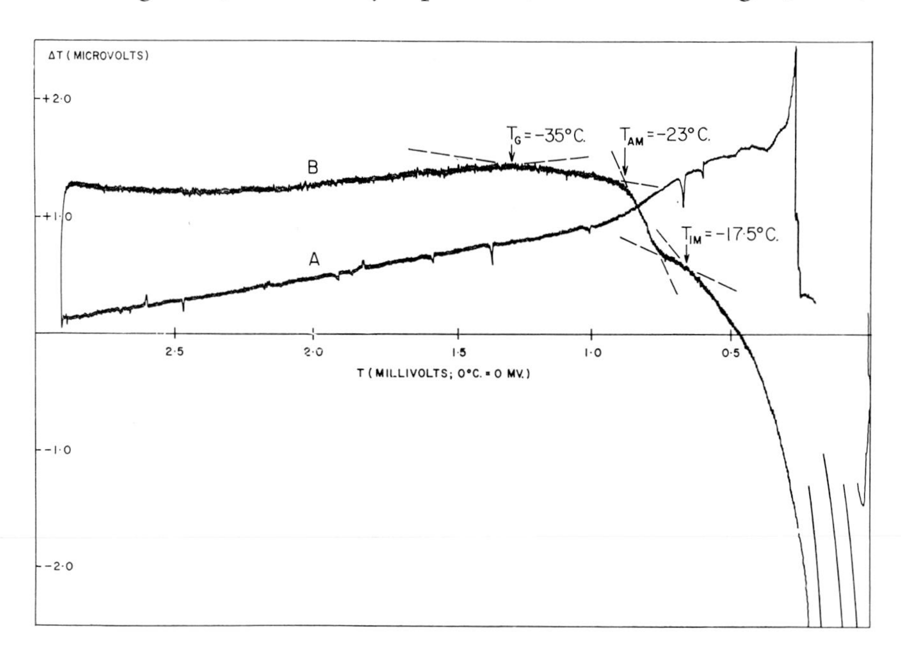

FIG. 3. Thermogram from a differential thermal analysis of *S. cerevisiae* in distilled water frozen by contact with ice at −3°C, cooled thereafter at 3°C/min (curve A), warmed at 4°C/min (curve B).

The cooling at 3°C/min in the one case was sufficiently similar to the 1°C/min employed by Mazur and Schmidt to yield a wholly extracellular freezing; the cooling in "freezing nitrogen", though not as rapid as that achieved on the microporous filters, was, despite the size of the samples, rapid enough to ensure a totally intracellular freezing.

In the analysis duplicating, very nearly, the slow cooling employed by Dr Mazur, slow warming caused the yeast cell suspensions to demonstrate

(1) a gradual exothermic transition at about −50°C, and (2) well-defined endothermic changes at −35, −23, and −17·5°C, the last of these being completed between −0·5 and 0°C. In the analysis in which we tried to simulate the very rapid cooling, the same slow warming caused the yeast suspension cooled by abrupt immersion in "freezing nitrogen" to undergo a first change on warming at −42°C, a second change at about −26°C, a third at −20°C, and a melting very close to 0°C.

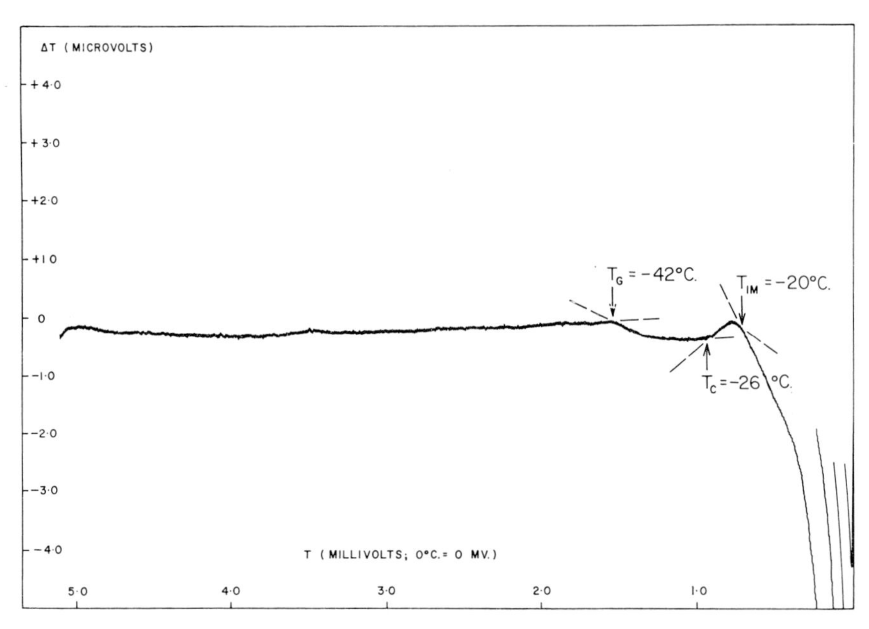

FIG. 4. Thermogram from a differential thermal analysis of *S. cerevisiae* in distilled water frozen by abrupt immersion in liquid nitrogen at its freezing point (−210°C), warmed at 4°C/min.

Comparisons with data gained from the study of various model systems (Rasmussen and MacKenzie, 1968; MacKenzie and Rasmussen, 1969) prompt us to ascribe the first of the transitions observed in Fig. 4 (at −42°C) to a glass transformation. Similarly, we attribute the second transition (at −26°C) to the recrystallization and/or the further crystallization of ice within and also, perhaps, outside the cells. In the same way, we explained the endothermic transitions observed in the slowly cooled materials (Fig. 3) in terms of a glass transformation (at −35°C), an "antemelting" (at −23°C) and an "incipient melting" (at −17·5°C). We cannot yet explain the gradual exothermic transformation observed at about −50°C.

Significantly, the glass transformation in the rapidly and the slowly cooled suspensions are observed to occur about 6°C apart, indicating, as we might have expected, the more extensive dehydration resulting from the less rapid freezing. The size of the peak at −20°C in Fig. 4, however, indicates the increase in the quantity of ice occurring at this temperature during slow warming. Presumably the cells enter the subsequent melting process dehydrated by intracellular freezing to a degree quite similar to that arising in the slowly frozen material in the course of the original extracellular freezing.

What, then, can we say of the processes by which yeast cells, unharmed by cooling, are injured during slow rewarming? Clearly, the rapidly frozen cells die during rewarming some 15 to 20°C before the crystallization/recrystallization processes are initiated. Clearly, too, the slowly frozen yeast dies during rewarming some 30 to 35°C before the cells soften to the point where ice present gains the freedom to recrystallize (though we did not detect any extracellular recrystallization by differential thermal analysis). One is tempted, in view of these observations, to discount the importance of any gross physical change involving ice and to seek the source of injury in a more subtle process.

If, however, we note the near-coincidence of the glass temperature (−42°C) and the lethal temperature (−40°C, rising) during the rewarming of the rapidly cooled material, we are forced to seek the explanation of the death of the slowly cooled material (beginning, it appears, below −50°C) some other way. That is, we are obliged to assume more than one subtle lethal process capable of acting during warming.

In the circumstances we should perhaps seek a single explanation, which must, if it applies to the slowly cooled material some 15°C or more below the glass temperature, be of a chemical nature, acting very much at the molecular level. We have noted the greater dehydration of the much less rapidly frozen specimens. If we now postulate a damaging event in the form of a "rate process", the term being used in its standard physicochemical context, the death of the drier cells at the lower temperature could be explained. We need only suppose the *rate constant* to increase with increasing temperature and the *rate* to increase with decrease in water content (or perhaps with water activity). The locus of the reaction is, of course, another matter.

To say that more rapid freezing results in survival, on warming to a higher temperature, is in any case, in the light of the studies presented, equivalent to attributing to water the role of a cryoprotective agent. What simpler substance could we hope to find!

Acknowledgements

I gratefully acknowledge the assistance of R. M. Albrecht and D. H. Rasmussen.

REFERENCES

MacKenzie, A., and Rasmussen, D. (1969). *Biophys. J.*, **9**, A-193.
Mazur, P., and Schmidt, J. (1968). *Cryobiology*, **5**, 1-17.
Rasmussen, D., and MacKenzie, A. (1968). *Nature, Lond.*, **220**, 1315-1317.

POSSIBLE RELATIONSHIPS BETWEEN THE PHYSICAL PROPERTIES OF SOLUTIONS AND CELL DAMAGE DURING FREEZING

J. FARRANT and A. E. WOOLGAR

Clinical Research Centre and National Institute for Medical Research, Mill Hill, London

DURING the process of freezing and thawing living cells are subjected to severe changes in their environment. In particular, the physical properties of the residual liquid phase surrounding the cells are altered both by the removal of some of the solvent water as ice and by lowering the temperature. Other physical changes may be caused by the addition of various protective additives. With some cells survival after freezing and thawing is maximal at a certain rate of cooling (Gehenio, Rapatz and Luyet, 1963; Mazur, 1963). At rates of freezing faster than that giving optimal recovery, increasing cell damage has been linked with the observed formation of intracellular ice (Mazur, 1960, 1961). It has been suggested that under these conditions the loss of intracellular water induced by the osmotic pressure gradient set up by extracellular freezing is not fast enough to prevent some of the residual cell water from reaching equilibrium by freezing (Mazur, 1963). The exact mechanism by which intracellular ice causes cell damage is not clear.

It has been suggested that at rates of freezing slower than that giving optimal recovery, the mechanism by which red blood cells are damaged is primarily a function of an increase in the concentration of extracellular solutes (Lovelock, 1953*a*). Our earlier paper (Mazur *et al.*, pp. 69–85) examined in some detail the effects of altering the cooling rate on the survival of certain mammalian cells, but the present work is concerned only with damage to cells cooled very slowly or held at constant temperatures below the freezing point. These conditions rule out the complications that could be caused either by thermal shock or by intracellular ice. Most of the experimental work investigating possible mechanisms of freezing damage has been done on red blood cells and the present work is no exception. Data have been obtained on three factors relevant to slow freezing:

(*a*) The effect of protective additives on the concentration of sodium chloride in the liquid phase during freezing.

(*b*) The possibility that changes in the osmotic coefficient of haemoglobin could explain the finding that during exposure to hypertonic conditions red cells shrink to a minimal volume.

(*c*) The effect of different electrolytes on the haemolysis of red blood cells during freezing to temperatures down to −10°C.

EFFECT OF PROTECTIVE ADDITIVES ON CONCENTRATIONS OF SODIUM CHLORIDE DURING FREEZING

Concentrations of sodium chloride during freezing were calculated from the measured freezing points of solutions of known weight in weight compositions.

Methods of Freezing Point Determinations

(*a*) *Differential thermal analysis*

Samples (about 0·3 ml) were placed in a small glass tube (25 × 6 mm) contained in a cylindrical aluminium block (60 mm high × 75 mm diameter). Temperatures were recorded from copper-constantan thermocouples in the sample. The thermocouples were calibrated using a platinum resistance thermometer calibrated at four fixed points. The reference substance (polyvinyl alcohol) was in an identical tube in the block. With the chopper-stabilized d.c. amplifiers used, the potentiometric recorder (Rikadenki) had a maximum full-scale deflection of 50 μV for both the differential channel and the temperature channel. Samples were frozen to about −20°C or until an exothermic response was observed during cooling. During slow rewarming (0·1°C/min) the melting point was taken to be the temperature at which the differential record showed the end of the endothermic melting phenomenon.

(*b*) *Seeding method for freezing points*

Samples (about 5 ml) were cooled in boiling tubes 0·2°C at a time in an insulated alcohol bath. At each temperature attempts were made to seed the sample with a small ice crystal in a Pasteur pipette. The mean temperature at which ice first began to form was taken as the freezing point. Temperatures were recorded by using a copper-constantan thermocouple and a potentiometric recorder.

(*c*) *Osmometer*

Samples (about 1 ml or 3 ml) were supercooled and frozen in a Fiske osmometer (Model G66). The instrument was calibrated by means of standard solutions of sodium chloride in water and the reading of an unknown sample was bracketed by readings of two of the standard solutions.

Solutions

One example of each of four types of protective additive was studied:

(i) permeant non-electrolyte; glycerol (mol.wt. 92),
(ii) impermeant non-electrolyte; sucrose (mol.wt. 392),
(iii) permeant electrolyte; ammonium acetate (mol.wt. 77),
(iv) impermeant polymer; polyvinylpyrrolidone (PVP) (mol.wt. 40 000).

Solutions were prepared both by volume and weight so that molar, molal and mole fraction concentrations could all be obtained.

Glycerol

The well-documented effect of glycerol in reducing the mole fraction of sodium chloride during freezing (Lovelock, 1953*b*) was included for comparison. Starting with a concentration of sodium chloride of 1 per cent w/w before freezing, a mole fraction of 0·01 was reached at −2°C in the absence of glycerol. In the presence of 10 and 20 per cent w/w glycerol, the same mole fraction of sodium chloride was not reached until −7·3°C and −17·1°C respectively. Lovelock showed that haemolysis first takes place during freezing at a temperature at which a mole fraction of sodium chloride of 0·014 is reached over a range in glycerol concentration of 0 to 3 M (Lovelock, 1953*b*).

Sucrose

Sucrose also protects some cells from damage during freezing and thawing (Florio, Stewart and Mugrage, 1943; Strumia, Colwell and Strumia, 1960; Doebbler and Rinfret, 1962; Mazur *et al.*, 1969), although unlike glycerol it does not seem to penetrate the red cell easily (Jacobs, 1950). It has been tacitly assumed that sucrose would also have the capacity to restrict the rise in sodium chloride concentration during freezing. Fig. 1 (upper curve) records the freezing point depressions of different (w/w) concentrations of sucrose in the presence of a single concentration of sodium chloride (1 per cent w/w). The lower curve

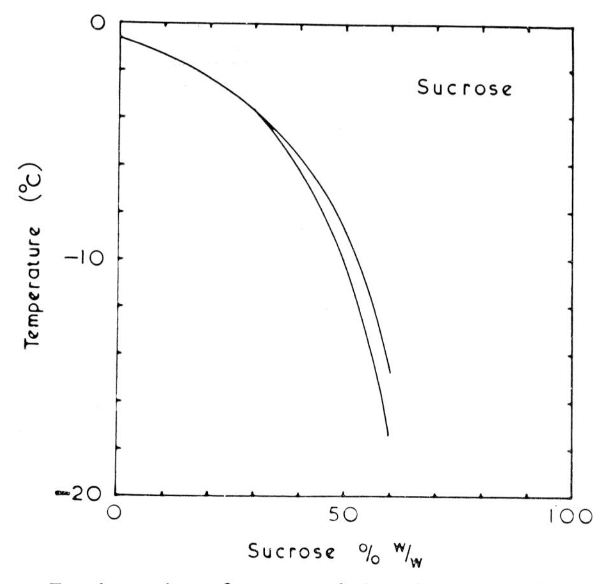

FIG. 1. Freezing points of sucrose solutions (per cent w/w) in the presence of (upper curve) sodium chloride (1 per cent w/w). The lower curve shows the temperature–concentration relationships during the freezing of a solution containing initially sucrose (30 per cent w/w) and sodium chloride (1 per cent w/w).

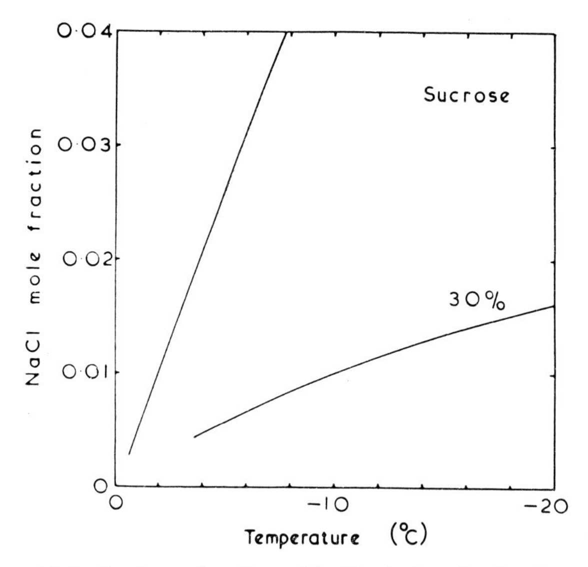

FIG. 2. Mole fractions of sodium chloride during the freezing of a solution without sucrose (upper curve) and in the presence of sucrose (30 per cent w/w). The concentration of sodium chloride initially was 1 per cent w/w.

shows the sucrose concentration–temperature relationships during the freezing of a solution containing sucrose (30 per cent w/w) and sodium chloride (1 per cent w/w) initially. As ice forms the concentrations of sucrose and sodium chloride proceed from 30/1 per cent to 45/1·5 per cent and then to 60/2 per cent respectively. Fig. 2 shows the sodium chloride mole fractions at different temperatures during the freezing of this solution. As with glycerol, the presence of the sucrose lowers the temperature at which a given mole fraction of sodium chloride is reached.

Ammonium acetate

Fig. 3 is a similar plot showing that the weak electrolyte ammonium acetate, reported by Meryman (1968) to protect red blood cells during freezing, also brings about a great reduction in the mole fraction of sodium chloride (initial concentration 1 per cent w/w) at given temperatures during freezing. The initial concentrations of ammonium acetate were 10 and 20 per cent w/w.

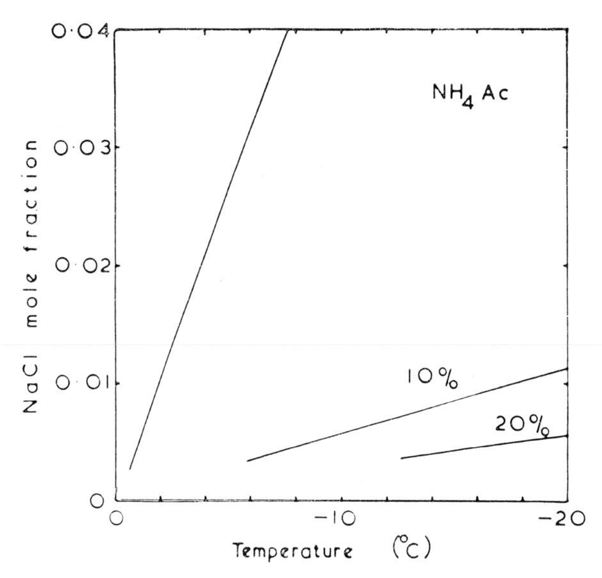

FIG. 3. Mole fractions of sodium chloride during the freezing of a solution without ammonium acetate (upper curve) and in the presence of 10 and 20 per cent w/w ammonium acetate. The concentration of sodium chloride initially was 1 per cent w/w.

Polyvinylpyrrolidone (PVP)

It has been thought that protective polymers like PVP have too high a molecular weight to exert similar effects to glycerol in reducing the temperatures at which given mole fractions of sodium chloride are reached during freezing (Sloviter, 1962; Mazur, 1966; Meryman, 1966; Nash, 1966). However the publication of two component PVP-water phase diagrams (Jellinek and Fok, 1967; Luyet and Rasmussen, 1967) suggested to us that the non-ideal nature of the system at high polymer concentrations could in fact allow sodium chloride concentrations to be modified during freezing. Fig. 4 (upper curve) shows the freezing point depressions

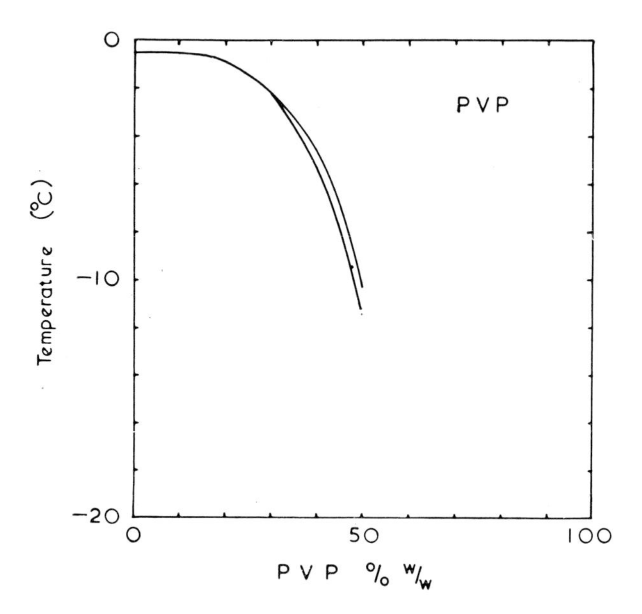

FIG. 4. Freezing points of PVP solutions (per cent w/w) in the presence (upper curve) of sodium chloride (1 per cent w/w). The lower curve shows the temperature–concentration relationships during the freezing of a solution containing initially PVP (30 per cent w/w) and sodium chloride (1 per cent w/w).

observed with varying concentrations of PVP in the presence of sodium chloride (1 per cent w/w). The lower curve shows the PVP concentration–temperature relationship during the freezing of a solution containing PVP (30 per cent w/w) and sodium chloride (1 per cent w/w) initially. Fig. 5 shows the sodium chloride (mole fraction)–temperature relationship during the freezing of this solution, and it can be seen that the presence of the PVP has, like glycerol, markedly lowered the temperature at which a particular mole fraction of sodium chloride is reached. Some of these results have been reported in a preliminary communication (Farrant, 1969).

Thus these four different examples of protective additives have the capacity to reduce the otherwise high concentration levels of a normally non-permeant extracellular solute, sodium chloride, during freezing.

Another important experimental fact is the observed dehydration of red blood cells to a virtual minimal volume at external concentrations of sodium chloride of about 0·7 M (Takei, 1921; Zade-Oppen, 1968*a*; Meryman, 1968). Several possible explanations of this phenomenon have

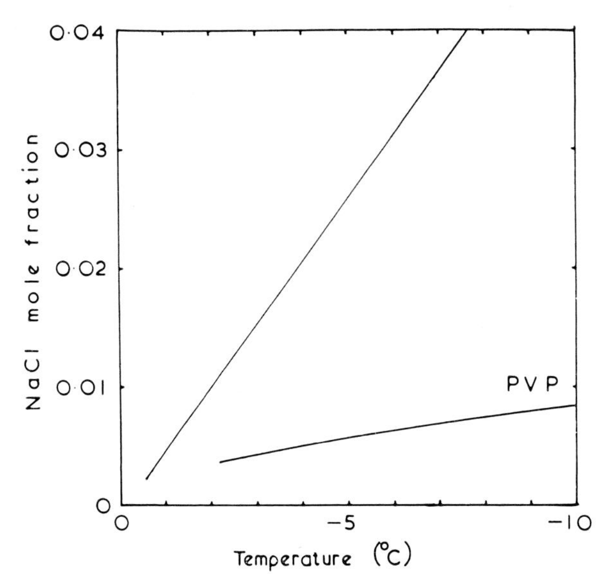

FIG. 5. Mole fractions of sodium chloride during the freezing of a solution without PVP (upper curve) and in the presence of PVP (30 per cent w/w). The concentration of sodium chloride initially was 1 per cent w/w.

been put forward by Ponder (1948) and Dick and Lowenstein (1958), including (*a*) rigidity of the cell membrane resisting changes in volume, (*b*) concentration dependence of the osmotic coefficient of haemoglobin, and (*c*) leakage of ions across the cell membrane. We decided to investigate the osmotic properties of haemoglobin by determining the freezing points of solutions of freeze-dried haemoglobin.

HAEMOGLOBIN

Human blood was collected by venipuncture using a syringe wetted with a solution of heparin (5000 units/ml). The blood was centrifuged for

15 minutes at 3000 *g*, the plasma was removed and the cells were washed three times with a solution of the following composition (mM): NaCl 135, KCl 4·4, $CaCl_2$ 1·2, $MgCl_2$ 0·5, Na_2HPO_4 1·7, NaH_2PO_4 4·2, $NaHCO_3$ 15. The solution was equilibrated with O_2 95 per cent and CO_2 5 per cent and its pH was 7·06 ± 0·04 with a freezing-point depression equivalent to 304 ± 1 m-osmolal. The washed packed cells were haemolysed by mixing with one volume of distilled water and half a volume of chloroform. After centrifugation the supernatant was dialysed in Visking dialysis tubing (diameter 15 mm) against distilled water. Finally the non-diffusible material was freeze-dried.

Weight in weight solutions of freeze-dried haemoglobin in water were prepared in small stoppered bottles, care being taken to minimize and record any gain or loss of water during the weighing and mixing procedure. Before freezing point measurements were made on any batch of haemoglobin, residual sodium and potassium was estimated by flame photometry. The freezing points obtained so far are preliminary since the residual water content of the freeze-dried haemoglobin has not yet been measured. This has however been estimated by Chai and Vogelhut (1967) to be about 5 per cent w/w. With the same batch of haemoglobin, freezing point depressions of −0·035°C (Fiske osmometer) and −0·4°C (differential thermal analysis) were obtained with concentrations of 31·5 per cent w/w and 54 per cent w/w. After allowing for the very small effect of the trace of sodium and potassium still present (0·06 mM-sodium and 0·03 mM-potassium in the 31·5 per cent w/w concentration), the above freezing points suggest osmotic coefficients of haemoglobin of 2·7 and 12 respectively.

HAEMOLYSIS EXPERIMENTS BETWEEN 0°C AND −10°C

Samples of packed cells (0·1 ml) were pipetted (Hamilton Co. Inc.) into round-bottom glass freezing tubes (100 × 12 mm) already containing 0·9 ml of the relevant test solution. The test solutions contained only the electrolyte of interest and not more than 0·012 mM total of NaH_2PO_4 and Na_2HPO_4 in the correct proportions to buffer at the required pH. The solutions were isotonic (303 ± 1 m-osmolal) and their pH was 7·30 ± 0·02. All concentrations were known in w/w terms as well as w/v. After the addition of the cells, the tubes were corked and the suspensions mixed. Solutions containing cells were kept at 0°C before freezing. The controls for complete haemolysis were similar except that cell aliquots were added to 0·9 ml of distilled water. Tubes were suspended on holders and placed in a constant temperature bath. This was a stirred insulated alcohol bath cooled by a submerged coil of tubing through which alcohol was passed from a cooling unit (Planer Ltd) at a temperature about 5°C below the desired temperature. The final temperature was achieved by the use of a 500-watt heater controlled from a resistance thermometer and a thermostatic control unit (Ether Ltd). Temperature fluctuations caused in the sample by the bath did not exceed ±0·1°C. After seven minutes the cell suspensions were seeded by touching the surface with a Pasteur pipette containing in its tip a very small volume of the appropriate frozen solution. After seeding, the tubes were left in the bath for 20 minutes. This time was sufficient to allow the sample temperature to return to the bath temperature for at least ten minutes after the dissipation of latent heat. Thawing was carried out by immersing the tubes in a water bath at +37°C until the last traces of ice had vanished. This took 45 to 50 seconds. After thawing the cell suspensions were transferred to 1·5 ml plastic centrifuge tubes and centrifuged for one minute at 15 000 ***g***. Haemoglobin was measured as cyanmethaemoglobin. Then 0·5 ml of the supernatant was pipetted into 4·5 ml of solution consisting of 2·5 ml of double strength Drabkin's reagent and 2 ml of water. (Single strength Drabkin's reagent contains $NaHCO_3$ 1 g, KCN 50 mg and $K_3Fe[CN]_6$ 200 mg/l.) The solution was well mixed and after at least 30 minutes was read at 540 nm (mμ) against a reagent blank in a Unicam SP 800 spectrophotometer with an added digital voltmeter readout. If necessary further dilutions of the more concentrated samples were made with single strength Drabkin's reagent. Haemolysis was expressed as a percentage of that observed for the cells diluted with water. For each experiment there were nine tubes in the control group and six tubes in each experimental group (for each test solution at each temperature). Also, in each experiment one of the test solutions was always sodium chloride to act as an internal standard.

Fig. 6 shows the percentage haemolysis of red blood cells after freezing at three temperatures in initially isotonic sodium chloride, potassium chloride and lithium chloride solutions. The extent of the haemolysis depends on the temperature but not on the cationic species present. A similar result was obtained with rubidium chloride.

Fig. 7 shows however that different anions do affect the amount of haemolysis. It can be seen that, compared with sodium chloride, more haemolysis was obtained at a given temperature with isotonic sodium iodide and less haemolysis with sodium acetate. All of these salts have eutectic temperatures with water well below −10°C.

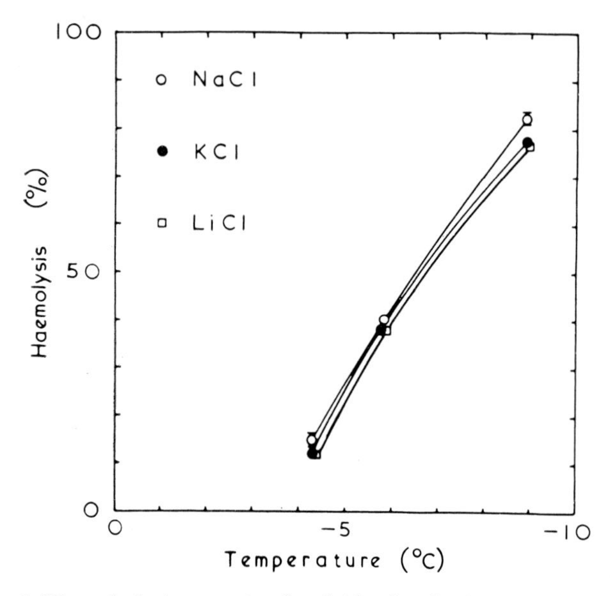

FIG. 6. Haemolysis (per cent) of red blood cells frozen at various temperatures in isotonic solutions of sodium chloride (○—○), potassium chloride (●—●) and lithium chloride (□—□). On this and following figures the bars denote ±1 standard error (bars are not shown if this range lies within the point).

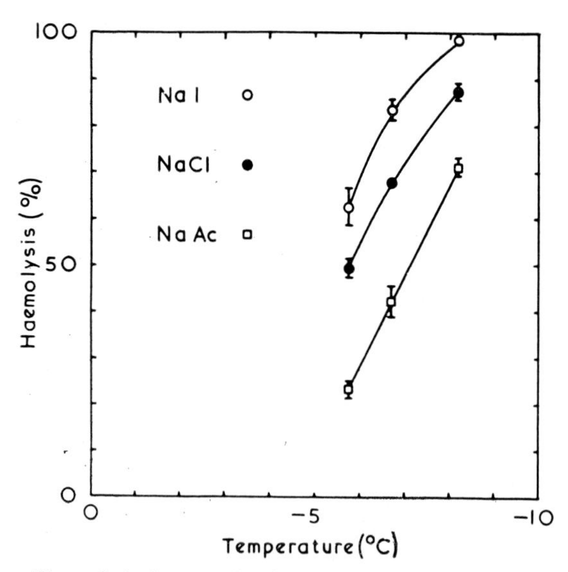

FIG. 7. Haemolysis (per cent) of red blood cells frozen at various temperatures in isotonic solutions of sodium iodide (○—○), sodium chloride (●—●) and sodium acetate (□—□). Bars: see legend to Fig. 6.

Fig. 8 records results with sodium sulphate and potassium sulphate. Both of these salts induce more haemolysis at a given temperature than sodium chloride. This finding is not in agreement with that of Lovelock (1953*a*), who showed that haemolysis in sodium sulphate was similar to that in sodium chloride. However, both of the sulphates have eutectic temperatures above −5°C, so it is possible that the separation of ice induced by seeding is sometimes accompanied by a spontaneous crystallization of potassium sulphate or, for sodium sulphate, of the hydrates $Na_2SO_4 \cdot 10H_2O$ or $Na_2SO_4 \cdot 7H_2O$.

Fig. 9 shows the results from a different type of experiment. Here sodium chloride was present at the same w/w concentration in all experimental groups. In the control group the solution was isotonic initially as before. The other two test solutions contained in addition polyvinylpyrrolidone at concentrations of 15 per cent w/w and 30 per cent w/w neutralized with sodium hydroxide to pH 7·22 and 7·26 respectively. The results show that polyvinylpyrrolidone is able to reduce markedly the extent of haemolysis after freezing.

DISCUSSION

The present results provide further evidence that the haemolysis of erythrocytes during slow freezing is a function of the raised concentrations of normally non-permeant extracellular solutes such as sodium chloride.

The varying proposals for the initial mechanism of damage to red blood cells following the extracellular hypertonic conditions caused by freezing and the redilution caused by thawing are similar to those proposed for haemolysis after hypertonic conditions and redilution in the absence of freezing. For example, Söderström (1944) suggested that after hypertonic conditions followed by a return to isotonicity the cell membranes are still able to provide a relative obstacle to the outward diffusion of those ions that may have penetrated due to an increased permeability in the hypertonic phase. He referred to the effect as paradoxical hypotonic haemolysis. In 1953 Lovelock proposed that during freezing in sodium chloride solutions the cells become permeable to sodium ions when concentrations greater than 0·8 M are reached. Then the cells tend to haemolyse when resuspended in "physiological" saline, possibly as a result of excessive internal osmotic pressure due to their burden of sodium ions (Lovelock, 1953*a*). More recently, Meryman (1968) has proposed that as the concentration of sodium chloride rises beyond 0·8 M during freezing, resistance to further shrinkage prevents the cells from achieving osmotic equilibrium with the suspending medium, resulting in a pressure gradient

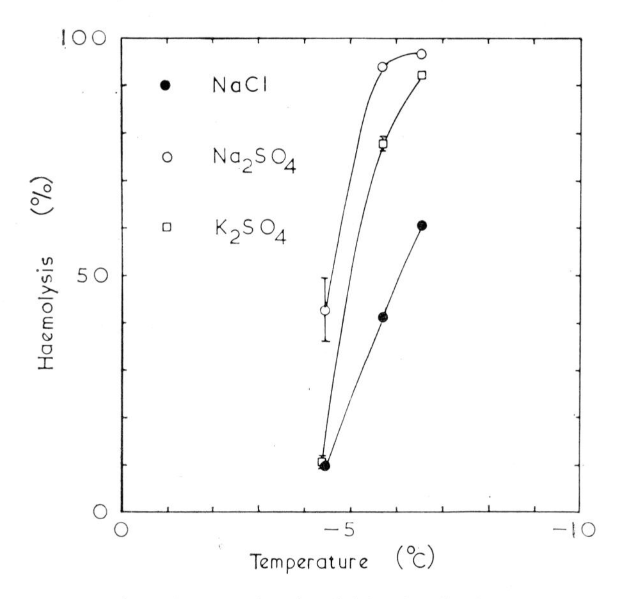

FIG. 8. Haemolysis (per cent) of red blood cells frozen at various temperatures in isotonic solutions of sodium chloride (●—●), sodium sulphate (○—○) and potassium sulphate (□—□). Bars: see legend to Fig. 6.

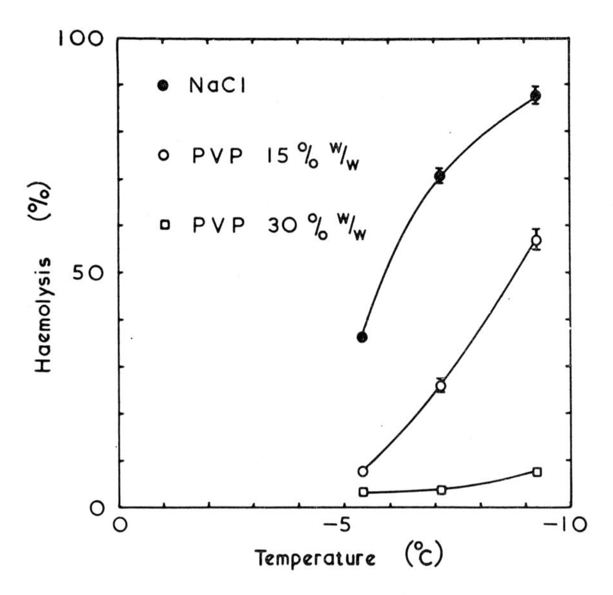

FIG. 9. Haemolysis (per cent) of red blood cells frozen at various temperatures in isotonic sodium chloride (●—●) and in sodium chloride solutions containing PVP, 15 per cent w/w (○—○) and PVP, 30 per cent w/w (□—□). Bars: see legend to Fig. 6.

across the membrane. He suggests that this pressure gradient damages the membrane, leading to an accelerated diffusion of low molecular weight solutes. Then a sudden return to an isotonic medium will result in osmotic haemolysis.

Perhaps the clearest statement on paradoxical hypotonic haemolysis has come from Zade-Oppen (1968*a*) in his consideration of hypertonic conditions without freezing. He writes that "... provided other events, such as cell damage, do not supervene, the (cell) volume will approach a limit asymptotically at high external concentrations. An important consequence of this virtual minimal volume is that there is also a virtual limit to which the internal salt concentration can be raised. At high external salt concentrations there will be thus, after cessation of net water movements, a gradient which can drive salt into the cell. When the cell is returned to an isotonic medium, water is again quickly taken up. If the critical volume is reached in this process then haemolysis occurs."

The present results show that four widely different types of protective additive can reduce the temperature at which a given concentration of sodium chloride is reached during freezing in equilibrium conditions. Thus the protective abilities of these substances with widely different properties (including penetrating and non-penetrating compounds, an electrolyte and non-electrolytes, low molecular weight solutes and a polymer) are all compatible with freezing damage being some function of the raised concentration during freezing of the normally non-permeant solute present (sodium chloride).

All of the proposed theories of freezing damage or post-hypertonic damage without freezing seem to have in common an increased influx of low molecular weight solutes during the hypertonic phase. Whether this increase in the influx of solutes is due merely to the increase in the concentration gradient once the virtual minimal cell volume has been reached, or whether it is also due to a concentration dependence of the permeability constant to the particular solute, is not clear. Zade-Oppen (1968*a*) provides evidence that under these conditions (but in the absence of freezing) the permeability constant for the entry of sodium ions is dependent on the external concentration of sodium chloride.

The suggestion by Meryman (1968) that the minimal volume phenomenon brings about a pressure gradient leading to membrane damage before the increased influx of solutes is worth further examination. The relative minimal cell volume under hypertonic conditions has been estimated to be 0·64 (Takei, 1921), 0·6 (Olmstead, 1960) and 0·53 (Zade-Oppen, 1968*a*). If the normal haemoglobin concentration is taken to be

33 g/100 ml cells and the water content of normal red blood cells to be 71·7 ml/100 ml of cells (Savitz, Sidel and Solomon, 1964), the concentration of haemoglobin in cell water (neglecting other solutes) is 31·5 per cent w/w. At a minimal volume of 0·53 the calculated concentration will have risen to 57·2 per cent w/w. Estimates of the minimal external concentration of sodium chloride leading to the smallest cell volume include 0·72 M (Takei, 1921), 0·64 M (Passow and Eggers, 1950) and 0·65 M (Zade-Oppen, 1968*a*). In addition, Olmstead (1960) showed by extrapolation that the minimal volume is first reached in sodium chloride solutions at a freezing point depression of -2°C.

In the present study, an osmotic coefficient of 2·7 was calculated from the freezing point of 31·5 per cent w/w freeze-dried haemoglobin. This compares with a value of about 2·6 found by Adair (1929) by osmotic pressure measurements for a similar concentration (i.e. that in the normal red blood cell). Agreement was also found between the present results and those of McConaghey and Maizels (1961) for 54 per cent w/w haemoglobin. The osmotic coefficient of 12 from our freezing point data lies in the range of 12–14 estimated indirectly by McConaghey and Maizels using red cells shrunken at 20° and 37°C. Dick and Lowenstein (1958) have considered that changes to the osmotic coefficient could explain the osmotic properties of the red blood cell. More recently Gary-Bobo and Solomon (1968) have suggested that when a shrinkage of the cell causes an increase in the concentration of haemoglobin a reduction in the charge on the haemoglobin molecules takes place, inducing an uptake of Cl^- into the cell together with some water to preserve osmotic equilibrium. Thus the cell tends to shrink less than it would otherwise. The combination of all these findings and the present results suggests that at high concentrations of haemoglobin, very small changes in its concentration would have a profound effect on the activity of the cell water, thus maintaining the cell interior in osmotic equilibrium with the external medium. This suggests that the proposal of Meryman (1968) that an osmotic imbalance is the prime cause of damage to the cell membrane may not be the case.

High concentrations of electrolytes can themselves cause the loss of lipoproteins from cell membranes (Lovelock, 1957) and this may be involved in the increased permeability to salts during freezing. However, post-hypertonic haemolysis is caused when the hypertonic conditions are also induced by normally non-penetrating non-electrolytes, such as sucrose, in the presence of an "isotonic" electrolyte (Valdivieso and Hunter, 1961; Zade-Oppen, 1966).

In the present study the haemolysis measured after freezing at constant temperatures above −10°C in different electrolytes indicates that, at least for the salts investigated, the nature of the anion is more important to cell survival than the cation. This finding is in agreement with the results of Doebbler, Greco and Rinfret (1960). If the initial mechanism of cell damage during freezing and thawing is that of paradoxical hypotonic haemolysis as described by Söderström (1944) and Zade-Oppen (1968*a*), it is possible that permeability differences could explain the different amounts of haemolysis in different salts. With each salt different concentrations will be reached at any temperature during freezing. This is sufficient to explain the difference in haemolysis in sodium chloride and sodium acetate. It is difficult to interpret these results, however, since part of the measured haemolysis could have taken place in the hypertonic conditions during freezing before the redilution on thawing. In fact Söderström (1944) showed that in hypertonic conditions alone, without redilution, sodium iodide began to cause haemolysis at a lower concentration level than did sodium chloride. If red blood cells are suspended in an isotonic solution of a penetrating substance like ammonium acetate or glycerol (without any sodium chloride), the cells haemolyse due to the inflow of water brought about by the osmotic pressure of the intracellular solutes, including the colloid osmotic pressure of the haemoglobin. Similarly, it has been suggested that in hypertonic conditions cells that have become permeable to sodium chloride may also be liable to colloid osmotic haemolysis (Lovelock, 1957; Zade-Oppen, 1968*b*). It is clear that damage is not solely a function of the water activity in the external medium since that is identical at any given temperature, providing the solution is in equilibrium with ice, whatever the nature of the solute.

The demonstration that PVP can protect red blood cells from haemolysis after freezing in a constant temperature bath above −10°C, together with the finding that PVP can also reduce the concentration levels of sodium chloride during freezing, brings this polymer into line with other protective substances of lower molecular weight as possibly acting via this concentration-reducing mechanism. Previously Rapatz and Luyet (1965) have shown some indications that concentrations of PVP up to 15 per cent can reduce the level of haemolysis when red blood cells are frozen above −20°C. Thus it is now not essential to invoke other mechanisms to explain the protective effects of PVP. However, it has been shown by Davies and co-workers (1968) that other macromolecules including dextran and polyethylene glycol can inhibit the release of haemoglobin during hypotonic haemolysis without the prior exposure to hypertonic conditions. The work of Meryman (1966) also suggests that more than one mechanism is involved. He showed that there was less haemolysis of cells frozen in a sodium chloride solution but thawed into a solution containing PVP (20 per cent) in addition, than of cells frozen and thawed in sodium chloride solutions alone.

If paradoxical hypotonic haemolysis and colloid osmotic haemolysis are involved in the haemolysis caused by slow freezing and thawing, it is possible to discuss the implications for the actions of different protective additives.

First, providing the cells remain impermeable to the protective substance, as would be the case with PVP, then colloid osmotic haemolysis during the hypertonic phase could be inhibited.

Secondly, in the presence of non-penetrating substances like sucrose and PVP the cells will still become dehydrated during freezing but the external salt will be less concentrated. This would tend to reduce the uptake of salt and consequently minimize the swelling on thawing. However, if the protective substance itself becomes more permeant at high concentrations then it too may induce haemolysis on thawing.

Thirdly, penetrating substances like glycerol will also modify the external concentration of electrolytes but, because they can penetrate, cellular dehydration will be less at any temperature during freezing. Thus on thawing there will be protection since (*a*) there will have been little entry of salt and (*b*) the glycerol is free to diffuse out of the cells.

In addition, polymers such as dextran and polyethylene glycol reduce the liberation of haemoglobin under hypotonic conditions (Davies *et al.*, 1968). This phenomenon may be related to colloid osmotic haemolysis.

Finally, with many types of cells improved survival is observed as the cooling rate is increasing, providing of course that intracellular ice does not form. It is now clear that one explanation of this phenomenon is that at higher cooling rates there is less time for the normally non-permeant solutes such as sodium chloride to enter the cells during the hypertonic conditions produced by freezing.

SUMMARY

Data have been obtained on several factors relevant to the mechanism of cell damage during slow freezing. It has been shown that different protective additives (glycerol, sucrose, ammonium acetate and poylvinylpyrrolidone) lower the temperature at which a particular mole fraction of sodium chloride is reached during freezing under equilibrium conditions. The freezing points of solutions of freeze-dried haemoglobin have been obtained at concentration levels similar to those expected in the normal and maximally shrunken red blood cell and the results indicate that the increase in the osmotic coefficient of haemoglobin at high concentrations could maintain the cell water in osmotic equilibrium with the external medium even in the shrunken cell. The haemolysis of red blood cells frozen at constant temperatures above $-10°C$ in the absence of protective substances in different isotonic electrolytes is shown to depend more on the nature of the anion than on the cation. Finally, polyvinylpyrrolidone is shown to reduce the haemolysis of red cells frozen in sodium chloride solutions at temperatures above $-10°C$.

Acknowledgement

We would like to thank Miss S. Byford for her excellent technical assistance.

REFERENCES

Adair, G. S. (1929). *Proc. R. Soc. A*, **126,** 16–24.
Chai, S. Y., and Vogelhut, P. O. (1967). *J. appl. Phys.*, **38,** 613–618.
Davies, H. G., Marsden, N. V. B., Ostling, S. G., and Zade-Oppen, A. M. M. (1968). *Acta physiol. scand.*, **74,** 577–593.
Dick, D. A. T., and Lowenstein, L. M. (1958). *Proc. R. Soc. B*, **148,** 241–256.
Doebbler, G. F., Greco, R. E., and Rinfret, A. P. (1960). *Fedn Proc. Fedn Am. Socs exp. Biol.*, **19,** 67.
Doebbler, G. F., and Rinfret, A. P. (1962). *Biochim. biophys. Acta*, **58,** 449–458.
Farrant, J. (1969). *Nature, Lond.*, **222,** 1175–1176.
Florio, L., Stewart, M., and Mugrage, E. R. (1943). *J. Lab. clin. Med.*, **28,** 1486–1490.
Gary-Bobo, C. M., and Solomon, A. K. (1968). *J. gen. Physiol.*, **52,** 825–853.
Gehenio, P. M., Rapatz, G. L., and Luyet, B. J. (1963). *Biodynamica*, **9,** 77–82.
Jacobs, M. H. (1950). *Ann. N.Y. Acad. Sci.*, **50,** 824–834.
Jellinek, H. H. G., and Fok, S. Y. (1967). *Kolloidzeitschrift.*, **220,** 122–133.
Lovelock, J. E. (1953*a*). *Biochim. biophys. Acta*, **10,** 414–426.
Lovelock, J. E. (1953*b*). *Biochim. biophys. Acta*, **11,** 28–36.
Lovelock, J. E. (1957). *Proc. R. Soc. B*, **147,** 427–433.
Luyet, B., and Rasmussen, D. (1967). *Biodynamica*, **10,** 137–147.
McConaghey, P. D., and Maizels, M. (1961). *J. Physiol., Lond.*, **155,** 28–45.
Mazur, P. (1960). *Ann. N.Y. Acad. Sci.*, **85,** 610–629.
Mazur, P. (1961). *J. Bact.*, **82,** 662–672.
Mazur, P. (1963). *J. gen. Physiol.*, **47,** 347–369.
Mazur, P. (1966). In *Cryobiology*, pp. 213–315, ed. Meryman, H. T. London & New York: Academic Press.
Mazur, P., Farrant, J., Leibo, S. P., and Chu, E. H. Y. (1969). *Cryobiology*, **6**, 1–9.
Meryman, H. T. (1966). In *Cryobiology*, pp. 1–114, ed. Meryman, H. T. London & New York: Academic Press.
Meryman, H. T. (1968). *Nature, Lond.*, **218,** 333–336.
Nash, T. (1966). In *Cryobiology*, pp. 179–211, ed. Meryman, H. T. London & New York: Academic Press.
Olmstead, E. G. (1960). *J. gen. Physiol.*, **43,** 707–712.
Passow, H., and Eggers, J. H. (1950). *Pflügers Arch. ges. Physiol.*, **252,** 609–619.

PONDER, E. (1948). *Hemolysis and Related Phenomena.* London: Churchill.
RAPATZ, G., and LUYET, B. (1965). *Biodynamica*, **9,** 333–350.
SAVITZ, D., SIDEL, V. W., and SOLOMON, A. K. (1964). *J. gen. Physiol.*, **48,** 79–94.
SLOVITER, H. A. (1962). *Nature, Lond.*, **193,** 884–885.
SÖDERSTRÖM, N. (1944). *Acta physiol. scand.*, **7,** 56–68.
STRUMIA, M. M., COLWELL, L. S., and STRUMIA, P. V. (1960). *J. Lab. clin. Med.*, **56,** 576–586.
TAKEI, T. (1921). *Biochem. Z.*, **123,** 104–127.
VALDIVIESO, D., and HUNTER, F. R. (1961). *J. appl. Physiol.*, **16,** 665–668.
ZADE-OPPEN, A. M. M. (1966). *Acta physiol. scand.*, **68,** suppl. 277, 224.
ZADE-OPPEN, A. M. M. (1968*a*). *Acta physiol. scand.*, **73,** 341–364.
ZADE-OPPEN, A. M. M. (1968*b*). *Acta physiol. scand.*, **74,** 195–206.

DISCUSSION

Meryman: May I comment first on your reference to Zade-Oppen's report (1968*b*) that sugars postpone the hypotonic haemolysis of cells? This I believe is explained by the findings I described in my paper (pp. 51–64), that is, that the sugars permit the cells to lose potassium rather than to haemolyse during hypotonic stress.

Secondly, you compared freezing in sodium chloride and in two sulphate solutions. Cells in sulphate are smaller than they are in isosmotic solutions of sodium chloride. For every sulphate that enters the cell, two chlorides are lost. This maintains electrical neutrality but represents a net loss of osmotically effective anion. If you were to compare haemolysis of these cells on a volume basis you might find that the responses were about the same. I have never looked at cell volume in sodium acetate. A volume measurement might be significant here. After months of examining cells in a wide variety of isosmotic and hypotonic solutions I feel that unless one also looks at cell volume and at intracellular potassium one never knows the whole story.

Farrant: We haven't looked at either of those things. I agree the sulphate information is the least satisfactory here, because we don't yet know what is happening to the solutions during freezing.

Mazur: You indicated that 15 per cent PVP did not produce any great protection of the red cell.

Farrant: In fact PVP did produce protection, but these cells were in frozen solutions above -10°C.

Mazur: In the experiments I reported earlier (pp. 69–85) where the hamster cells were reasonably well protected by 15 per cent PVP, the osmotic coefficient of the unfrozen PVP was not significantly different from 1·0. Do you think your concept will still account for the observed protection?

Farrant: We took the hamster cells right down to liquid nitrogen temperatures. The red cell experiments show that on freezing at temperatures above $-10°C$ there is some correlation between modifying the salt concentration and the percentage haemolysis. If we froze red cells in 15 per cent PVP to $-196°C$ very slowly they would probably be quite severely damaged, but we haven't yet done that.

Mazur: Why should they be damaged? As the temperature drops further and further the PVP would concentrate, and its osmotic coefficient would rise to a high value.

Farrant: Although 15 per cent PVP reduces the salt concentration during freezing it does not have as marked an effect as does 30 per cent PVP. We are still getting an increased gradient to salt across the cell membrane and perhaps an increased permeability to salts, and PVP in an initial concentration of 15 per cent is clearly not enough.

Rinfret: Do you maintain the same rate of heat removal with 15 per cent PVP all the way down to $-196°C$, or are you considering this in the context of cooling the specimen in a liquid nitrogen bath?

Farrant: The experiments on hamster cells that Dr Mazur talked about were done by putting a lagged vessel in a bath of liquid nitrogen. Of course the cooling curve was exponential and we measured the rate between $-10°C$ and $-65°C$. In contrast the red cells were taken to a particular temperature above $-10°C$ and then seeded. Latent heat dissipation brought the temperature up and then the cells returned to the bath temperature. They were then left at that temperature and thawed rapidly. No cooling rate was involved other than that induced by the latent heat itself.

Rinfret: If you had cooled an equal volume of blood in 15 per cent PVP rapidly, that is by immersion in nitrogen, you would have recovered 96 or 97 per cent of the red cells after rapid thawing. After 50 per cent of the freezable water is frozen out in any given volume of the specimen at perhaps -2 or $-3°C$, the concentration of PVP in the residual unfrozen fluid would have doubled. During the ice formation process the PVP concentration would thus reach high levels.

Farrant: If you start with 1 per cent NaCl and 1 per cent PVP both concentrations will go up during freezing.

Mazur: If a macromolecular additive is non-permeating, and even if it is able to reduce the external salt concentration by virtue of possessing an anomalously high osmotic coefficient, how is it going to prevent the cell from dehydrating to the point where the intracellular electrolyte concentration rises above mole fraction 0·014?

Farrant: The volume of the cell is diminishing by dehydration, so once it gets to the minimum volume the concentration of electrolytes inside cannot go up any more by the loss of intracellular water.

Mazur: Isn't the minimum volume sufficient to bring the concentration up to this critical level? What is the minimum volume?

Farrant: There are various figures, but the minimum volume is about 53 per cent of the original volume (Zade-Oppen, 1968*a*).

Mazur: But much of the cell volume is taken up by haemoglobin so the water volume is actually going to decrease a great deal more than that.

Farrant: The concentration of haemoglobin goes up from about 31 to about 57 per cent (weight in cell water).

Meryman: In fact, you don't know how far the cell can shrink because if you shrink it beyond a certain point it starts to leak solute instead.

Farrant: But why does it get leaky? That is the point.

Meryman: To say that the haemoglobin is incompressible beyond a certain point because you can't shrink the cell beyond a point is to beg the question, because the reason you can't shrink it is because something happens. If the membrane didn't get leaky maybe the cell could shrink more.

Farrant: I am not saying it is incompressible. I am just saying that the concentration of electrolytes inside the cell goes up at first because the cell shrinks, but once the minimal volume is reached the concentration can't go up any more by that mechanism, so it must go up by extracellular salts coming in.

Mazur: The critical thing we need to know is the molar fraction of potassium chloride in the red cell when it is at its minimal volume before this leak occurs. Is it above 0·014?

Meryman: I have calculated that a normal cell in 1200 m-osmolal salt has a volume about 62 per cent of normal. Assuming 65 per cent of the original cell to have been water, this is a reduction in water content to about 40 per cent of the original, meaning that the molar concentration of cell solutes has been increased by a factor of 2·5. Since KCl is initially present at about 100 m-equiv./l, this means that the concentration at 1200 m-osmolal will be about 250 mM for a mole fraction of 0·005.

Farrant: The minimum volume seems to be reached at around the same temperature at which haemolysis begins.

In the presence both of salt and of salt plus PVP the driving force for dehydration is going to be the activity of water, in both situations. For example, before the red blood cell becomes permeable to low molecular weight solutes the dehydrating force is going to be the same with PVP present as in its absence, and yet we get more damage in one situation than in the other (Fig. 9, p. 108).

Meryman: Have you seen the sudden change in the osmotic coefficient of haemoglobin that I showed in my paper (pp. 51–64)? McConaghey and Maizels (1961) only went to about 600 m-osmolal but I found that the osmotic coefficient suddenly fell again at about 1000 m-osmolal.

Farrant: As I showed, when we did direct freezing point measurements on haemoglobin solutions our data lay close to McConaghey and Maizels' transposed data. When we went up to beyond the calculated concentration of haemoglobin in a maximally shrunken cell (about 57 per cent w/w cell water) the freezing point depression continued to increase.

Meryman: That means that there is some discrepancy between haemoglobin in the cell and out of the cell.

Farrant: McConaghey and Maizels' data were obtained in the cell.

Meryman: But they only went to about 800 m-osmolal and at 1000 or 1200 m-osmolal cell volume decreases suddenly.

Farrant: But our freezing point determinations are made under virtually salt-free conditions. I agree it is only about $-0\cdot5$°C at the lowest freezing point. We haven't yet done it with a salt present as well, but McConaghey and Maizels (1961) got the same information on the osmotic coefficients of haemoglobin when they dehydrated their cells with solutions of lactose in high concentration which caused low-salt cell interiors as when they dehydrated with hypertonic salt solutions which produced salt-rich cell interiors.

Elford: Were they sure that extracellular solute (lactose) wasn't penetrating the cell membrane?

Farrant: They made the assumption that it was not.

Elford: There is another way to achieve osmotic equilibrium between the inside of the red blood cell and extracellular space. In 1946 Perutz showed that a certain amount of water associated with a haemoglobin crystal ($0\cdot3$ g H_2O/g protein) was not able to dissolve mobile ions; there was a partition of ions between a bathing solution in which a crystal was suspended and some of the water in the hydrated crystal. In the normal red blood cell, according to Perutz in a later paper (1948), the concentration of haemoglobin is about as much as can exist in solution within the cells as a close-packed lattice of freely rotating molecules. It seems to me that, whereas all the water in the red blood cell in the isotonic state is available to dissolve electrolytes, as the cell is dehydrated more water within the cell may become unavailable to dissolve electrolytes if the lattice of haemoglobin takes on the form of a hydrated crystal. In the haemoglobin crystal there is 0·3 g of non-solvent water/g protein and as the concentration of haemoglobin in the maximally shrunken cell is the same as that in the hydrated crystal it seems possible that the same amount of water in a dehydrated cell is also unable to dissolve ions; this will have a more marked effect on the osmotic properties of the red cell than the non-ideal behaviour of haemoglobin in high concentration but it should still be possible to achieve osmotic equilibrium between the inside of a maximally shrunken cell and its environment.

Farrant: There is strong evidence that labelled non-electrolytes can penetrate all of the cell water (Gary-Bobo and Solomon, 1968).

Elford: Yes, but it is mainly the intracellular electrolytes that are achieving osmotic equilibrium in increasing hypertonic conditions.

Mazur: Isn't this value of 0·3 g bound water/g protein the usual value for the hydration of the proteins?

Meryman: In the cell at its minimal volume about 60 to 65 per cent of the water has been removed, so the concentration is approaching a point where, on the basis of that figure, there is almost nothing but osmotically unavailable water left.

Mazur: Then that would explain why the cell stops shrinking; i.e. it is in fact in osmotic equilibrium.

Meryman: Our measurements of cell volume show a plateau at about 1000 m-osmolal. With further concentration the volume suddenly decreases about 10 per cent. This occurs both in saline and in plasma. Since there is no exchange of cation across the membrane, the only way we can account for this volume change is by a change in osmotic coefficient inside the cell. The only explanation I have been able to think of is that, initially, the osmotic coefficient is going up as the haemoglobin is concentrated and therefore one gets more and more of a departure from ideal behaviour. Then suddenly the haemoglobin is precipitated or crystallizes, the bound water is rejected and cell volume is reduced.

Farrant: We don't see that in concentrated solutions of haemoglobin.

Meryman: Then it must be something associated with the intact cell.

Davies: I can confirm the effect described by Dr Meryman as we also find a similar plateau at about 2·5 times isotonic followed by a sudden drop in cell volume at about three times isotonic.

Mazur: Would haemoglobin be a good protective agent?

Farrant: If you could get enough of it perhaps it would be.

MacKenzie: Weinstein and Merk (1967) showed by freeze-cleavage replication that haemoglobin crystallized in sheep erythrocytes frozen in a medium containing glycerol. But it was doubtful, in that instance, that any significant quantities of glycerol were present within the cells. Are any other demonstrations of intracellular crystallization of haemoglobin available?

Meryman: We looked for it in freeze-etched cells and so far have been unable to demonstrate it.

Farrant: Haemoglobin can be crystallized but only with difficulty (Perutz, 1946).

REFERENCES

Gary-Bobo, C. M., and Solomon, A. K. (1968). *J. gen. Physiol.*, **52,** 825–853.
McConaghey, P. D., and Maizels, M. (1961). *J. Physiol., Lond.*, **155,** 28–45.
Perutz, M. F. (1946). *Trans. Faraday Soc.*, **42B,** 187.
Perutz, M. F. (1948). *Nature, Lond.*, **161,** 204.
Weinstein, R. S., and Merk, F. B. (1967). *Proc. Soc. exp. Biol. Med.*, **125,** 38–40.
Zade-Oppen, A. M. M. (1968*a*). *Acta physiol. scand.*, **73,** 341–364.
Zade-Oppen, A. M. M. (1968*b*). *Acta physiol. scand.*, **74,** 195–206.

GENERAL DISCUSSION

Huggins: Multiple causes of freezing injury and differences in vulnerability between one cell type and another are a problem when we contemplate preservation of whole organs. All cells must function if a transplanted kidney, heart, or liver is to benefit a sick patient. It seems to me that the only possible approach to organ preservation at this time is to freeze without freezing. That is—some ice must solidify the tissue to prevent uncontrolled mixing of intra- and extracellular solute. However, the extent of ice formation must not be so great that solute concentrates to toxic levels. The concept of degrees of freezing raises an important question for this symposium: "How do we define freezing?"

Meryman: You have answered your own question. One certainly has to maintain the solute concentration at a point where it is not injurious. Any method which involves special rates of freezing and thawing is going to be entirely impracticable for organs, since ultimately we are talking about organs of such sizes that cooling or warming rates in excess of a few degrees per minute will be impossible. Therefore we have to talk about essentially equilibrium conditions, in which case we are talking about the effects of solute concentration. I would generalize this by stating that the limit for non-penetrating solute concentration should be set at about 1000 m-osmolal. I think most tissues will tolerate this degree of concentration.

Huggins: I agree. One possible approach is to analyse the problems of freezing; the other—and it seems to me the more productive—is to avoid freezing. In our experience most cells, including perfused canine kidney cells, can tolerate osmotic gradients of 1100 to 1500 m-osmolal/kg (Huggins, 1965; Huggins and Miura, 1968). Recently, however, we have found that human granulocytes when suspended for 30 minutes in 0·4 M-NaCl lose their ability to phagocytize *Staphylococcus albus*. Human granulocytes appear to be much more sensitive to concentrated salt than the red blood cell.

Meryman: If you provide the granulocyte with a little bit of potassium and the other things that it might need to maintain a membrane potential, would you find the same thing? Are you talking about a straight sodium chloride suspension?

Huggins: The granulocytes were suspended in hypertonic sodium chloride solutions. Other conditions might have been better tolerated by the granulocytes. However, the observation that one type of cell can be destroyed at salt concentrations that do not bother another type leads me to conclude that we must avoid concentrating solute during freezing. This in turn implies that we may have to avoid freezing. Once again: "what is freezing?"

Meryman: Some of these cells can lose intracellular cations at a tremendous rate when the Donnan relationship is not maintained.

Mazur: Not all cells are killed when the electrolyte concentration reaches a value of 0·014 mole fraction. Yeast cells can be dried to 8 per cent water content without damage. Furthermore, they will survive immersion in 2 M-NaCl for an hour.

Leibo: During the experiments on the freezing of marrow cells Dr Mazur reported, we assayed the population of marrow cells in two ways. One was for the stem cell activity which, as Dr Mazur mentioned, represents only one out of about 50 000 cells in the total population. At the same time we ran a dye exclusion test to get some measure of the viability or survival of the major portion of the population. With some additives the optimum cooling rate was very different for the marrow population as a whole and for the stem cell population. On the other hand, in 0·35 M-sucrose the optimum rate was almost the same for both the marrow population as a whole and for the stem cell population. This agrees with what Dr Huggins said about whole organ preservation; that is, the preservation of mixed cell populations is liable to be a very difficult problem. Only when we know the interaction of cooling rate, warming rate and additive concentration for more cells will we be able to make any sort of prediction for the preservation of organs or other mixed cell populations.

Elford: Dydyńska and Wilkie in 1963 found that striated muscle fibres from the frog sartorius were damaged when placed in hypertonic sucrose-Ringer solutions having osmolar concentrations greater than about four times that of isotonic Ringer. They found that with up to four times the osmolar concentration of Ringer there was no irreversible damage to a muscle, which resumed a normal contractile response on reimmersion in isotonic Ringer. However, above this concentration there was irreversible damage and sucrose penetrated into the fibres. Dydyńska (1961) has also found that when 75 per cent or more of the water in sartorius muscle was removed, using silica gel, irreversible damage occurred and nucleotides and phosphate started leaking out of the cells. There seems to be some correlation between damage to a cell and its minimal volume. A high internal electrolyte concentration need not necessarily be causing the damage, because in hypertonic sucrose-Ringer solutions intracellular electrolytes tend to leak out of the cells.

Mazur: We seem to agree that when a cell freezes, the solutes are concentrated so as to reduce the chemical potential of intracellular water to that of extracellular ice. However, there are actually two ways to reduce the chemical potential of water: either by increasing solute concentration or by applying hydrostatic pressure (in this particular case, a negative pressure, or tension). I believe that what Dr Meryman refers to as an osmotic pressure gradient across a cell membrane is in fact a difference in hydrostatic pressure (or tension).

Elford: Aren't they the same?

Mazur: Yes and no. I am talking here about osmotic pressure simply in terms of water activity (i.e. as defined by the equation $\pi = \frac{-RT}{V_o} \ln a_w$, where π = osmotic pressure, R = gas constant, T = absolute temperature, V_o = partial molar volume of water, and a_w = activity of water).

In the plant cell, the rigid cell wall permits a hydrostatic pressure difference to be maintained. The red cell membrane, I believe, has practically no elasticity and no rigidity.

Meryman: Hydrostatic pressure *per se* is going to do nothing. We have taken red cells up to 304×10^6 Newtons per square metre (3000 atmospheres) without any ill-effects whatsoever, as long as the pressure is uniformly distributed. There must be a gradient across the membrane to create injury.

Mazur: If a yeast or plant cell with $2 \cdot 03 \times 10^6$ N m^{-2} (20 atmospheres) osmotic pressure inside is suspended in distilled water, how can the chemical potential of the water outside be the same as the chemical potential of the water inside?

Meryman: It seems to me that there are two ways. One is to have an extraordinarily strong cell wall simply to withstand the pressure; the other is to have such an impenetrable cell wall that the activity inside doesn't know what the activity outside is.

Mazur: But we know that the cell wall is permeable to water. An unfrozen yeast cell in distilled water remains perfectly viable, just like the plant cell. I think it is the counter-hydrostatic pressure exerted by the cell wall that is responsible for raising the chemical potential of the internal water to make it equal to that of the external water.

Levitt: Isn't that what Dr Meryman means when he talks about a hydrostatic pressure gradient?

Mazur: Yes, but I am saying that a semi-rigid or rigid cell wall is needed to establish this hydrostatic pressure gradient. It seems to me more reasonable that the red cell ceases to shrink not because there is a hydrostatic pressure gradient but because the chemical potential has been made equal inside by a mechanism such as Mr Woolgar and Dr Farrant suggested; namely, that the haemoglobin is acting non-ideally in some way so as to reduce the chemical potential of the intracellular water to the equilibrium value.

Meryman: Then there is no reason for the membrane to disintegrate unless we then postulate some biochemical effect. This is really our dilemma: we have to decide whether dehydration or the volume reduction of the cell is responsible for the injury. The only direct experiments we have done in this regard are those on potassium-depleted cells which underwent haemolysis at an identical volume but at lower osmolality.

Mazur: Dr Farrant says that perhaps those results can be accounted for in terms of the anomalous osmotic coefficient of haemoglobin.

Farrant: The cell with a lower potassium content had a lower volume to start with. So the concentration of haemoglobin will be higher than normal, and therefore the haemoglobin within the cell will have a higher osmotic coefficient than in the situation with normal potassium levels.

Meryman: Yes. The cells succumbed at the same volume and presumably at the same haemoglobin concentration regardless of the extracellular osmolality, but the intracellular potassium concentration was less.

Levitt: So you are left with only the biochemical change.

Meryman: Not necessarily.

MacKenzie: Are you supposing that the pressure which acts primarily at about the same time as damage sets in, acts from the inside out, or from the outside in?

Meryman: My analogy would be that of a plastic bag filled with water, with little pieces of sponge rubber floating around in it. As the water is extracted the bag gets smaller and flaccid, until finally only the water-soaked sponge rubber is left. At this point if more water is removed there starts to be a resistance to continued shrinkage of the bag. This means that there can be a pressure gradient across the surface of the bag. If the interior of the cell is not absolutely molecularly uniform these pressure gradients across the membrane could lead to the breakdown of some metastable structures in the cell. This, of course, is sheer speculation. Lovelock has reported the appearance of cholesterol and phospholipids in the suspending solution when this breakdown happens. Some of the models of membranes show cholesterol as penetrating the membrane and being involved in the ion transport system. Conceivably these structures are just simply blown out. I find it intuitively acceptable, but there is absolutely no experimental or theoretical evidence that such a thing can happen.

MacKenzie: But is the excess pressure on the inside?

Meryman: No, the pressure is on the outside. The cell can no longer shrink. The extracellular concentration continues to rise, and if the cell cannot shrink the intracellular concentration remains the same. So now there is a higher osmotic pressure on the outside than on the inside.

Brandts: Can you talk about osmotic pressure here in the same way as you are talking about hydrostatic pressure, Dr Meryman?

Klotz: Osmotic pressure is not a pressure.

Meryman: The fact remains that a semipermeable membrane with solutions of different concentration on both sides has a pressure gradient across it.

Klotz: No; only at equilibrium would there be a pressure gradient across it.

Meryman: But if the volume of the solutions on the two sides is not allowed to change, there must be a pressure gradient.

Mazur: Either the membrane itself somehow resists shrinkage, as the cell wall of the plant does, or the haemoglobin has to resist shrinkage. You have to have something to resist shrinkage in order to establish a pressure gradient.

Meryman: That is right. I have no evidence for this. I am just proposing the idea.

Elford: (*Note added in proof*) Many authors have discussed the relationship between hydrostatic pressure and osmotic pressure (for example see Dainty, 1965). The chemical potential of water, μ, in a solution, e.g. the medium surrounding red cells, can be written:

$$\mu_1 = \mu_1^\circ + \bar{V}_1 P + RT \ln a_1$$

where μ_1° = the chemical potential of water in its standard state, $\bar{V}_1$ = partial molar volume of water (considered to be independent of pressure), R and T = gas constant and absolute temperature, P = hydrostatic pressure in excess of 1 atmosphere ($101 \cdot 3 \times 10^3$ N m^{-2}), and a_1 = the activity of the solvent (water). The corresponding relationship for conditions the other side of a semipermeable membrane, e.g. inside the red cell, may be written as:

$$\mu_1' = \mu_1^\circ + \bar{V}_1 P' + RT \ln a_1'$$

For conditions of zero solvent flow such as appear to hold experimentally when the virtual minimum volume of the red cell is reached during exposure to hypertonic conditions in the presence or absence of freezing, then the chemical potential of water is the same on either side of the membrane, i.e.:

$$\mu_1 = \mu_1'$$

If the external concentration of solutes (electrolytes) is increased after the minimum volume is reached then a_1 will decrease, leading to a reduction in μ_1, as P_1 is constant. μ_1' must also decrease and this can be brought about by (*a*) an entry of solutes into the cell leading to a reduction in a_1', or (*b*) a very small increase in the concentration of intracellular haemoglobin due to an immeasurable change in cell water leading to a progressive reduction in a_1', as suggested by Dr Farrant and Mr Woolgar (pp. 97–114), or (*c*) a reduction in the hydrostatic pressure term P' producing a hydrostatic pressure gradient across the cell membrane as suggested by Dr Meryman. This is the osmotic pressure gradient across the membrane.

Combinations of the alternatives are also possible.

Meryman: Hypotonic haemolysis of the cell is certainly an example of the development of a pressure inside the cell which swells it up and bursts it, due to the difference in concentration between inside and outside.

Klotz: The term "osmotic pressure" is causing trouble because it is being used in two different ways. The activity of water is lowered when one puts a solute in the water. One can increase the activity of the water by imposing a real physical hydrostatic pressure to compensate for the lowering of the activity by the added solute. We use the words "osmotic pressure" unfortunately for either this compensating pressure, or as an alternative measure of the solute concentration even in the absence of the compensating hydrostatic pressure.

Did the use of PVP as a protective agent occur to people because it is a highly water-soluble polymer? Have dextran and other water-soluble starches been tried too, and are they equally effective?

Levitt: PVP has a much higher solubility, and therefore produces a much higher osmotic value than other polymers. But dextran is not very cheap anyway.

Klotz: How about polyvinyl alcohol?

Levitt: That doesn't have anywhere near the solubility of the PVP.

Rinfret: There are differences in cryoprotection between polymers. The only one that appears to give as much protection to the red cell as PVP is hydroxyethyl starch.

Klotz: There are a number of different water-soluble starches.

Rinfret: We would be very anxious for you to study them in this context of cryoprotection.

Levitt: They can easily be broken down, enzymically.

Klotz: I would like to ask another question, as an outsider to the field. I was particularly impressed by the discussions of rates of warming and subsequent thawing, and rates of cooling. In regard to explaining damage to the cell, all the attention seems to be paid to the rates of cooling. Is it possible that the warming is doing the damage, and that in both warming and cooling one really has a very asymmetric temperature gradient because heat is applied only from the outside? Is there a way of warming in which one wouldn't have a steep asymmetric gradient, for example, with high-frequency irradiation? Does it make any difference whether you warm by, let's say, high-frequency irradiation, instead of by an external heat source?

Meryman: The big problem there is that the absorption of energy by ice is negligible and absorption by the liquid portion is very high, so there is a positive feedback problem. The minute any ice begins to thaw its energy absorption skyrockets and the specimen may explode.

Rinfret: This was pretty thoroughly studied in the frozen food industry. For example, steam explosions have been observed when strawberries were thawed by radiation.

Mazur: Another point is that although there can be a large thermal gradient in the sample as a whole, it is probably small in a cell. Crude calculations (Mazur, 1963) indicate that the difference in temperature between the edge and centre of a 50-μm sphere of water cooled at 10^4 °C/min will only be 0·08°C.

Rinfret: In the surgical situation the distances are over 2 or 3 mm.

Mazur: The thermal gradients are very large in such a situation.

Rinfret: In terms of a given cell the gradient can be of the order of hundredths of a degree.

Mazur: Lovelock in his original paper (1953) studied the relation between haemolysis and the time taken for thawing. When he froze red cells slowly without a protective additive he found much more haemolysis after slow thawing than after rapid thawing. I would like to know how Dr Farrant and Dr Meryman would fit that in to their hypotheses.

Farrant: Presumably there was a longer exposure to hypertonic conditions, even during thawing. Have you got any figures?

Mazur: When Lovelock cooled cells slowly (9 seconds between −3°C and −40°C) he found that slow warming gave 93 per cent haemolysis, while rapid warming only gave 35 per cent haemolysis.

Meryman: That is not a very slow cooling rate.

Mazur: No. Is there any other information on the relation between the rates of cooling and thawing and the amount of haemolysis?

Meryman: There is quite a close relationship between the proportion of cells which succumb to hypertonic stress and the time available. If we dump red cells into strongly hypertonic solution and get them right out again, quite a large proportion of them are still intact. The rate at which they succumb appears to be proportional both to the magnitude of stress and the time permitted. So I would suggest that there can be insufficient time for all cells to succumb to hypertonic stress during cooling, but during slow warming additional opportunity is provided.

Mazur: The easiest explanation for Lovelock's data is to assume that the damage was proportional to the total time the red cells spent between −3°C and −40°C.

Huggins: A number of effects may occur. When results from many laboratories that preserve red blood cells are compared, they all report very good, but not perfect protection against haemolysis. These observations imply the presence of two damaging factors, one explaining 98·5 per cent of injury, the other 1·5 per cent. I suspect that injury from concentrated solute explains 98·5 per cent, and mechanical damage from ice explains the other 1·5 per cent.

Rinfret: Have we any reason to regard a population of cells such as erythrocytes as a simple entity on which these factors are operating uniformly? There is at least a chronological division here: 0·8 per cent of any population is at a given day of age, so to speak.

Elford: How long does it take for the red cell to reach its minimal volume?

Rinfret: With glycerol-treated bovine erythocytes there is very slow penetration. Exposure of this type of cell to glycerol should have produced a minimum volume very rapidly. If one returns such a cell to an isotonic medium without freezing, without interposing any other step, would this cell of minimal volume disintegrate?

Luyet: Jacobs and associates, who studied the changes in volume undergone by cells of various sorts in solutions of glycerol and ethylene glycol of various concentrations, generally reported an initial shrinking followed by swelling. Thus, with fertilized sea urchin eggs, Stewart and Jacobs (1932) observed that, in 0·5 M-ethylene glycol, the volume decreased to a minimum of about 96 per cent of its original value in one minute and returned to the original value in nine more minutes. In similar experiments with small pieces of chick embryo heart (about 1 mm^3), we found that, at glycerol concentrations extending from 0·5 to 2 M, the volume may decrease to a minimum of some 75 per cent of its original value in 30 to 60 seconds, and then may increase to as much as 25 per cent of the original value in four more minutes (Luyet and Keane, 1953). Jacobs (1928) determined the time necessary to produce haemolysis when red cells were immersed in glycerol solutions of various concentrations. In half-molar and molar solutions, haemolysis was "obtained with human corpuscles in a few minutes" (Jacobs, 1928, p. 154). Since haemolysis occurs as a result of swelling, the time required for the previous shrinking should be of the order of a fraction of a minute.

Rinfret: Yes, but does this cell then disintegrate on returning to an isotonic situation, after having reached a minimum volume?

Luyet: In the experiments described, the cells are destroyed when they swell above a certain volume; what happens if they are removed from the glycerol solution earlier remains to be established.

Rinfret: But glycerol would not enter into the bovine cell immediately. One might expect a reduction in volume to some minimum size fairly quickly. Would the attainment of minimum volume have a lethal effect? Wouldn't the hypothesis of a minimum volume require that it have such an effect, quite independently of any freezing operation?

Luyet: Jacobs (1928) reported that bovine red cells take several hours to haemolyse in conditions in which human cells haemolyse in a few minutes.

REFERENCES

DAINTY, J. (1965). *Symp. Soc. exp. Biol.*, **19**, 75.
DYDYŃSKA, M. (1961). *Bull. Acad. pol. Sci. Cl. II Sér. Sci. biol.*, **9**, 57–60.
DYDYŃSKA, M., and WILKIE, D. R. (1963). *J. Physiol., Lond.*, **169**, 312–329.
HUGGINS, C. E. (1965). *Fedn Proc. Fedn Am. Socs exp. Biol.*, **24**, 190–195.
HUGGINS, C, and MIURA, T. (1968). In *Organ Perfusion and Preservation*, pp. 71–76, ed. Norman, J. C. New York: Appleton-Century-Crofts.
JACOBS, M. H. (1928). *Harvey Lect.*, **22**, 146–164.
LOVELOCK, J. (1953). *Biochim. biophys. Acta*, **10**, 414–426.
LUYET, B., and KEANE, J. F., Jr. (1953). *Biodynamica*, **7**, 141–155.
MAZUR, P. (1963). *J. gen. Physiol.*, **47**, 347–369.
STEWART, D. R., and JACOBS, M. H. (1932). *J. cell. comp. Physiol.*, **1**, 83–92.

MECHANISM OF HAEMOLYSIS OF ERYTHROCYTES BY FREEZING, WITH SPECIAL REFERENCE TO FREEZING AT NEAR-ZERO TEMPERATURES

TOKIO NEI

The Institute of Low Temperature Science, Hokkaido University, Sapporo, Japan

IN spite of the numerous reports by various investigators on the freezing of erythrocytes, only a few fundamental investigations have been made, especially as regards the mode of action of haemolysis by freezing.

Lovelock (1953) emphasized salt injury as the most important cause of haemolysis by freezing; he stated that the concentration of electrolytes induced by ice formation in the extra- and intracellular fluid affects constituents of the cell membrane, resulting in an increase of cell permeability and subsequent haemolysis. Since then, his salt injury theory has generally been accepted as the explanation of the mechanism of cell damage by freezing in animal cells, especially in mammalian cells.

Gehenio, Rapatz and Luyet (1963) and Nei, Kojima and Hanafusa (1964) presented typical haemolysis curves obtained by freezing mammalian erythrocytes at various temperatures ranging from freezing point to −150°C. They (Rapatz, Nath and Luyet, 1963; Nei, Kojima and Hanafusa, 1964) also made morphological observations with an electron microscope on the frozen state of the specimens, and proposed that intracellular ice formation and salt concentration are causal factors in haemolysis by freezing.

Recently, Meryman (1967) reported the effect of dehydration on the haemolysis of frozen erythrocytes. He attempted to explain the mechanism of haemolysis on the basis of removal of cellular water.

In the present paper, after a consideration of general freezing over a wide temperature range, special emphasis is placed on haemolysis by freezing at near-zero temperatures.

FREEZING OF ERYTHROCYTES AT VARIOUS TEMPERATURES

Freezing of rabbit erythrocytes at various temperatures was investigated, with measurement of haemolysis and morphological observations (Nei,

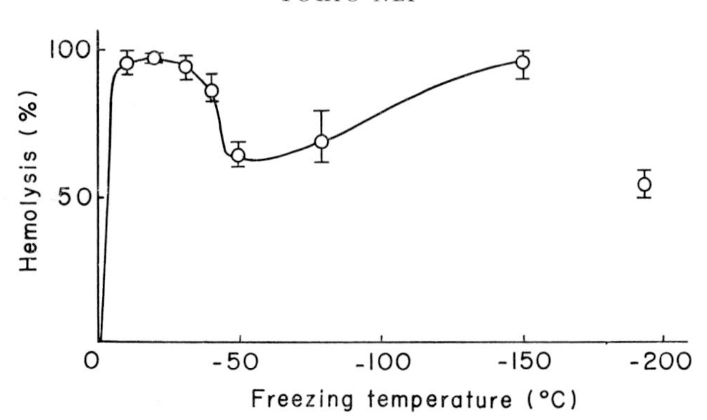

FIG. 1. Haemolysis curve: rabbit blood frozen at various temperatures.

Kojima and Hanafusa, 1964). Thin-layered blood specimens without any additives, placed between two sheets of aluminium foil, or small quantities of the same specimens in small test tubes, were frozen in cooling baths kept at temperatures ranging from freezing point to -150°C.

Haemolysis was measured spectrophotometrically after freeze-thawing. Fig. 1 shows a haemolysis curve in which complete haemolysis was found at two points around -30°C and -150°C.

The state of these frozen erythrocytes was studied in the electron microscope after freeze-drying. Cells frozen at -30°C lost their original size and shape, showing a remarkable shrinkage or distortion, whereas

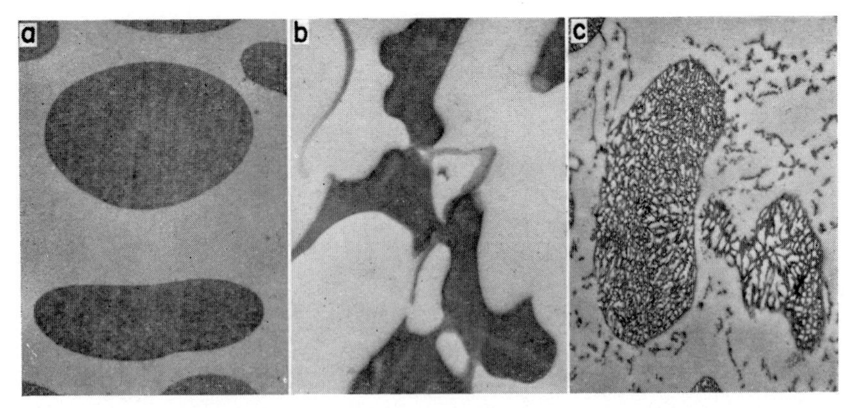

FIG. 2. Electron micrographs of frozen cells in the thin-sectioned specimens prepared by freeze-drying (× 6000).

(*a*) Unfrozen normal control.
(*b*) Irregularly shrunken cells frozen at -30°C.
(*c*) Cells frozen at -150°C. Note numerous intracellular cavities, probably left after sublimation of ice crystals.

cells frozen at −150°C kept their round shape but had numerous intracellular cavities (Fig. 2). The complete haemolysis observed at higher temperatures may therefore have been brought about by the concentration of solutes accompanying slow cooling, while the complete haemolysis at very low temperatures might be due to intracellular freezing induced by rapid cooling.

POSSIBLE FACTORS AFFECTING HAEMOLYSIS BY FREEZING AT HIGHER TEMPERATURES

As described above, two temperature-dependent factors are responsible for the haemolysis which occurs during freezing of erythrocytes. Subsequent experiments on the mechanism of those two factors (Nei, 1967, 1968) showed that, at temperatures from the freezing point to −10°C, haemolysis was affected to a certain extent by factors such as cell concentration or solute concentration. When haemolysis was compared in suspensions of normal and one-tenth cell concentration, the two haemolysis curves crossed each other at certain temperatures, as illustrated in Fig. 3. At higher temperatures around −4°C or over there was more haemolysis in cell suspensions of one-tenth concentration than in normal cell concentrations. However, at temperatures below −4°C the results were reversed.

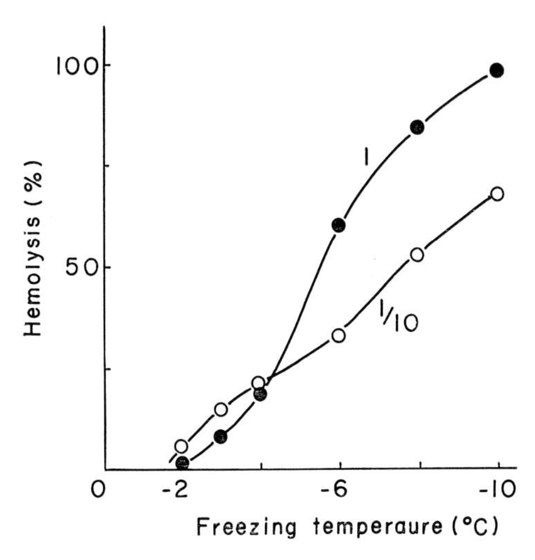

Fig. 3. Comparison of haemolysis curves of erythrocytes in 0·15 M-NaCl suspensions of normal and one-tenth cell concentrations during freezing ranging from the freezing point to −10°C.

Fig. 4 shows haemolysis curves plotted for specimens of varying cell concentrations during freezing at different temperatures. The extent of haemolysis decreased at lower temperatures and increased at higher temperatures with a reduction of cell concentration from normal to one tenth, but no change was observed in specimens at lower cell concentrations.

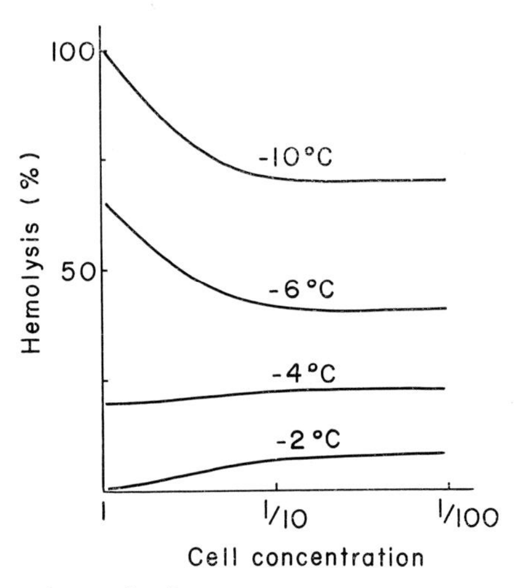

FIG. 4. Comparison of cell concentrations affecting haemolysis by freezing at −2, −4, −6 and −10°C.

MECHANISM OF HAEMOLYSIS BY FREEZING AT NEAR-ZERO TEMPERATURES

In general, the solute in the specimen is concentrated as freezing proceeds and the concentration reaches a maximum value depending upon the final temperature. It is unlikely, however, that the solute concentration is affected by cell concentration. In slow freezing down to −10°C, almost all the cells are frozen extracellularly. Therefore, some factor other than solute concentration or intracellular ice formation should be considered. From the evidence so far obtained, mechanical damage is assumed to be the most probable factor affecting haemolysis by freezing at near-zero temperatures (Nei, 1968).

FREEZING PATTERN PRODUCED BY SLOW FREEZING

When the specimens are frozen at near-zero temperatures and cooled slowly to lower temperatures, ice columns grow and advance gradually, leaving enclosed unfrozen channels. Although a few cells are individually surrounded by ice columns or are collected in lines between the ice columns, most are displaced and confined in the channels, as shown in Fig. 5. As the temperature is lowered, the channels of unfrozen solution are narrowed and the confined cells become tightly packed. Some of the cells thus packed together are haemolysed during the freezing process down to −10°C and most of the remaining cells are also haemolysed during thawing from −10° to 0°C. Cells inserted between ice columns

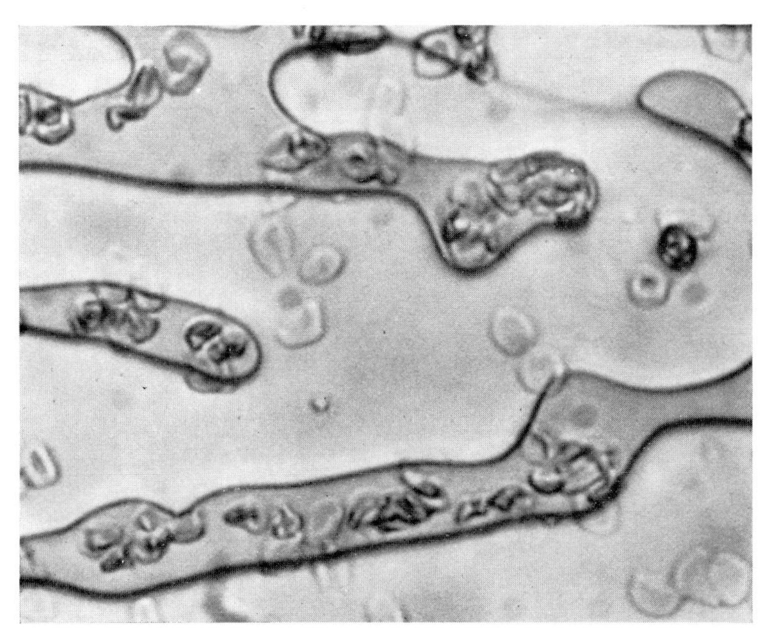

FIG. 5. Photomicrograph of blood frozen by inoculation at −1·0°C and cooled slowly to −2°C. A few cells are disseminated in ice crystals, but most are confined in channels (× 800).

at the moment when freezing begins show no morphological change during further cooling to −10°C, but they haemolyse during the thawing process at temperatures near to 0°C. In slow freezing of this kind, all cells shrink because of extracellular ice formation and no intracellularly frozen cells can be seen.

AN ASSUMED CLASSIFICATION OF FROZEN CELLS

In order to interpret the characteristics of the haemolysis curves mentioned above, the frozen cell was classified into two types, on the basis of the findings so far. A cross-section of the frozen specimens, as schematically drawn in Fig. 6, indicates that one type (A) consists of cells in a single row on the borderline of ice crystals, and the other type (B) of cell masses collected in channels of unfrozen solution. From the morphological data obtained, it was surmised that cells of type A, which are in the minority, appear in the early stage of freezing at higher temperatures and come directly in contact with ice crystals, while cells of type B, which constitute the majority of the cells, are surrounded by ice crystals and confined in the unfrozen channels, where they become more tightly packed as the temperature decreases.

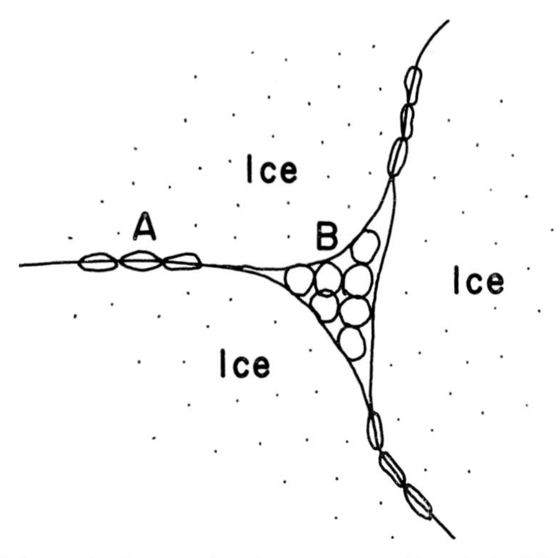

FIG. 6. Schematic diagram showing two types (A and B) of frozen cells.

HAEMOLYSIS AT HIGHER TEMPERATURES ABOVE −4°C

Provided that the same experimental conditions, such as inoculation of ice at the same temperature or the same rate of cooling, are used for slow freezing, different specimens usually show almost the same shape of freezing pattern. The ice crystals formed should therefore fall within certain limits of length or area. The total number of cells of type A on the ice crystal boundaries (Fig. 6) is also limited. If the number of cells haemolysed, possibly by the compression of ice crystals, is expressed as a and the total number of cells in suspension as t, the percentage of haemolysis induced may be indicated as $a/t \times 100$. This value becomes larger when the concentration of cells (t) becomes smaller, because a is limited to a certain extent as described above. The finding that at temperatures between the freezing point and about −4°C the extent of haemolysis was greater at lower cell concentrations supports such an interpretation.

HAEMOLYSIS AT LOWER TEMPERATURES DOWN TO $-10°C$

A marked morphological change observed on further cooling down to $-10°C$ was the narrowing of channels produced by the growth of ice and tight packing of the enclosed cells. As mentioned above, the cells which had already been singly packed between ice columns in the initial stage of freezing showed no change during this freezing process. The mode of action of haemolysis by freezing at lower temperatures down to $-10°C$ will be discussed from various points of view.

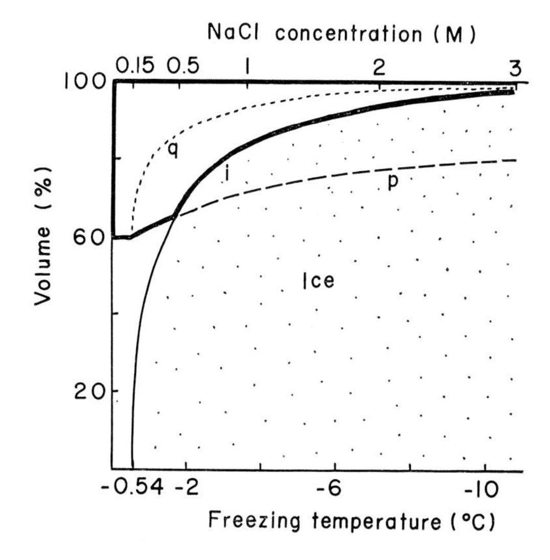

FIG. 7. Diagram showing the volume of ice and concentrated solution and also cells in the frozen state at various temperatures between freezing point and $-10°C$.

i: volume of ice occupying specimens at each temperature.
p: total volume of cells in the concentrated NaCl solutions, indicated as M on the upper abscissa, packed by centrifugation.
q: total volume of cells in the unfrozen solution estimated on the assumption that ideal osmotic shrinkage occurs.

(1) *Volume of ice formed during the freezing process*

The volume of ice formed in the specimen in 0·15 M-NaCl solution at temperatures between the freezing point and $-10°C$ was estimated by the freezing point depression and the concentration of NaCl. As illustrated in Fig. 7, for example, about 85 per cent of the water in the 0·15 M solution changes to ice at $-4°C$ (NaCl, 1·1 M).

(2) *Total cell volume in the specimen, estimated on the basis of osmotic shrinkage*

The original cell concentration contained in the 0·15 M-NaCl solution used in this experiment was 40 per cent (cell volume/total volume × 100), which is almost the same as that in normal whole blood. As the amount of ice increases during the freezing process, the concentration of NaCl in the channels increases. It is conjectured that, with the temperature drop, the volume of each erythrocyte decreases in parallel with the solute concentration. If the cells can be shrunk to a greater extent in exact correspondence with the osmotic pressure in the surrounding media during the freezing process—in other words, if theoretically ideal shrinkage proceeds as illustrated in line q in Fig. 7—some space must be left between the cells in the channels in specimens at any temperature. However, data obtained from the sedimentation of cells suspended in varying concentrations of NaCl showed higher total volumes than that estimated, as will be described next.

(3) *Total volume of packed cells in concentrated salt solution, measured by centrifugation*

The total volumes of cells suspended in concentrated salt solutions of 1, 2 and 3 M and packed by sedimentation by the haematocrit method were 77, 71 and 63 per cent, respectively, of the normal cell control in

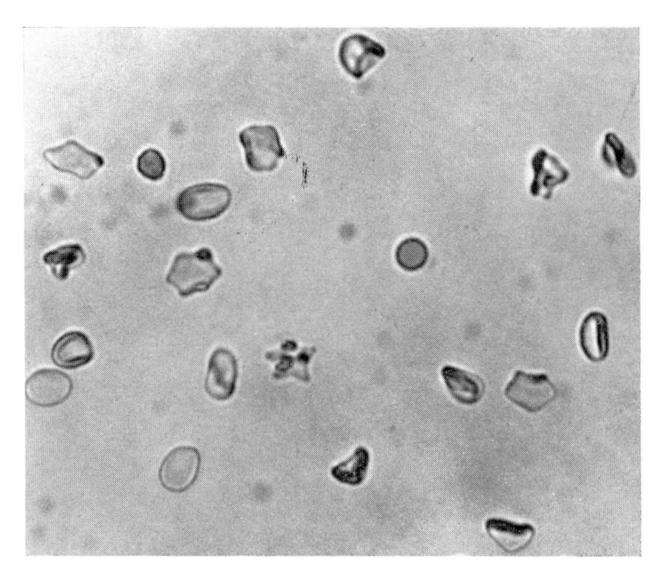

FIG. 8. Polymorphous figures of erythrocytes suspended in 1 M-NaCl solution (× 800).

the 0·15 M-NaCl solution. These values, plotted as line *p* in Fig. 7, exceed not only the total quantity of cells as estimated by theoretical osmotic shrinkage (*q*), but also the amount of unfrozen solution (*i*), as shown in Fig. 7.

The fact that the total volumes of packed cells in concentrated solutions were greater than the expected values may be explained by the irregular morphology of the cells in these solutions, characterized by the extreme crenation or shrinkage shown in Fig. 8. In frozen specimens, therefore, the total volume of packed cells confined in unfrozen solutions should be reduced to the area indicated by heavy lines in Fig. 7, as estimated from the amount of ice formed. This means that line *p* should recede to line *i*.

The haematocrit values of erythrocyte ghosts produced by hypotonic haemolysis with water was about one-tenth that of the normal intact cells. If some of the cells confined in the unfrozen solutions are haemolysed during the freezing process, the total volume of packed cells will be reduced by the increase in ghosts produced until it is finally balanced to the volume of the unfrozen solution (*i*).

(4) *Haemolysis during freezing*

Morphological observations also showed that haemolysis occurs during freezing as well as during thawing. In particular, there was a high rate of haemolysis in the specimens containing high concentrations of cells which

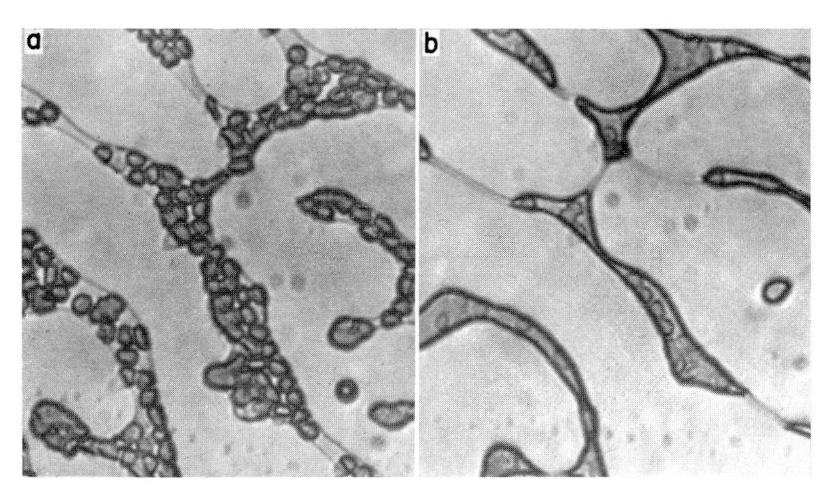

FIG. 9. Haemolysis of packed cells confined in channels occurring during the freezing process (× 600).

(*a*) Cells frozen at −2°C.

(*b*) Cells cooled to −10°C. Note that most of the cells are haemolysed.

become tightly packed in the freezing process, as illustrated in Fig. 9. This suggests that the increased cell concentration in suspensions leads to tighter packing in the channels and subsequently to increased haemolysis.

(5) *Haemolysis in concentrated solutions*

The assumption that, in addition to the effect of salt concentration, haemolysis during freezing down to $-10°C$ might be due to mechanical damage caused by tight packing of cells in the narrow channels was also supported by the finding that even if the solute concentration in cell suspensions is raised by dehydration produced by evacuation, very few cells are haemolysed; most cells are haemolysed only when they are compressed in a narrow space, as shown in Fig. 10.

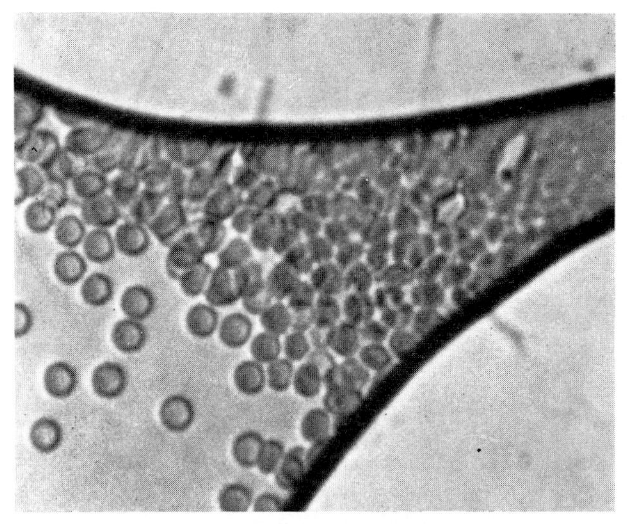

Fig. 10. Haemolysis of cells in solutions concentrated by dehydration produced by evacuation. The two empty areas are dehydrated parts. Only cell masses compressed in the narrow space are haemolysed (× 800).

CONCLUSION

From the results so far, and the assumption based on them, it is concluded that mechanical damage may well be one of the main causes of haemolysis during freezing at near-zero temperatures. The mechanical stress which causes haemolysis may be the compression of the surrounding ice against cells and the tight packing or mutual contact of cells confined in the narrow space of unfrozen solution left during the freezing process. In any event, one or more of these factors may play an important role in haemolysis. The importance of salt injury is not denied, and indeed this still seems to be essential, to some extent, in freezing injury to mammalian cells. The role of mechanical damage introduced here has not so far attracted any attention, but this work confirms that it has some share in freezing injury, just as salt injury does.

SUMMARY

Measurements of released haemoglobin and morphological observations of erythrocytes frozen at various temperatures suggest that haemolysis by slow freezing at near-zero temperatures might be caused by the concentration of solutes accompanying extracellular freezing, and that haemolysis by rapid freezing at lower temperatures might be due to intracellular ice formation.

During slow freezing at near-zero temperatures the cell concentration considerably affected the degree of haemolysis. This indicates that mechanical stress due to extracellular ice formation might be one of the causes of haemolysis. Suspensions containing lower cell concentrations showed more haemolysis at high temperatures and less at low temperatures than did normal cell suspensions, and accordingly these haemolysis curves crossed each other at a certain temperature. To interpret this finding, the assumption was made that there are two types of freezing pattern for erythrocytes cooled to −10°C, one consisting of monolayered frozen cells directly surrounded by ice columns at the early stage of freezing, and the other of multi-layered cells confined in the unfrozen solution and packed tightly as the temperature drops.

Morphological findings and quantitative measurement of haemolysis during freezing support this assumption.

Mechanical damage therefore may well be one of the dominant factors, in addition to salt injury, affecting haemolysis by freezing at near-zero temperatures.

Acknowledgement

I am indebted to Mr K. Tanno, The Institute of Low Temperature Science, for kindly suggesting the classification of two types of freezing pattern, especially for haemolysis at higher temperatures.

REFERENCES

GEHENIO, P. P., RAPATZ, G. L., and LUYET, B. J. (1963). *Biodynamica*, **9**, 77–82.
LOVELOCK, J. E. (1953). *Biochim. biophys. Acta*, **10**, 414–426.

MERYMAN, H. T. (1967). In *Cellular Injury and Resistance in Freezing Organisms*, pp. 231–244, ed. Asahina, E. Sapporo: Institute of Low Temperature Science, Hokkaido University.
NEI, T. (1967). *Cryobiology*, **4**, 153–156.
NEI, T. (1968). *Cryobiology*, **4**, 303–308.
NEI, T., KOJIMA, Y., and HANAFUSA, N. (1964). *Contr. Inst. low Temp. Sci. Hokkaido Univ. B*, **13**, 1–6.
RAPATZ, G., NATH, J., and LUYET, B. (1963). *Biodynamica*, **9**, 83–94.

DISCUSSION

Huggins: Professor Nei's beautiful experiments with red blood cells at high subfreezing temperatures appear similar to a series of experiments that we did with cryophylactic agents at much lower temperatures (Huggins, 1963).

In the first group of experiments blood cells at room temperature were damaged whenever sodium chloride concentrations exceeded 0·8 M-NaCl. In the second group, erythrocytes were frozen slowly to −85°C, held at this temperature for 24 hours and thawed. In these experiments salt concentration was kept constant at 0·15 M-NaCl. DMSO concentration in the different aliquots of blood ranged from 1·0 to 8·0 M-DMSO. When haemolysis was measured after thawing no protection was found until the 3·0 M-DMSO concentration had been reached. Between 3·0 and 6·0 M-DMSO the degree of protection increased progressively. Almost complete (98·6 per cent) protection against haemolysis during the slowest rates of freezing and thawing was found when the DMSO concentration was between 5·0 and 5·7 M.

If excessive concentrations of solute were the cause of slow freezing injury, and if DMSO acted to reduce the degree to which solutes concentrated in the frozen state, we postulated that differences in the prefreezing salt concentration of blood cells in 5·0 M-DMSO should affect haemolysis after freezing and thawing. An experiment was designed to test this hypothesis. Minimal haemolysis was found with cells frozen in 5·0 M-DMSO/0·1 M-NaCl. Total haemolysis, however, was found with other aliquots of blood frozen in DMSO/0·5 M-NaCl. This series of three experiments supports the hypothesis that excessive concentrations of solute, or other effects related to sequestration of solvent water into ice, are a major cause of damage during slow freezing and thawing.

Of great importance, however, was the subsequent finding that blood cells frozen in 5·0 M-DMSO/0·1 M-NaCl underwent progressive destruction during storage at −85°C. Aliquots that showed less than 1 per cent haemolysis when stored for 24 hours showed 50 per cent haemolysis when stored for one week. Numerous other experiments have shown that the cells swollen by a hypotonic prefreezing environment undergo progressive destruction during storage. We interpret these findings as mechanical damage to the swollen cryophylactic agent containing cells at low temperatures. The results of these experiments appear similar to those that Professor Nei has just shown us with untreated cells at high subfreezing temperatures. I agree strongly with Professor Nei that mechanical damage from ice crystals can be an important cause of cellular injury during freezing and thawing.

Meryman: If you started with, say, a 3 or 4 per cent salt solution, a high concentration of salt would be reached before the volume of the unfrozen portion had been reduced much. Would you then see more, or less, or the same amount of haemolysis on freezing?

Nei: In my previous experiment, I examined the effect of the initial concentration of solute (NaCl) in cell suspension upon haemolysis by freezing at −2 to −10°C. As the salt concentration was increased from 0·15 M to 0·6 M, the extent of haemolysis was reduced. Beyond this range, however, haemolysis was increased and it was higher in 1 M-NaCl than in 0·15 M-NaCl. The volume of unfrozen channels might be greater in 0·6 M-NaCl than in 0·15 M-NaCl at all freezing temperatures studied and this was confirmed by morphological observation. The increase in haemolysis observed in 1 M-NaCl may be due to thermal shock, which occurs in highly concentrated solutions, as mentioned by Lovelock (see Nei, 1968, Fig. 4).

Farrant: If this experiment is done starting off with less than isotonic salt, the minimum cell volume will still be reached at the same concentration of salt. The only thing that would be different is the amount of ice in the system. So one can't distinguish between the effects of the reduction in cell volume and of the increase in salt concentration.

Meryman: If there is twice the absolute amount of salt, there will be twice the amount of liquid volume when the same temperature is reached.

Farrant: The freezing point will be lowered but then the concentrations will increase as ice forms, to the same levels at the same temperature.

Mazur: The concentrations will be the same but the fraction unfrozen will not be the same.

Meryman: The space available for the cells will be larger if one starts with a higher salt concentration.

Farrant: But the salt concentration during dehydration will be the same at any given temperature whatever the cell concentration.

Mazur: Therefore the fact that differences in survival are seen with different sodium chloride concentrations can't be explained in terms of a concentration difference at those temperatures. The concentration is the same. But survival could be related to differences in the fraction unfrozen.

When you froze the red cells, what percentage haemolysed after they were frozen and before they were thawed?

Nei: It is very difficult to determine quantitatively the extent of haemolysis during freezing and before thawing. It was estimated by morphological observation to be approximately 50 per cent of total cells haemolysed. Some haemolysis of erythrocytes occurs on exposure to highly concentrated solutions and damage increases extensively when the cells are returned to isotonic solution. If it is assumed that the cells haemolyse during freezing only because of the salt concentration, the extent of haemolysis during freezing should be the same as that obtained by exposure of the cells to a corresponding concentration of solute. In order to determine quantitatively the extent of haemolysis in the frozen state, the following experiment was carried out. The frozen specimen was dissolved by the addition of unfrozen concentrated NaCl solution maintained at given temperatures to prevent the diluting effect of thawing, and the extent of haemolysis was measured after solution was complete. This kind of experiment took a long time and there might be some technical error. But it was found that haemolysis was much higher than expected—nearly the same value as observed under ordinary freeze-thawing conditions. Such a difference in percentage of haemolysis may be due to some factor other than solute concentration, probably mechanical damage (see Nei, 1968, Fig. 6).

Rinfret: With the thin films you used for observing the state of the formation of ice and the relative position of the cells within the channels, is there not grossly visible an increase in haemolysis with time?

Nei: Increase in haemolysis was observed with time and temperature drop.

Meryman: The answer also depends partly on the interplay between thermal shock and the irreversible leak due to hypertonic stress. As Lovelock showed in his paper on thermal shock (1955), the proportion of cells that are shockable is reduced with time over a period of an hour as the cells are suspended in hypertonic solution. As we interpret this in the light of our work, the cell which is suspended in a solution sufficiently concentrated to be stressful has two equally unattractive alternatives. It may spontaneously break down and become irreversibly leaky, or it may be haemolysed by thermal shock due to a temperature reduction. Cells which are haemolysed during freezing are presumably drawn only from those cells which have not leaked spontaneously. The cells which have spontaneously leaked are no longer under osmotic stress and will not haemolyse from thermal shock. However they have taken in an excess amount of solute and, on thawing and return to an isotonic solution, they will be destroyed by hypotonic haemolysis.

Mazur: If Dr Nei's estimate that about 50 per cent of the cells haemolysed after freezing and before thawing is correct, it would suggest that thermal shock is as important in haemolysis as osmotic shock.

Meryman: It depends on how slowly you cool the cells. If the cells are exposed for a long time to a hypertonic solution with almost no temperature drop, a larger proportion of them will leak and haemolyse during thawing. If the freezing is more rapid, a larger proportion of them will not leak and are susceptible to thermal shock. Professor Nei has demonstrated a third factor operative during freezing which is probably not thermal shock because the temperature is being reduced so slowly.

Mazur: Does thermal shock depend on the rate of cooling, or just on a temperature change?

Meryman: Lovelock showed a relationship between cooling rate and haemolysis, but his data are not conclusive because he failed to consider that, during very slow cooling, more cells would leak and therefore fewer cells would be candidates for thermal shock, thus producing an apparent reduction in thermal shock at low cooling rates. Thermal shock is definitely not related to absolute temperature, but to a reduction in temperature. If the cells are mixed with the hypertonic solution at any temperature down to −30°C in the presence of cryoprotective agent, there is no haemolysis. At borderline osmolalities the cells can go back and forth between isotonic and hypertonic solutions without any haemolysis as long as there is no temperature change. The haemolysis occurs with a downward change in temperature, never an upward, and only at temperatures below +25°C.

Mazur: I thought going from isotonic to hypertonic was what produced haemolysis.

Meryman: You have to reduce the temperature in the presence of a hypertonic stress in order to get thermal shock haemolysis.

Leibo: How hypertonic?

Meryman: If the solution is too hypertonic it will cause the membrane to leak.

Mazur: Lovelock's papers (1953*a*,*b*) indicate that he observed haemolysis when cells in an isotonic solution at, say, 0°C were transferred to a hypertonic solution at 0°C and then transferred back to isotonic at 0°C; i.e. the temperature remained constant. Lovelock's data seem to indicate that when the concentration of NaCl got above about 1·5 M, subsequent transfer to isotonic sodium chloride at the same temperature produced considerable haemolysis.

Huggins: Experience suggests that many factors can damage cells during freezing and thawing. The great challenge of low temperature biology is to preserve cells under conditions that avoid *all* causes of damage.

Levitt: There may still be one common denominator by means of which they all act. There may be a chemical effect on the membrane as well as a physical effect, but in both cases it may be a membrane effect.

Huggins: I agree that membrane damage is the final common pathway.

MacKenzie: It should be possible to determine whether or not the cells designated by Dr Nei as "type A" suffer from mechanical injury at these rather high subzero temperatures. Where a cell is caught in a pocket, so to speak, formed by the pairing of two cup-shaped cavities, one in each of two abutting ice crystal surfaces, the tendencies of the two crystals to reduce their surface-to-volume ratios can be stated quantitatively. So also could the resistance of the cell membrane to an excess internal pressure (though we would probably have to get this from data obtained at higher temperatures, by extrapolation). If indeed there is a pressure exerted on a type A red blood cell from two sides, it should be possible to calculate the tendency of the cell to burst around the rim at various high subzero temperatures.

Mazur: But as long as the cell is free to shrink, that is as long as it is above the minimum volume, the cell ought just to shrink osmotically without the appearance of any hydrostatic pressure.

MacKenzie: I am discussing the situation at a constant temperature at which it has stopped shrinking. A cell could undergo a gradual flattening without a change in temperature. It should be possible to estimate the magnitude of the forces involved and then decide whether such mechanical destruction is an important mechanism.

Huggins: High pressures alone can damage blood cells. Dr Edward Finch at the Massachusetts General Hospital has recently constructed a pressure chamber in which blood cells at room temperature were kept at 300 atmospheres ($30{\cdot}4 \times 10^6$ N m^{-2}) for 48 hours. Total haemolysis was found after release of pressure, whereas none was present in identical control aliquots stored at atmospheric pressure for the same time, at room temperature (unpublished data).

Luyet: In regard to Professor Nei's observation of the combined effects on haemolysis of the cell concentration and the freezing rate, it may be of interest to note that, when the cell concentration in a suspension is high, the amount of extracellular fluid is proportionally low, and that the time during which the temperature stays at the freezing plateau as well as the time required for the cryodehydration of the cells are thereby modified. These latter quantities, in turn, are controlled by the rate of heat removal and the rate of the passage of water through the membranes. The final outcome will, thus, be the result of a complex interplay of factors.

This reminds me that Dr Meryman (personal communication) said that the amount of extracellular water in living tissues is generally small, and that this may explain why, in frostbite, tissues may be frozen to quite low temperatures without suffering much damage.

Mazur: One would expect that haemolysis would be less in a more concentrated cell suspension, but Dr Nei found the opposite: there was more haemolysis in more concentrated cell suspensions at low temperatures. Can one be certain whether this effect is due to differences in temperature or to differences in cooling velocity?

Nei: The cooling rate used was slow. The specimens were frozen by inoculation of ice at about −1·5°C and then cooled to −10°C at a rate of approximately 1°C/min.

Huggins: We have seen increased haemolysis after thawing when blood samples with packed cell volumes greater than 80 per cent were frozen to −85°C in the presence of glycerol.

REFERENCES

HUGGINS, C. E. (1963). *Transfusion*, **3**, 483.
LOVELOCK, J. E. (1953*a*). *Biophys. biochim. Acta*, **10**, 414.
LOVELOCK, J. E. (1953*b*). *Biophys. biochim. Acta*, **11**, 28.
LOVELOCK, J. E. (1955). *Br. J. Haemat.*, **1**, 117–129.
NEI, T. (1968). *Cryobiology*, **4**, 303–308.

THE ROLE OF MEMBRANE PROTEINS IN FREEZING INJURY AND RESISTANCE

J. LEVITT and JOHN DEAR

Department of Botany, University of Missouri, Columbia, Missouri

SCIENTIFIC concepts often undergo cyclical changes. The mechanism of freezing injury and resistance is a case in point. In 1912, Maximov proposed that freezing injury was due to injury to the plasma membrane. During the following 25 years, so many investigators produced evidence of a precipitation of soluble proteins in plant juices, due to freezing, that Maximov's concept was dropped. Twenty-four years after Maximov's proposal, Levitt and Scarth (1936) showed that the increase in freezing resistance of plants which occurs during hardening is accompanied by an increase in cell permeability. This pointed once again to the plasma membrane as the cell organelle involved in freezing injury. Again, this seemed to be an over-simplification when Siminovitch and Briggs (1949) discovered that soluble proteins increase markedly during the hardening of plants. Recently, however, many investigations have shown that the quality of the soluble proteins may be largely unaffected either by freezing injury or by freezing resistance (Gerloff, Stahmann and Smith, 1967; McCown, Hall and Beck, 1969), and once again many investigators are turning to the plasma membrane for the answer. In a recent symposium on environmental cryobiology, all five participants were forced to consider this possibility (Smith, 1968; Sakai and Yoshida, 1968; Modlibowska, 1968; Heber, 1968; Siminovitch *et al.*, 1968). It is, therefore, high time to evaluate all the available evidence.

THE RESPONSE OF PLANTS TO FREEZING

When plants are exposed to freezing temperatures, whether or not they are injured depends not only on the actual freezing conditions but also on the nature and previous treatment of the plant. The tender plant cannot harden, the hardy plant can (Table I). Different hardy plants, of

Table I

KILLING TEMPERATURES (°C) FOR SOME TENDER AND HARDY PLANTS

Growing temperatures (°C)	*Tender plants (e.g. sunflower)*	*Hardy plants (e.g. cabbage)*
20–30° ("unhardened")	−2	−2
5° for 1 week ("hardened")	−2	−7
0° for 2nd week ("hardened")	−2	−12

course, respond differently to the hardening treatment. Plants may, therefore, be classified according to the maximum hardening they can achieve:

	Tender	*Slightly hardy*	*Moderately hardy*	*Very hardy*
Killing temperature (°C) ..	> −5	−5 to −10	−10 to −20	< −20

Freezing injury, therefore, poses four basic questions. (1) What is the mechanism of injury? (2) What happens during the hardening process to prevent injury during subsequent freezing? (3) Why are tender plants incapable of hardening? (4) Why are there all degrees of hardening? To these four questions on the behaviour of the plant under natural conditions can be added a fifth for artificial conditions. (5) Why do some substances act as cryoprotectants against freezing injury?

FREEZING INJURY

Effects of freezing on proteins

Since the proteins are the largest component of the dry matter of protoplasm, it is not surprising that they are usually assumed to be the key to freezing injury (Levitt, 1956). Recent work with pure proteins now provides the basic information for testing this assumption. Thermodynamic considerations, which have been fully supported experimentally, indicate that proteins can be expected to denature (reversibly) when their temperature is lowered sufficiently (Brandts, 1967). The actual temperature at which this occurs depends on the nature of the bonds responsible for the tertiary structure. Since it is the hydrophobic bonds which are weakened by the low temperatures, the higher the proportion of hydrophobic bonds, the greater the danger of unfolding, and the higher the low temperature which is capable of producing this denaturation. Although the denaturation is reversible ($N \rightleftharpoons D$), it can be converted to an irreversible aggregation ($D \rightarrow A$). This aggregation may occur during freezing, as a result of extracellular ice formation which dehydrates the protoplasm progressively until the proteins approach close enough for bonds to form between them. Thus Thiogel, a denatured thiolated protein, rapidly forms intermolecular SS bonds on freezing "extragellularly" at -3 to -8°C (Levitt, 1965). Bovine serum albumin, an SS protein in the native state, fails to aggregate appreciably when similarly frozen (Goodin, 1969). However, if it is first denatured by unfolding in urea and reducing the SS bonds to SH groups, the repurified protein will aggregate on freezing. The aggregate has a lower SH content, and it can be completely solubilized by 8 M-urea only if the urea solution contains a thiol (e.g. mercaptoethanol). The aggregation on freezing can be prevented by tying up the SH groups of the denatured protein with *N*-ethylmaleimide. This evidence therefore points to SS bridges as the major bonds responsible for aggregation of the denatured protein on freezing. It is, however, also possible for the *N*-ethylmaleimide form of bovine serum albumin to become aggregated as a result of freezing, for instance if the dehydrated protein is allowed to warm up to room temperature before rehydration. This is presumably due to hydrophobic bond formation and readily explains the dangers of rapid thawing in extracellularly frozen plants.

Like native bovine serum albumin, many enzymes remain native, soluble and active after freezing, and some in fact are often stored in the frozen state in order to preserve their activity for a longer time. The rapid freezing of enzyme solutions (or of tissues before extraction of enzymes), however, is not comparable to the normal extracellular freezing of cells or to the "extragellular" freezing used in the above experiments, since the ice crystals form between the protein molecules and may therefore prevent aggregation. Nevertheless, some proteins which are known to have many SH groups are inactivated by freezing and may be protected in the frozen state by inclusion of small-molecule thiols (Levitt, 1966*a*; McCown, Hall and Beck, 1969). On the other hand, even enzymes which are inactivated by freezing in the pure state may be protected from such inactivation by the presence of substrate, other proteins, or other impurities. Since such other substances are always present in the living cell, it seems likely that even these more sensitive enzymes would not be inactivated in frozen plants. It is in fact obvious that a plant normally killed by extracellular freezing retains active enzymes, since oxidative and autolytic reactions occur rapidly in the thawed plant after a fatal freeze. This conclusion is supported by the lack of any detectable decrease in extractable (soluble) protein from cabbage leaves killed by frost (Morton, 1969). There is also no decrease in enzyme activity in carnation tissues killed by freezing (McCown, Hall and Beck, 1969). We must therefore conclude that the irreversible aggregation which occurs on freezing of Thiogel and denatured bovine serum albumin does not apply to these soluble enzymes, or to the soluble proteins of the cell in general. These soluble proteins, therefore, either do not denature sufficiently at freezing temperatures, or else the reversible denaturation is not converted to an irreversible aggregation on freezing. Since the hydrophobic groups are present in minimal quantity in the soluble globular proteins, and in maximal quantity in the structural membrane proteins (Hatch and Bruce, 1968), the reversible denaturation at low temperature will occur much more readily in the latter. These membrane proteins must, in fact, unfold on cooling to a greater degree than any of the other protoplasmic proteins. They must therefore be the most vulnerable of all the cell's proteins to irreversible aggregation on freezing. Five physical tests made three to five decades ago, and one observation made even earlier, support this conclusion.

(1) Freezing injury is commonly instantaneous and therefore is not simply due to a disruption of metabolism by enzyme inactivation. There is apparently an immediate loss of semipermeability, for the thawed plant has its intercellular spaces infiltrated with liquid which the flaccid cells cannot reabsorb (Levitt, 1956). (2) Maximov (1912) showed that non-penetrating solutes (e.g. 0·5 M or higher sucrose) prevent freezing injury to tender plants. Unprotected cells which were all killed at −5°C survived freezing at as low as −30°C when protected by such solutions. Since these protective solutes do not penetrate the semipermeable plasma membrane (and in fact produce their maximum effect immediately), Maximov concluded that they must exert their effect on the surface of the protoplasm, and that injury to the plasma membrane must be the cause of death by freezing. He suggested that the solutes protect the membrane by preventing pressure from external ice crystals. Later observations (Iljin, 1933) showed that the cell contracts markedly as a result of this extracellular freezing, producing a tension on the protoplasmic surface instead of the pressure postulated by Maximov. The non-penetrating solute prevents contraction of the cell wall and the consequent tension by inducing plasmolysis, which occurs as the protective solution becomes more concentrated due to freezing and penetrates between the cell wall and the protoplasm (Åkerman, 1927). (3) When plants are hardened at low temperatures, they become resistant to freezing injury and their cell permeability simultaneously increases (Levitt and Scarth, 1936). This indicates that resistance requires a change in the protoplasmic surface—

i.e. in the plasma membrane. (4) Plasmolysed cells of non-hardy plants undergo a stiffening of their protoplast surface usually within about six hours, depending on the degree of plasmolysis. As a result, when transferred to water or hypotonic solutions they rupture and disgorge their contents before deplasmolysis can be completed. Cells which are fully frost-tolerant do not show this stiffening and can be deplasmolysed successfully after many hours of the severest plasmolysis. It has in fact been possible to determine the degree of freezing tolerance of a series of plants by the concentration of the plasmolysing solution and the length of plasmolysis time tolerated (Siminovitch and Levitt, 1941). (5) If the cells are severely plasmolysed, it is possible to detect a rupture of the cytoplasmic strands which connect the protoplast surface to the cell wall. This rupture occurs in unhardened but not in hardened cells, demonstrating that the dehydrated protoplast surface becomes stiffened in the former but not in the latter (Siminovitch and Levitt, 1941). (6) Finally, micrurgical tests proved that the rigid layer which forms on plasmolysis of unhardened cells is at the surface of the protoplast (Levitt and Siminovitch, 1940).

Two important facts emerge from all the above observations. (1) Dehydration of the cell surface may induce a rigidity which is not reversed when the cell reabsorbs water. (2) Extracellular freezing dehydrates and therefore contracts the cell, producing a tension on the protoplast surface. The second fact is self-explanatory. Recent biochemical information has now supplied an explanation for the first, as follows: (*i*) An increase in protein SS has been detected as a result of freezing, drought and heat injury (Levitt, 1962). It has now been shown that this increase occurs in the structural proteins and not in the soluble proteins (Gaff, 1966). (*ii*) The soluble proteins of cabbage leaves possess SH groups, but no measurable quantity of SS groups. This is not surprising, since up to 70 per cent of the soluble leaf proteins consist of Fraction I protein which contains about 100 SH groups for a molecular weight of 500 000, but no SS groups (Sugiyama and Akazawa, 1967), and since intracellular enzymes are in general SH and not SS proteins. The soluble proteins can therefore form SS bridges only by SH $\rightleftharpoons$ SS oxidation. The insoluble (including the membrane) proteins on the other hand possess both, the SS groups in cabbage accounting for upwards of 20 per cent of the total. They are therefore capable of SS bonding by the additional mechanisms of SH $\rightleftharpoons$ SS interchange and SS $\rightleftharpoons$ SS interchange (see below). (*iii*) Maximov's method of cryoprotection can be overcome by including a thiol (10^{-2} M-mercaptoethanol) in the protective solution (Krull, 1966). The mercaptoethanol presumably reduces SS groups to SH, leading to a marked unfolding of SS proteins, which was shown above to lead to irreversible aggregation of bovine serum albumin on freezing. (*iv*) If sections of tissue are allowed to react with SH reagents (*p*-chloromercuribenzoate or iodoacetate) before being frozen, the cells show a marked decrease in freezing resistance (Levitt, 1969*a*). Since the reagents are equally effective after reacting for ten minutes or one hour at 0°C (after which the excess is washed away), the effect must be primarily on the cell surface.

To the six lines of older biophysical evidence, and the four lines of more recent biochemical evidence, can be added one direct test. Though the plasma membranes have not been isolated, the chloroplast lamellar system has. These lamellae are capable of photophosphorylation. Freezing, however, inactivates them (Heber, 1968). Heber searched unsuccessfully for an enzyme sufficiently sensitive to freezing to explain this drastic loss of phosphorylating activity. Since the process is membrane-bound and requires a continuous membrane system with special permeability properties, Heber concludes that the freezing damages the membranes by altering their permeability. Thus the unfrozen membrane vesicles function as osmometers, but the frozen ones completely lose this property. From these results, and from evidence that photosynthesis is more sensitive to freezing than are other cell activities, Heber concludes that the inactivation of chloroplast membranes is causally related to freezing injury. This explanation, of course, cannot apply to non-photosynthesizing cells. Root cells, for instance, are far more readily injured by freezing than are green cells of leaves or twigs (Levitt, 1956). It must also be pointed out that Heber freezes the chloroplast lamellar system in an aqueous medium by exposure to −20°C. This is, of course, analogous to intracellular and not to the intercellular freezing which occurs in plants during winter. Therefore, the injury he produces is presumably due to ice formation within the membranes. Finally, until it is shown that the chloroplast lamellae lose their property of semipermeability whenever leaves are killed by normal extracellular freezing, there is no reason to consider them as a factor in freezing injury. On the other hand, it is known that the plasma membrane loses its property of semipermeability whenever a cell is killed by freezing, and indeed this fact is made use of when determining freezing injury quantitatively. Nevertheless, categorical proof of direct injury to the plasma membrane during extracellular freezing is still lacking.

Can the biochemical observations explain the above-described biophysical changes associated with extracellular freezing and other kinds of cell dehydration? With bovine serum albumin (Goodin, 1969) the proteins must be in the denatured state before dehydration by freezing can induce an irreversible aggregation. As mentioned above, the membrane proteins should denature more readily at the low temperature, due to their greater hydrophobicity. But this denaturation may not be a necessary precursor of aggregation. Whether the membrane proteins are oriented with their polar groups adjacent to the polar groups of the membrane lipids (Davson and Danielli, 1943), or with their hydrophobic groups adjacent to those of the membrane lipids (Green and MacLennan, 1969), in either case they must be in at least a partially extended form; for both of these kinds of bonds are weak, and there would have to be many points of contact between a single protein molecule and the adjacent lipid layer, in order for the bonding to be strong. Therefore, they may be as available for intermolecular bond formation when in the native state as are the globular proteins in the denatured state. This would explain the well-known correlation between drought and freezing resistance (Levitt, 1956). For if freezing injury required a preliminary reversible denaturation produced by low temperature before the irreversible aggregation could occur, there would be no reason for a correlation with drought injury, which occurs at normal temperatures and therefore in the absence of reversible denaturation.

Even osmotic dehydration due to plasmolysis at room temperature may be expected to induce intermolecular SS bonding between the membrane proteins, for the surface area of the protoplast shrinks, bringing the dehydrated membrane proteins close enough together for covalent bond formation. With Thiogel this bonding results in a rigid framework which prevents melting of the gel. The bonded membrane proteins must therefore form a similar rigid framework. When the plasmolysed cells are then transferred to a hypotonic solution, the absorption of water will produce a hydrostatic pressure within the protoplast, leading to rupture of the rigid membrane. With Thiogel, each molecule is surrounded on all sides by similar molecules and the bonding occurs rapidly. In contrast to Thiogel, the rigidity produced by plasmolysis develops slowly, presumably because SS bonding can occur only laterally, between adjacent molecules occurring in a monomolecular layer of proteins on either side of the lipid layer. It is, in fact, because of the slowness in development of rigidity that plasmolysis actually protects cells against freezing injury. However, if the cells frozen in protective solutions of non-penetrating solutes are kept in the resulting plasmolysed (and frozen) state for too long a period, they too will rupture on reabsorption of water, and the protective effect will be destroyed (Levitt, 1956). On the other hand if, after the protoplasts have developed their rigid surface, they are not allowed to reabsorb water, they remain alive in the plasmolysed state and do not lose their cell contents.

Normal freezing, in the absence of protective solutions, produces a tension on the protoplast surface. It is only when this tension is combined with the tendency for the surface to become rigid that freezing injury occurs instantly. The tension must therefore accelerate the process leading to rigidity of the cell surface. On the other hand, the stretched protoplast surface is supported by the much stronger cell wall to which it adheres, so it cannot be ruptured as happens during deplasmolysis. Why then should these two changes kill the cell? The following hypothesis provides a logical explanation.

Membrane-hole hypothesis

All the above evidence can be explained if freezing injury is due to the formation of permeable non-lipid "holes" in the lipid layer of the plasma membrane. Injury due to intracellular freezing has long been explained in this way. If ice crystals form in, or protrude into, the plasma membrane, they must leave permeable holes in the membrane on thawing. This, of course, would explain the need for both rapid freezing (small crystals, not protruding through the membrane) and rapid thawing (prevention of growth of crystals), for cell survival of intracellular freezing.

But plants are injured in nature by extracellular freezing. Therefore, the external ice crystals cannot produce holes in the membrane, which is separated from them by a cell wall. The collapse of the cell due to extracellular freezing produces a tension on the thin plasma membrane. If the collapse and the consequent tension become sufficiently severe, a break will occur; and it will occur sooner in the thin, bimolecular, apolar lipid layer, due to its smaller cohesive force, than in the adjacent protein layers, which due to their much higher polarity must possess a greater cohesive force.

Support for this concept is provided by experiments with free, spherical protoplasts (Levitt, Scarth and Gibbs, 1936). When allowed to expand in a hypotonic solution, their permeability to water is two to three times greater than when they contract through the same volume changes in a hypertonic solution. This is presumably due to a temporary, excessive separation of the lipid molecules when the plasma membrane is stretching. A similar stretch occurs when the protoplast adheres to the cell wall during cell collapse. This postulated break in the lipid layer will be accompanied by contact between the two protein membrane layers on either side of the lipid layer. If this contact between the two dehydrated protein layers results in covalent bond formation between them the hole becomes irreversible, leading to instant death on thawing, when the cell contents will diffuse out through the hole now filled with hydrated protein. If covalent bonds do not form between the two membrane layers of proteins, the lipid layer may perhaps be able to become continuous again on thawing, by flowing together between the two rehydrated protein layers as the tension on the membrane decreases during reabsorption of thaw water by the cell.

What is the nature of these covalent bonds which are presumed to result in irreversible "holes" in the membrane? It has been shown that tension greatly lowers the temperature at which SS $\rightleftharpoons$ SS interchange occurs between adjacent protein molecules in keratin (Feughelmann, 1966). Tension presumably has the same effect on membrane proteins. Intermolecular SS bonds will also result more readily from SH $\rightleftharpoons$ SS interchange and from 2SH $\rightleftharpoons$ SS oxidation since the tension will tend to unfold the protein and therefore to unmask these groups. Besides forming between the two layers of membrane proteins, SS bridges may also form between soluble proteins and adjacent membrane proteins, for instance if the holes continue through one or both layers of membrane proteins. Such mixed aggregates are apparently responsible for the loss in semipermeability and injury in abnormal red blood cells containing Heinz bodies (Jacob *et al.*, 1968). Labelling of SH groups with [^{14}C]-*p*-chloromercuribenzoate has led to evidence of a similar mixed aggregation of proteins as a result of freezing injury to cabbage leaves (Morton, 1969).

It is now understandable why tension markedly speeds up the development of rigidity in the protoplast surface, as compared with the plasmolysed, unstressed surface. In the plasmolysed cell, intermolecular SS bonds can form only between adjacent proteins in what may be a monomolecular layer of membrane proteins on each side of the lipid layer. Tension produced on the protoplast by freezing must speed up this process by (1) inducing an additional contact between the two layers of membrane proteins via holes in the lipid layers, and (2) accelerating SS $\rightleftharpoons$ SS and probably SH $\rightleftharpoons$ SS interchange between adjacent stressed protein molecules. Maximov's method protects because instead of the membrane being stretched on freezing, it contracts. Therefore no holes can form, and the cells survive uninjured.

It must be emphasized that any change in the surface area of the protoplast, whether an increase or a decrease, may be injurious if it leads to stiffening. This is proved by plasmolysis injury. Consequently, cells which differ from the typical plant cell in not possessing a rigid wall may conceivably suffer a decrease in protoplast surface as a result of extracellular freezing. In *Arbacia* (sea urchin) eggs both changes may occur, depending on whether or not the cells have been fertilized, and both types may suffer freezing injury (Asahina, 1967).

There is one basic question, however, which must be answered before accepting the membrane-hole concept as a working hypothesis. Do the membrane proteins contain a sufficient number of SH and SS groups to fix the holes and induce rigidity in the membrane? Heber (1968) points

TABLE II

TOTAL POTENTIAL SH GROUPS (SH + 2SS) IN MEMBRANE PROTEINS

Protein	*SH groups/10 000 mol.wt.*	*Reference*
Thiogel	0·6–0·8	
Chloroplast structural:		
Chlorella pyrenoidosa	1·2	Weber, 1962
Allium porrum	0·3	Weber, 1962
Antirrhinum majus	0·32	Weber, 1962
Spinacea oleracea	0·44	Weber, 1962
	1·2	Criddle, 1966
Beta vulgaris	0·5–0·8	Bailey, Thornber and Whybom, 1966
Mitochondrial structural:		
Neurospora	2·0–2·5	Woodward and Munkres, 1966
Yeast	5·0	Woodward and Munkres, 1966
Beef heart	4·6	Criddle *et al.*, 1962
Beef heart	2·5	Woodward and Munkres, 1966
Beef heart	1·0	Lenaz *et al.*, 1968
Erythrocyte membrane	1·0	Morgan and Hanahan, 1966
Erythrocyte membrane	1·0	Mazia and Ruby, 1968
Liver membrane (Eigen)	0·0	Neville, 1969

out that analyses of chloroplast lamellar protein have revealed a very low SH content (0·11 mol per cent cystine according to Weber, 1963). Such a low value would, indeed, seem to eliminate SH and SS groups as a factor in membrane structure, though it seems incompatible with Heber's (and other investigators') evidence of the need to add mercaptans in order to maintain the activity of the membrane system. Furthermore, several of the published results yield values just as high as, or higher, than those for Thiogel (Table II), and therefore fully adequate for protein aggregation. In any case, chloroplast lamellar proteins occur only in green cells, whereas all plant cells are susceptible to freezing injury. Consequently, if membrane proteins are a factor, they must be found in all cells. Furthermore, the proposed mechanism can apply only to plasma membranes. Unfortunately, little information is available as to the amino acid content of plasma membranes. The only reliable values at present available are those for erythrocytes (Morgan and Hanahan, 1966). These values are particularly pertinent since the erythrocyte does not contain any sub-cellular structures to complicate analyses. The haemoglobin-free stroma quantitatively retains the cellular lipid and is therefore considered to represent the plasma membrane. The lipoprotein isolated from this erythrocyte membrane was found to contain one half-cystine and six cysteic acid residues for a total of seven SH groups for a molecular weight of 58 000. This is about double the content of Thiogel (6–8 SH groups for a mol. wt. of 100 000). Since Thiogel so readily forms a rigid structure due to intermolecular SS formation on freezing, the erythrocyte membrane should do so four times more readily, for the opportunity to form SS bonds rises geometrically with the number of SH groups. That these high SH values are characteristic of membrane proteins in general is indicated by analyses of mitochondrial structural proteins (Woodward and Munkres, 1966), which are believed to be the membrane proteins: 3·7–5·0 SH groups per mol. wt. of 22 000 in six strains of *Neurospora*, five SH groups per mol. wt. of 23 000 in beef heart, and double this amount in yeast (mol. wt. not given). These values are three to 20 times as high as in Thiogel (Table II).

Evidence with SH reagents has also proved the existence and importance of large numbers of SH and SS groups in the membrane of red blood cells. Approximately 7 per cent of the SH groups are readily reactive with *N*-ethylmaleimide, chlormerodrin, and $HgCl_2$, up to 25 per cent react with chlormerodrin and $HgCl_2$, and the remaining 75 per cent react only with $HgCl_2$ (Rothstein and Weed, 1963). Furthermore, both the passive and active uptake of substances by red blood cells are affected by SH reagents (Webb, 1966).

On the basis of the above evidence, membrane proteins must unquestionably be exposed to the danger of intermolecular SS formation when dehydrated. The proposed mechanism is summarized in Table III.

FREEZING TOLERANCE

The above hypothesis provides an answer to the first question posed: "What is the mechanism of freezing injury?" The second and third questions presented by the existence of freezing tolerance are: (1) "Why are tender plants incapable of becoming tolerant (i.e. of hardening)?" and (2) "What happens to hardy plants during the hardening process, which prevents injury during subsequent freezing?"

TABLE III

PROPOSED MECHANISM OF FREEZING INJURY

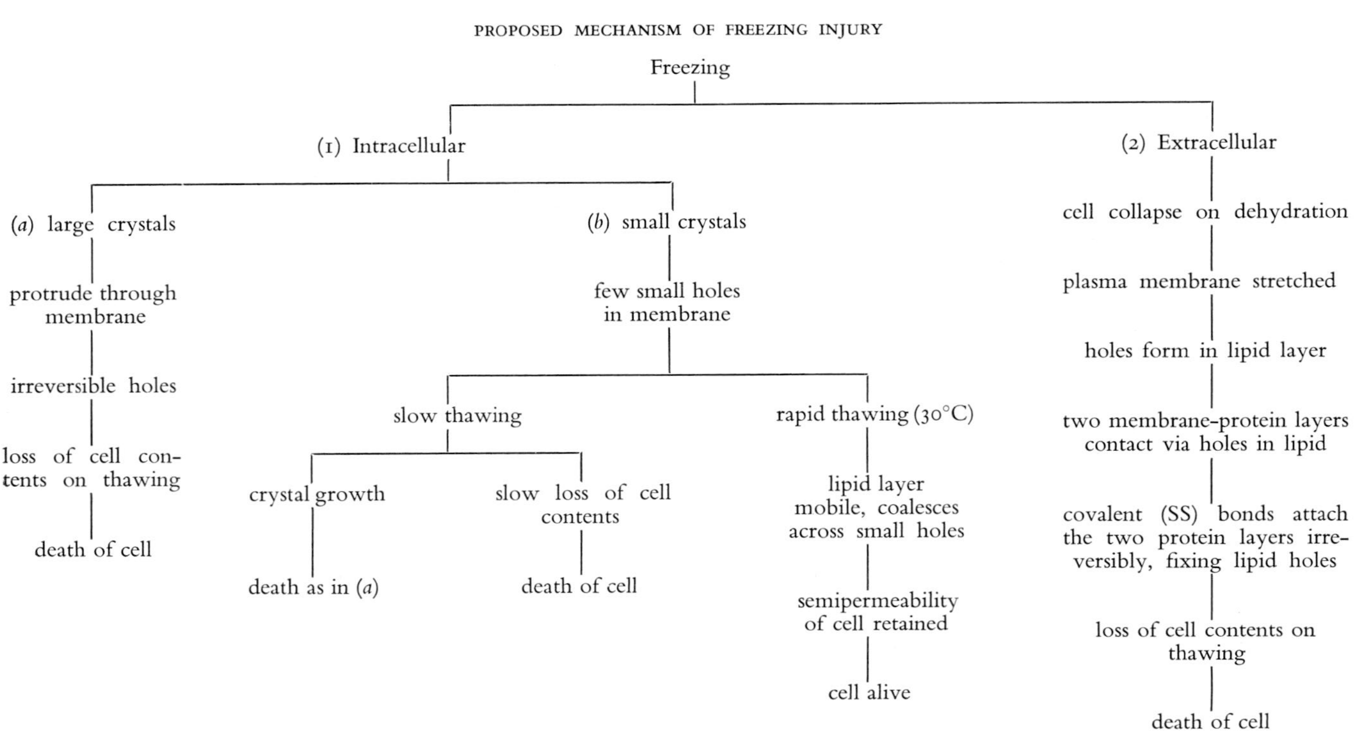

Tender and hardy plants

Even plants which survive the severest freezes of winter possess little if any freezing tolerance in spring and summer. They must be exposed to a period of low (but above freezing) temperature in the fall, during which they "harden" progressively—i.e. they become progressively tolerant of more and more severe freezes. Some plants are "tender" and are unable to develop freezing tolerance by hardening, in contrast to the "hardy" plants. The hardy plants when in the unhardened state may be just as intolerant of freezing as the tender plants; but in the hardened state they can survive freezing without injury, the degree of freezing survived depending on (*a*) the hardening potential and (*b*) the amount of this hardening potential actually induced.

What then distinguishes hardy from tender plants? On the basis of the above concept of freezing injury, we may define tender plants as those with proteins which undergo a sufficient degree of reversible denaturation ($N \rightleftharpoons D$) at hardening temperatures (e.g. 5°C) to inactivate reversibly the enzymes required for the metabolic processes leading to hardening. Hardy plants would, then, possess enzymes which remain in the native state at these hardening temperatures. This would permit the hardy plants to continue their normal metabolism at the low (hardening) temperature, which would lead to a gradual increase in freezing tolerance. Indirect evidence in favour of this explanation is obtained from growth measurements. Sunflower (a tender plant) grows rapidly at 15–25°C. When cooled to 5°C, its growth is drastically reduced. If returned to 15–25°C, its growth rate immediately returns to the original value. Cabbage (a hardy plant) also grows rapidly at 15–25°C. When cooled to 5°C, its growth is reduced less than in sunflower, and its Q_{10} for the change in rate of growth is 2·0–2·3 (Cox and Levitt, 1969), which is characteristic of simple chemical reactions. These results agree with the hypothesis that the enzymes required for growth of the tender plant (sunflower) are partially inactivated (i.e. denatured) reversibly at 5°C, while those of cabbage retain the activity expected of the fully native enzyme.

What can account for this apparent difference in enzyme response to low temperature? If the enzymes of the tender plants possessed a higher content of hydrophobic bonds, they would unfold (denature) more readily at low temperature than would the enzymes of the hardy plant. Again, indirect evidence is in agreement with this concept. Hydrophobic bonds become stronger with rise in temperature. Therefore, if the enzymes from tender plants have more hydrophobic bonds, they should require a *higher* temperature for *heat* denaturation than would the enzymes from the hardy plants. This expectation is fulfilled, judging from the higher maximum temperatures for growth in the tender plants (Levitt, 1969*b*).

The hardening process

(*a*) *Changes expected from membrane hole hypothesis.* On the basis of the above hypothesis, the following three changes, which would confer increased freezing resistance on the plant, may conceivably occur during the hardening process:

(1) A change may occur in the properties of the lipid layer in the plasma membrane which would prevent the formation of holes during freezing or repair them during thawing. Indirect evidence of repair is provided by the conductivity method of determining freezing resistance. This is a technique for measuring the freezing injury by the amount of electrolyte which diffuses out of the thawed tissue. Surprisingly, it is capable of distinguishing between plants which suffer no injury at the freezing temperature used—e.g. a plant which must be frozen at −15°C in order to injure it will show a greater loss of electrolyte after freezing at −8°C than will a plant which must be frozen at −20°C in order to injure it. This must mean that these plants suffer some membrane damage at −8°C, but this is slight enough to be completely repaired on thawing. On the other hand, it also indicates that the lipid layer cannot develop the necessary cohesive force to prevent the formation of holes, and the plant must protect itself by repairing them. The fact that hardened plants have no resistance to intracellular freezing injury is further evidence that their lipids cannot prevent the formation of holes. This is reasonable since all lipids are apolar and therefore their molecules attract each other with a very small force. It has, however, been known for some time that the lipids are more fluid, due to greater unsaturation, when synthesized by the plant at low temperature (Levitt, 1956; Gerloff, Richardson and Stahmann, 1966). This fluidity may perhaps play a role by permitting the membrane to coalesce across the holes on thawing. But coalescence would be possible only if the holes are reversible. Consequently, this kind of change in the lipids during hardening can be of secondary importance only.

Siminovitch and co-workers (1968) have suggested another way in which lipids could increase hardiness. They found that total lipids increased during hardening in the fall by 30 to 40 per cent but that phospholipids increased over 100 per cent. The increase in the polar phospholipids was therefore at least partially at the expense of the non-polar or neutral lipids. This increase corresponded to the increase in membrane content of the cells since the organelles and the cytoplasm as a whole also showed a 100 per cent increase. On the other hand, starved cells increased markedly in hardiness and in phospholipid content without any increase in organelles or cytoplasm. On the basis of modern concepts of membrane structure, the lipid content of a membrane is fixed (e.g. as a bimolecular leaflet) and no increase per unit membrane area would be expected. Even if a membrane with permeable "holes" is a true picture, such an increase in lipid content would decrease the permeability of cells to polar substances, whereas an increase actually occurs. The only possible way in which an increase in membrane lipids could occur, as Siminovitch and co-workers (1968) suggest, is by a folding of the protoplast surface. If this occurred, it would of course be expected to prevent the tension on the protoplast surface, and therefore the injury during freezing. But this mechanism requires proof of the existence of such folds in the plasma membrane.

(2) The tension on the protoplast surface during freezing may decrease, preventing the formation of holes. This effect could be brought about osmotically by an increase in cell sap concentration. It has long been known that such an increase, primarily in carbohydrates or sugar alcohols, occurs during hardening. But it must be realized that this osmotic factor is of little value by itself. If the unhardened plant is killed by freezing at −2°C, doubling its cell sap concentration would lower the killing point only to −4°C, since the same degree of dehydration (and therefore of cell collapse) in the hardened plant would occur at −4°C as in the unhardened plant at −2°C. This osmotic factor is, therefore, also secondary in importance.

(3) A change may occur in the properties of the membrane proteins which would prevent stiffening during extracellular freezing. Since the other two changes are secondary factors in freezing resistance, this third change must be the primary, most fundamental factor developed during hardening. That the proteins do show this change as a result of the hardening process has been proven by micrurgical and plasmolytic tests (Levitt, 1956).

(*b*) *Mechanism of hardening*. The above three changes are expected on the membrane hole hypothesis, and have been shown to take place experimentally. But how are these changes brought about by the hardening process? We shall first consider the last of the above changes, since this is the primary one, without which no marked hardening can occur. There are two conceivable mechanisms for preventing the stiffening of the membrane proteins.

(1) The first is the accumulation of protective substances. It has long been known that both denaturation and intermolecular SS bonding can be prevented by the presence of certain substances, for instance the substrate of an enzyme. Also the intermolecular SS bonding which occurs in Thiogel on freezing can be prevented by the presence of solutes such as sugars, glycerol and DMSO (Andrews and Levitt, 1967). This mechanism could easily explain the protection of soluble proteins and of some enzymes attached to mitochondrial or chloroplast membranes. But in order to prevent the formation of the postulated irreversible holes in the plasma membrane, these protective substances would have to be present both within the membrane and at both surfaces. While it is conceivable that the substances are in contact with the internal surface of the membrane, it is impossible for them to be present external to it, since there are essentially no solutes external to the protoplast (i.e. in the cell wall). It is also impossible for such solutes to occur within the membrane, for this would destroy the semipermeable properties of the plasma membrane if present in sufficient quantity to protect all the membrane proteins.

(2) Changes in properties of membrane proteins may prevent covalent bond formation between adjacent molecules. There are two possible mechanisms for producing such a change. (*a*) The proteins may be converted at the low temperature to another form—i.e. a conformational change may occur. Thus, it has been shown that an enzyme (L-amino acid oxidase) can be converted from one form to another by a change in temperature from 37°C to 0°C (Wellner and Hayes, 1968), provided that other factors such as pH are also changed. The forms apparently differ only in conformation. The best method for determining a conformational change is by a change in reactivity of an amino acid residue (Yankeelov and Koshland, 1965), and the first evidence appears to be increased reactivity of the cysteine residue. Direct amperometric measurements of free (reactive) and masked (unreactive) protein SH of cabbage leaves failed to detect any conformational change during a two-day hardening period (Table IV). This amount of hardening was shown in other tests to lower the killing temperature from −2° for the unhardened to −7° to −10°C for the hardened leaves. Furthermore, during a hardening period that was continued for 75 days, during which the killing temperature dropped to −20°C, the percentage of free SH remained at about the same level, fluctuating irregularly between 65 and 85 per cent. These results for the soluble proteins were essentially confirmed for the in-

TABLE IV

FREE (UNMASKED) SH GROUPS AS A PERCENTAGE OF TOTAL (MASKED PLUS UNMASKED) SH IN SOLUBLE PROTEINS OF CABBAGE LEAVES BEFORE AND DURING HARDENING

	(a)		(b)	
Time (hr)	*Hardened series*	*Unhardened series*	*Hardened series*	*Unhardened series*
− 47	72	68	73	70
− 23	71	70	72	68
− 2	72		69	
beginning of hardening				
+ 1 to 3	68	70	71	72
+ 13	66		68	
+ 26	67	64	67	74
+ 36	72		72	
+ 50	73	73	72	76

(*a*) Determined in whole supernatant, (*b*) proteins separated from non-proteins by Sephadex filtration. − = before hardening, + = after beginning of hardening. Average of three experiments.

soluble proteins, but the experimental error was much larger due to difficulties in titrating insoluble substances. (*b*) The proteins may be broken down and resynthesized at the low temperatures during hardening. This may lead to the replacement of freezing-sensitive by freezing-tolerant proteins. The fact that hardening is a gradual process, requiring days or weeks at low temperature for completion, favours this explanation. The necessary metabolic changes must occur slowly, due to the low temperature, and full hardiness would not be attained until all or nearly all the proteins had been broken down and resynthesized.

Several lines of evidence favour this mechanism. In overwintering plants both RNA and protein synthesis are accelerated in the fall during the hardening period, leading to a net increase (Siminovitch, Rhéaume and Sahar, 1967; Siminovitch, Gfeller and Rhéaume, 1967; Jung, Shih and Shelton, 1967; Li and Weiser, 1967). In cabbage plants, artificial hardening can be induced readily by exposure to low temperatures (0 to 5°C) and the rate of hardening of individual leaves on the plant parallels their growth rates (Cox and Levitt, 1969). Since growth rate is dependent on protein synthesis, this again points to a parallel between ability to harden and rate of protein synthesis. In favour of this conclusion, the cabbage leaves show a marked increase in protein content as a result of the artificial hardening, just as the overwintering plants do (Morton, 1969). Plants unable to grow at low temperatures (e.g. tomato, corn, millet, etc.) are also unable to harden. Presumably they are unable to synthesize proteins rapidly enough to support growth and therefore to support hardening. In some of these plants (e.g. tomato), in fact, the low temperature results in a net hydrolysis of proteins, presumably because little or no resynthesis occurs (Levitt, 1969*b*). Similarly, even the hardy plants are unable to harden in spring or early summer when growing. Again, this is probably because the growing parts of the plant (e.g. the leaves, flowers, fruit), like the above tender plants, are unable to grow and therefore to synthesize proteins actively at low temperatures. Even the non-growing portions of the plant would not be able to harden during spring and summer (as they do in the fall), because they have been depleted of proteins and substrates by translocation to the growing parts.

Although the measured increase in protein synthesis during hardening has been mainly due to soluble proteins, some measurements have also revealed an increase in the insoluble proteins (Levitt, 1966*b*). Furthermore, in some cases an obvious increase in total protoplasm per cell can be recognized by direct observation (Siminovitch *et al.*, 1968). And even though the soluble proteins do not appear to be irreversibly aggregated on freezing, it may be just as important for the freezing resistance of the plant to resynthesize them as to resynthesize the membrane proteins. This is because mixed disulphides between membrane and soluble proteins may be just as effective in fixing the membrane "holes" irreversibly as SS bonds are between the membrane proteins themselves.

The increase in quantity of protein *per se* is not the cause of the hardening, for both the RNA and soluble protein content remain at their maximum level throughout the winter and even in early spring when hardiness decreases (Sakai and Yoshida, 1968; Siminovitch *et al.*, 1968). It is only when the plant has lost all its hardiness in spring that the content of these two groups of substances falls to the summer level. Therefore it must be the protein quality that controls hardiness. In agreement with this conclusion, Siminovitch and co-workers were able to induce considerable hardening in starved cells in the absence of an increase in protein quantity. Presumably the rapid turnover of the proteins at the low temperature in the fall leads to a replacement of proteins characteristic of non-hardened cells by proteins characteristic of hardened cells. In the spring, due to the retention of the high RNA content, there is once again a rapid turnover of the proteins, this time at higher temperatures, which therefore regenerates the original proteins characteristic of the unhardened cells. Thus, we must conclude that the mere resynthesis of the cell's proteins would not by itself confer hardening on the plant, unless the properties of the newly formed proteins differed in some way from the properties of the proteins which they replace. On the basis of the membrane hole hypothesis, the newly-formed proteins must be able to prevent intermolecular SS bonding during freezing. Two lines of evidence support this prediction.

(1) It has been shown that Thiogel of high molecular weight (10^5) increases its melting point on freezing, but Thiogel of low molecular weight (10^4) does not (Levitt, 1965). This result is easily understood, since the protein contains six to eight SH groups per mol. wt. of 100 000. Since each protein molecule of 10 000 mol. wt. contains not more than one SH group, the proteins can aggregate to dimers but not to any larger aggregates. Such dimers would not be able to form a rigid framework in the membrane, and furthermore their denaturation would still be reversible since the unattached end of each molecule would be free to refold. This would also explain why the small (10 000–20 000 mol. wt.) globular soluble enzymes (e.g. RNase, papain, trypsin, etc.) are usually not inactivated irreversibly by freezing. Similarly, it may explain the role of the increase in soluble proteins during hardening, reported by so many investigators (Levitt, 1966*b*). This increase may be due to a dissociation of oligomers into monomers or dimers. This would be expected from the fact that the association of small protein units into multimers is commonly due to hydrophobic bonds between the units. The weakening of these bonds by low temperature would lead to the formation of the smaller units. This method has actually been used to dissociate isozymes into their subunits (Vesell, 1968), which then recombine on warming, forming hybrids.

(2) A decrease in content of protein SH groups has been found to occur during hardening, usually after an initial increase (Kohn and Levitt, 1966). It has already been shown (Table III) that no change occurs in reactivity of the SH groups, therefore these results must represent a true change in SH content. In support of this conclusion, an increase in non-protein SH accompanies the decrease in protein SH, at least during part of the hardening period. Unfortunately, translocation of amino acids into and out of the leaf prevents the preparation of a true balance sheet. All these results have since been corroborated; but a similar decrease in protein SH has been found during the normal ageing of leaves without an increase in freezing tolerance. Furthermore, these measurements were made on the soluble proteins and therefore do not necessarily apply to the membrane proteins, which have so far not yielded sufficiently accurate quantitative determinations to reveal the changes in SH content during hardening. This is unfortunate, since a change in SH content of the membrane would explain the increase in cell permeability which is

known to occur on hardening. It has long been known that both Cu^{++} and Hg^{++} inhibit the entry of glycerol into red blood cells. Since the amount of Hg^{++} required to alter the membrane properties is of the same order of magnitude as the estimated SH content, Webb (1966) suggests that the change in permeability is due to reaction of the Hg with SH groups (forming S–Hg–S bridges) in and around the membrane pores, thus impeding the passage of substances across the membrane. A similar effect could be produced by SS bridges between adjacent membrane proteins. Consequently, a decrease in SH (and SS) content of the membrane proteins could conceivably account for the increase in permeability. Such a decrease in protein SH would also prevent intermolecular SS formation between osmotically dehydrated membrane proteins, and would explain the fluidity of the protoplast surface in hardened cells, as opposed to the rigidity developed in unhardened cells during plasmolysis. Finally, it would explain the increase in freezing tolerance, since the lower the protein SH content of the membrane, the less chance to fix the "holes" irreversibly by intermolecular SS formation.

(*c*) *Metabolic control.* The mechanism of hardening cannot be understood unless it is known how the metabolic controls of the plant are altered by exposure to the hardening temperature, in such a way as to induce (1) the replacement of non-resistant proteins with resistant proteins, (2) an increase in osmotically effective solutes, and (3) an increase in fluidity of lipids.

The first effect of the low temperature is to slow down all the metabolic processes. This retardation will not be uniform for all reactions, because of differences in the temperature (or Arrhenius) coefficients. Some reactions may even be stopped altogether, due to lack of substrate (Selwyn, 1966). Both growth rate and respiration rate are, in fact, decreased to a greater degree in overwintering annuals than photosynthesis is (Levitt, 1967). As a result, carbohydrates accumulate to a greater degree than at normal growing temperatures. Since the carbohydrates are primarily in the soluble form in hardened plants, their accumulation may be sufficient eventually to inhibit assimilation of carbon. Even without such inhibition, assimilation of carbon would be slowed down to a greater degree than the light reactions of photosynthesis, since light-dependent reactions have smaller temperature coefficients than ordinary dark chemical reactions. As a result, most of the ATP and NADPH formed in the light reactions would not be used in assimilation of carbon and would be available for other metabolic paths (Levitt, 1967). This availability of ATP, NADPH, and excess carbohydrate is undoubtedly one reason for the increased RNA and protein synthesis during hardening at low temperature. In the winter perennials the difference is even greater because growth inhibitors accumulate in late summer and early fall, all growth stops and the plant becomes dormant. But although growth stops, metabolism continues; and since the assimilates are not utilized for growth, they accumulate to an even greater degree than in the winter annuals which continue to grow at a decreased rate. The winter perennials, therefore, attain an even higher degree of freezing resistance than do the winter annuals. The accumulation of substances may also protect proteins from denaturation or aggregation, since enzymes combined with substrate may be protected from denaturation under conditions which would denature and aggregate the pure enzyme. The accumulation of soluble carbohydrates also leads to osmotic protection.

It is conceivable that this metabolic shift at low temperature may also account for other changes during the hardening. The NADPH may lead to an increase in the ratio of SH to SS bonds in the membrane proteins, and this may conceivably be the cause of the increase in cell permeability.

The answer to questions (2) and (3) posed at the beginning of this paper is therefore provided by the following proposed mechanism (Table V):

TABLE V

MECHANISM OF HARDENING

(1) Temperature lowered to 5°C (and successively lower temperatures)

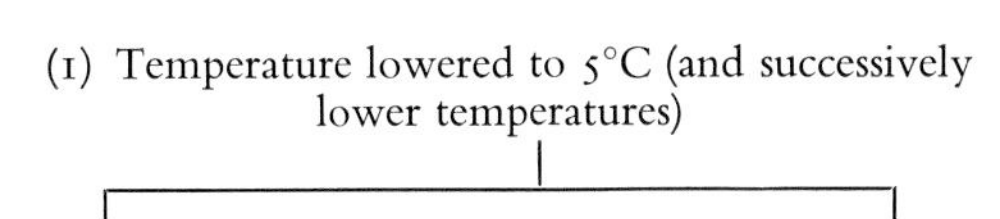

Tender plants	*Hardy plants*
(Proteins with high hydrophobicity)	(Proteins with low hydrophobicity)
(2) N → D	(2) Little N → D
(3) Some enzymes reversibly inactivated	(3) No appreciable enzyme inactivation and both growth and metabolism retarded by an amount dependent on the Q_{10} values
(4) Protein synthesis retarded and growth stopped	(4) Photosynthesis *less* retarded than respiration and growth; therefore, sugars accumulate
(5) No hardening	(5) Light reactions *less* retarded than assimilation of C; ATP and NADPH not used up
	(6) Due to (4) and (5), RNA and protein synthesis enhanced
	(7) Membrane proteins newly formed at 5°C less liable to intermolecular SS formation on freezing
	(8) Lowering of freezing temperature necessary to cause injury

Why, then, should plants differ in the degree of hardening which they can undergo at hardening temperatures? This may also depend on the degree of hydrophobicity of their proteins. Hardening is a progressive process, which increases only up to a certain point with the length of time at a single hardening temperature. Maximum hardening requires exposure for successive periods of time to 5, 0, and -3°C. The lower this hardening temperature, however, the greater the degree of denaturation (N → D) and, therefore, of inactivation of the enzymes. Consequently, slightly hardy plants may perhaps be able to harden only at 5°C, moderately hardy plants may continue to harden at 0°C, and very hardy plants may continue even at -3°C. Thus, white clover was the hardiest of three varieties tested and it continued to accumulate sugars at a later date in the fall (and therefore at a lower temperature) than did the others (Smith, 1968). Ladino, the least hardy, stopped its accumulation at the earliest date (and therefore at the highest temperature). Presumably, the degree of hardening achieved at each of these temperatures would depend on the degree of denaturation and inactivation of the enzymes, and therefore on their hydrophobicity. This answers the fourth question proposed at the beginning of this paper.

The fifth question, on the mechanism of artificial cryoprotection, has already been answered in the first part of this paper. Maximov's method of cryoprotection would be due to prevention of tension on the protoplast membrane, and the effectiveness of penetrating cryoprotectants would be due to prevention of intermolecular bonding between membrane proteins.

SUMMARY

Five basic questions must be answered in order to understand freezing injury and resistance. The following are the questions and the postulated answers.

Question 1: What is the mechanism of freezing injury?
Answer: "Holes" are formed in the semipermeable plasma membrane, leading to loss of semipermeability, efflux of the cell solution, and consequent death. The "holes" are produced (*a*) by ice crystals penetrating the membrane during intracellular freezing, (*b*) by tension on the membrane due to cell collapse during extracellular freezing. The "holes" are fixed irreversibly by intermolecular SS bonding of the membrane proteins.

Question 2: What happens during the hardening process which prevents injury during subsequent freezing?
Answer: There are three changes: (1) An increase in fluidity of the protoplasmic lipids due to an increase in unsaturation. This tends to

prevent and to repair "holes". (2) An increase in cell solutes (e.g. sugars). This decreases the extracellular ice formation and therefore the tension at any one freezing temperature. (3) A decrease in the chance of intermolecular SS bonding between the membrane proteins, due perhaps to a decrease in SH content of the membrane proteins, which are resynthesized during hardening. All three changes are due to a change in metabolic balance at the low (hardening) temperature, leading to an accumulation of sugars, ATP, NADPH, etc., which would otherwise be used up by respiration, growth, etc.

Question 3: Why are tender plants incapable of hardening?

Answer: Due to the high hydrophobicity of their proteins, they unfold reversibly when exposed to hardening temperatures. This inactivates certain enzymes required for protein resynthesis, and therefore prevents the above three processes.

Question 4: Why are there all degrees of hardiness?

Answer: The three proposed changes are quantitative, so may yield different degrees of hardening. Similarly, protein hydrophobicity is quantitative. It may be low enough to permit protein resynthesis at 5°C and therefore the first stage of hardening, but not low enough to prevent unfolding and inactivation of the enzymes at 0°C, the temperature required for the second stage of hardening. Finally, the last stage of hardening requires exposure to about −3°C, at which temperature only the least hydrophobic of proteins would remain native and therefore active.

Question 5: Why are cryoprotectants able to prevent freezing injury?

Answer: (*a*) Maximov's (non-penetrating) cryoprotectants protect by preventing the development of tensions on the protoplast surface. This protection, however, is limited in time of effectiveness, for plasmolysis occurs and the protoplast surface eventually stiffens, due to intermolecular SS formation, and is ruptured on deplasmolysis during thawing. (*b*) Penetrating cryoprotectants protect by acting as barriers between the protein molecules, preventing intermolecular SS formation.

REFERENCES

ÅKERMAN, Å. (1927). *Studien über den Kältetod und die Kälteresistenz der Pflanzen*, pp. 1–232. Lund: Berlingska Boktryckeriet.

ANDREWS, S., and LEVITT, J. (1967). *Cryobiology*, **4**, 85–89.

ASAHINA, E. (1967). In *Cellular Injury and Resistance in Freezing Organisms*, pp. 211–230, ed. Asahina, E. Sapporo: Institute of Low Temperature Science, Hokkaido University.

BAILEY, J. L., THORNBER, J. P., and WHYBOM, A. G. (1966). In *Biochemistry of Chloroplasts*, pp. 243–255, ed. Goodwin, T. W. London: Academic Press.
BRANDTS, J. F. (1967). In *Thermobiology*, pp. 25-72, ed. Rose, A. H. New York: Academic Press.
COX, W., and LEVITT, J. (1969). *Pl. Physiol., Lancaster*, **44**, 923–928.
CRIDDLE, R. S. (1966). In *Biochemistry of Chloroplasts*, pp. 203–231, ed. Goodwin, T. W. London: Academic Press.
CRIDDLE, R. S., BOCK, R. M., GREEN, D. E., and TISDALE, H. (1962). *Biochemistry*, **1**, 827–842.
DAVSON, H., and DANIELLI, J. E. (1943). *The Permeability of Natural Membranes*. New York: Cambridge University Press.
FEUGHELMANN, M. (1966). *Nature, Lond.*, **211**, 1259–1260.
GAFF, D. F. (1966). *Aust. J. biol. Sci.*, **19**, 291–299.
GERLOFF, E. D., RICHARDSON, T., and STAHMANN, M. A. (1966). *Pl. Physiol., Lancaster*, **41**, 1280–1284.
GERLOFF, E. D., STAHMANN, M. A., and SMITH, D. (1967). *Pl. Physiol., Lancaster*, **42**, 895–899.
GOODIN, R. (1969). *On the Cryoaggregation of Bovine Serum Albumin.* M.Sc. Thesis, University of Missouri, Columbia.
GREEN, D. E., and MACLENNAN, D. H. (1969). *BioScience*, **19**, 213–222.
HATCH, F. T., and BRUCE, A. L. (1968). *Nature, Lond.*, **218**, 1166.
HEBER, U. (1968). *Cryobiology*, **5**, 188–201.
ILJIN, W. S. (1933). *Protoplasma*, **20**, 105–124.
JACOB, H. S., BRAIN, M. C., DACIE, J. V., CARRELL, R. W., and LEHMANN, H. (1968). *Nature, Lond.*, **218**, 1214.
JUNG, G. A., SHIH, S. C., and SHELTON, D. C. (1967). *Cryobiology*, **4**, 11–16.
KOHN, H., and LEVITT, J. (1966). *Pl. Physiol., Lancaster*, **41**, 792–796.
KRULL, E. (1966). *Investigations of the Frost Hardiness of Cabbage in Relation to the Sulfhydryl Hypothesis.* Ph.D. Thesis, University of Missouri.
LENAZ, G., HAARD, N. F., SILMAN, H. I., and GREEN, D. E. (1968). *Archs Biochem. Biophys.*, **128**, 293–303.
LEVITT, J. (1956). *The Hardiness of Plants*, pp. 278. New York: Academic Press.
LEVITT, J. (1962). *J. theor. Biol.*, **3**, 355–391.
LEVITT, J. (1965). *Cryobiology*, **1**, 312–316.
LEVITT, J. (1966*a*). *Cryobiology*, **3**, 243–251.
LEVITT, J. (1966*b*). In *Cryobiology*, pp. 495–564, ed. Meryman, H. T. New York and London: Academic Press.
LEVITT, J. (1967). In *Cellular Injury and Resistance in Freezing Organisms*, pp. 51–61, ed. Asahina, E. Sapporo: Institute of Low Temperature Science, Hokkaido University.
LEVITT, J. (1969*a*). *Cryobiology*, **5**, 278–280.
LEVITT, J. (1969*b*). *Symp. Soc. exp. Biol.*, **23**, 395–448.
LEVITT, J., and SCARTH, G. W. (1936). *Can. J. Res. C*, **14**, 267–284.
LEVITT, J., SCARTH, G. W., and GIBBS, R. D. (1936). *Protoplasma*, **26**, 237–248.
LEVITT, J., and SIMINOVITCH, D. (1940). *Can. J. Res. C*, **18**, 550–561.
LI, P. H., and WEISER, C. J. (1967). *Proc. Am. Soc. hort. Sci.*, **91**, 716–727.
MCCOWN, B. H., HALL, T. C., and BECK, G. E. (1969). *Pl. Physiol., Lancaster*, **44**, 210–216.
MAXIMOV, N. A. (1912). *Ber. dt. bot. Ges.*, **30**, 52–65, 293–305, 504–516.
MAZIA, D., and RUBY, A. (1968). *Proc. natn. Acad. Sci., U.S.A.*, **61**, 1005–1012.
MODLIBOWSKA, J. (1968). *Cryobiology*, **5**, 175–187.
MORGAN, T. E., and HANAHAN, D. J. (1966). *Biochemistry, N.Y.*, **5**, 1050–1059.
MORTON, W. M. (1969). *Pl. Physiol., Lancaster*, **44**, 168–172.

NEVILLE, D. M. (1969). *Biochem. biophys. Res. Commun.*, **34,** 60–64.
ROTHSTEIN, A., and WEED, R. I. (1963). *The Functional Significance of Sulfhydryl Groups in the Cell Membrane*, 35 pp. Springfield, Va.: AEC Research and Development Report UR-633.
SAKAI, A., and YOSHIDA, A. (1968). *Cryobiology*, **5,** 160–174.
SELWYN, M. J. (1966). *Biochim. biophys. Acta*, **126,** 214–224.
SIMINOVITCH, D., and BRIGGS, D. R. (1949). *Archs Biochem.*, **23,** 8–17.
SIMINOVITCH, D., GFELLER, F., and RHÉAUME, B. (1967). In *Cellular Injury and Resistance in Freezing Organisms*, pp. 93–118, ed. Asahina, E. Sapporo: Institute of Low Temperature Science, Hokkaido University.
SIMINOVITCH, D., and LEVITT, J. (1941). *Can. J. Res. C*, **19,** 9–20.
SIMINOVITCH, D., RHÉAUME, B., POMEROY, K., and LEPAGE, M. (1968). *Cryobiology*, **5,** 202–225.
SIMINOVITCH, D., RHÉAUME, B., and SAHAR, R. (1967). In *Molecular Mechanisms of Temperature Adaptation*, pp. 3–40, ed. Prosser, C. L. Washington, D.C.: AAAS.
SMITH, D. (1968). *Cryobiology*, **5,** 148–159.
SUGIYAMA, T., and AKAZAWA, T. (1967). *J. Biochem., Tokyo*, **62,** 474–482.
VESELL, E. S. (ed.) (1968). *Ann. N.Y. Acad. Sci.*, **151,** 1–689.
WEBB, J. L. (1966). *Enzyme and Metabolic Inhibitors*, vol. 3, 892 pp. New York: Academic Press.
WEBER, P. (1962). *Z. Naturf.*, **17B,** 683–688.
WEBER, P. (1963). *Z. Naturf.*, **18B,** 1105–1110.
WELLNER, D., and HAYES, M. B. (1968). *Ann. N.Y. Acad. Sci.*, **151,** 118–132.
WOODWARD, D. O., and MUNKRES, K. D. (1966). *Proc. natn. Acad. Sci., U.S.A.*, **55,** 872–880.
YANKEELOV, J. P., and KOSHLAND, D. E. (1965). *J. biol. Chem.*, **240,** 1593–1602.

DISCUSSION

Lee: What changes in the protein biosynthetic pathway, or in its direction, are responsible for the reduction in the thiol content of the membrane proteins?

Levitt: All we know is that there is a very marked change in the metabolism of the plant during the hardening. There is an increase in RNA and there is an increase in protein, but what specific proteins are involved has not yet been investigated.

Meryman: How do you fit the protection provided by sucrose into this model?

Levitt: There is a reduced tension on the surface, with less ice formed.

Klotz: What you say suggests that a mercaptan such as cysteine ethyl ester should be able to interfere with the disulphide link.

Levitt: Maximov's protection with sugars can be overcome by mercaptoethanol, and similarly one can actually decrease the frost resistance of moderately resistant plants by parachloromercuribenzoate (PCMB) and iodoacetate. In other words we have succeeded in getting a negative effect but not a positive effect with thiol reagents.

Klotz: But shouldn't you get a positive effect with thiol reagents, if I understand the mechanism correctly?

Levitt: The trouble is that the thiol can produce two effects: it can protect the SH groups that are already there but it can also split disulphides. Splitting the disulphides will perhaps increase the injury because about 20 per cent of the total sulphur is in the form of SS in the insoluble proteins. If this is true in the membrane protein then splitting the disulphides may produce a lethal increase in permeability of the membrane. Many experiments, for instance those with blood cell ghosts (Webb, 1966), have shown that SH and SS groups in the membrane itself can be affected by SH reagents. The decrease in permeability produced by copper has been postulated as being due to a sulphur-copper-sulphur bond formation, and similarly that by mercury. All kinds of effects of this sort have been found but nobody yet knows the exact role of these SH or SS groups in the semipermeability of the plasma membrane.

Klotz: So the intolerant plant has greater rigidity of the protoplasm because of the disulphide linkages which are formed, whereas the plant that can tolerate freezing doesn't have these disulphide linkages. Is that right?

Levitt: This is the hypothesis, though it may not necessarily be that simple.

Klotz: You show the formation of an SS cross-link as the basis for providing a locus for injury. To me this implies that the conditions for susceptibility to injury could be avoided by having a mercaptan present.

Levitt: Yes, provided it did the right thing. If, for instance, the mercaptan breaks one of the disulphide bonds that is already present and converts it into two SH groups, these may then become more available for combining with another SH group and producing this bond.

Klotz: As long as you have an excess of mercaptan you should be able to maintain the reduced state.

Levitt: Yes, but 20 per cent of the sulphur is in the disulphide form and presumably this is essential for the intactness of the protein layer in some way. Presumably if one breaks these disulphides the integrity of the membrane will be destroyed. There are two opposite effects and it is difficult to know what changes each will produce because we haven't any idea really as to the actual role of these proteins in the plasma membrane.

REFERENCE

Webb, J. L. (1966). *Enzyme and Metabolic Inhibitors*, vol. 3, 892 pp. New York: Academic Press.

PROTEINS CAPABLE OF PROTECTING CHLOROPLAST MEMBRANES AGAINST FREEZING

U. Heber

Institute of Botany, University of Düsseldorf

The available evidence indicates that the ability of hardy cells to withstand freezing is not the result of a special "condition" of their protoplasm but rather the consequence of the presence of cryoprotective compounds. To acquire natural resistance against freezing a cell has to produce and to distribute within its protoplasm compounds capable of protecting its sensitive parts against the dehydration accompanying freezing. In the plant kingdom and probably also elsewhere, low molecular weight sugars serve as cryoprotective substances (Sakai, 1960, 1962; and others; see also Levitt, 1966). In order to be effective their concentration has to be relatively high (about 2 to 3 per cent w/v; Heber and Santarius, 1964). Even higher concentrations are required of agents such as glycerol or dimethylsulphoxide which are commonly used in cryopreservation of non-resistant mammalian cells (Doebbler, Rowe and Rinfret, 1966). They are assumed to exert their protective influence on a colligative basis. Only recently has evidence become available on the existence of another class of cryoprotective compounds (Heber and Ernst, 1967; Heber, 1968). This report describes a sensitive test system consisting of isolated protoplasmic membranes, which is useful for measuring the effectiveness of cryoprotective compounds, and some properties of protective protein fractions isolated from hardy plant material. Protection by these fractions appears to be remarkable from two points of view. The proteins have a molecular weight which is much higher than that of other cryoprotective agents. They presumably cannot act on a colligative basis. In addition, unusually low concentrations are required for protection.

TEST SYSTEM

The test system used consists of isolated chloroplast thylakoids (Heber, 1967). These are vesicular membrane systems, obtained by osmotic shock from isolated spinach chloroplasts. The membranes are separated from soluble chloroplast components including stroma proteins by repeated high speed centrifugation. It appears established that one of the main effects of freezing on cells and tissues, and probably the predominant one, is the alteration of biological membranes. Isolated thylakoid membranes are biologically active. In the light they support ATP formation at rates of up to 1000 µmoles/mg chlorophyll per hour. After freezing ATP formation is largely inactivated, or, when freezing occurs in the presence of salts, fully inactivated. The inactivation is irreversible. It is accompanied by the loss of the proton-pumping activity of the membranes and could be traced back to changes in their osmotic properties. ATP formation and pumping activities require semipermeable membranes *in vivo*. Unfrozen thylakoid vesicles function as osmometers. After freezing the vesicles appear collapsed. They no longer swell in hypotonic and shrink in hypertonic solutions. Freezing under "mild" conditions (low levels of salts), which leads to inactivation of phosphorylation and of pumping activities, does not inactivate individual enzymic reactions of the membranes and it appears that loss of osmotic properties is the first and probably main aspect of damage.

Thylakoid membranes are sensitive to freezing not only *in vitro*. There are also good indications that the dehydration accompanying freezing causes direct damage to the membrane systems of chloroplasts in frost-sensitive leaves *in vivo* (Heber and Santarius, 1964, and unpublished experiments). This damage and that to other membranes results in cell death.

Several compounds added before freezing protect the thylakoid system against freezing injury, as shown by the survival of photophosphorylation and of other membrane activities. Examples of such compounds are glycerol (Santarius, 1969) and dimethylsulphoxide (Heber and Ernst, 1967). The same compounds are active in the cryopreservation of intact cells such as red cells or spermatozoa. The concentrations required for the efficient protection of salt-depleted thylakoids are lower than those needed for intact cells by a factor of perhaps 4. This may be due to different stabilities of thylakoids and of the membrane systems of mammalian cells, but more likely so to differences in the electrolyte levels. The concentrations of salts in the test system determine the amount of protective agent needed for protection (Heber and Santarius, 1964). If sucrose is used to preserve phosphorylation of thylakoids a molar ratio close to 1 is needed to overcome the deleterious effects of a salt like sodium chloride.

Thylakoids can without damage be suspended in a medium of low ionic strength. Apart from the insensitivity to hypotonic conditions there are other advantages over test systems employing intact cells. As compared with intact cells, thylakoids are simple systems and open to experimental analysis. As an important feature, they lack permeability barriers, which limit the entrance of potentially active compounds to intact cells, and thus permit a clear decision on whether or not a compound is cryoprotective. It appears that only the outer surface of the membrane vesicles has to be protected against freezing injury, as compounds such as sucrose and protective proteins, which are unable to penetrate the membranes at significant rates, protect against freeze inactivation.

ISOLATION OF HARDINESS FACTORS OF HIGH MOLECULAR WEIGHT

During the winters of 1965, 1966 and 1967 several hardiness factors of high molecular weight were isolated from leaves of field-grown *Spinacia oleracea*, *Secale cereale* and *Valerianella olitoria*, and from the bark of *Populus nigra*. Attempts to isolate active material from spinach in the summer failed. Leaves or bark were ground in an isotonic salt buffer at pH 7·8 and particulate material was removed by filtration and high speed centrifugation. After adjustment to pH 4·6, brief heating to 95°C, and subsequent centrifugation, ammonium sulphate was added to the supernatant and the resulting precipitate collected. It contained active material and was further fractionated on Sephadex columns. On Sephadex G-25 five fractions were obtained. Only the first two protected thylakoid membranes against freeze inactivation. Since the second fraction was small, only the first one was collected routinely and then subjected to filtration through Sephadex G-50. There it separated into two large fractions. The slower one is designated protein factor II, while the faster one separates on Sephadex G-75 into two components, protein factor I and a faster fraction which is largely inactive in the chloroplast membrane test. Yields of protein factors I and II and the ratios of factor I to factor II varied considerably. From 100 grams of spinach leaves about 10 to 15 milligrams of the protein factors can be obtained. The actual content of the factors in the leaves may be higher.

In the leaf cells the protein factors are located in the protoplasm, as shown by the fact that they can be extracted also from isolated chloroplasts. In fact, they protect chloroplasts isolated from hardy leaves against freeze inactivation. The chloroplast membranes become sensitive to freezing only after their removal by washing. Chloroplasts isolated from non-hardy leaves do not require washing with water to become inactivated by freezing.

PHYSICAL PROPERTIES OF THE FACTORS

The protein factors I and II dissolve easily in water. They contain, possibly as a contamination, a low percentage of bound glucose and pentoses (2 to 3 per cent). Their nitrogen content is high. On hydrolysis with strong acid they yield amino acids. As calculated from the nitrogen data, about 100 per cent of protein factor I and about 80 per cent of protein factor II are made up of protein (or polypeptide). The ultraviolet spectra have maxima at 275 nm. In most preparations a shoulder at 340 nm indicates a material different from the protein component and this is assumed to be a contamination. Some (10 to 15 per cent) of the dry weight of protein factor II is unaccounted for. The protein factors are heat-stable. They are precipitated by trichloroacetic acid and the precipitate dissolves in an excess of trichloroacetic acid. Trypsin causes inactivation. Both protective fractions separated at pH 8·2 in the electrical field (microelectrophoresis in the liquid phase) in two negatively charged components. The faster component was somewhat smaller than the slower one. In the ultracentrifuge no separation into individual fractions occurred. Average molecular weights calculated from sedimentation and diffusion measurements were about 10 000 for protein factor II and 16 000 for protein factor I. (The ultracentrifugal analysis was carried out by Dr Kempfle, Institute of Physiological Chemistry, University of Bonn.)

PROTECTIVE ACTION

The protein factors are highly protective in the membrane test. The activity of protein factor I is comparable to that of protein factor II. The best fractions exceed, on a unit weight basis, the protection afforded by sucrose, dimethylsulphoxide or glycerol by a factor of 100. Protection follows a saturation curve. With average preparations maximum protection is reached at a concentration of about 0·1 per cent (w/v). This compares with about 2 or 3 per cent for sucrose, dimethylsulphoxide or glycerol. However, even at saturation levels the membranes may not be completely protected by the protein factors, but only to about 80 per cent of the unfrozen controls. It is thought that the failure to protect completely, which is often but not always observed, is due to diffusion problems. Thylakoid membrane vesicles often occur not as individual "bubbles" but in stacks. The high molecular weight factors are assumed to be unable to reach all membrane sites requiring protection. In fact, preincubation has been observed to increase the effectiveness of the protein factors (Heber and Ernst, 1967). Sucrose and other low molecular weight protective additives protect completely at saturation, but very high concentrations are required as compared with those needed for the protein

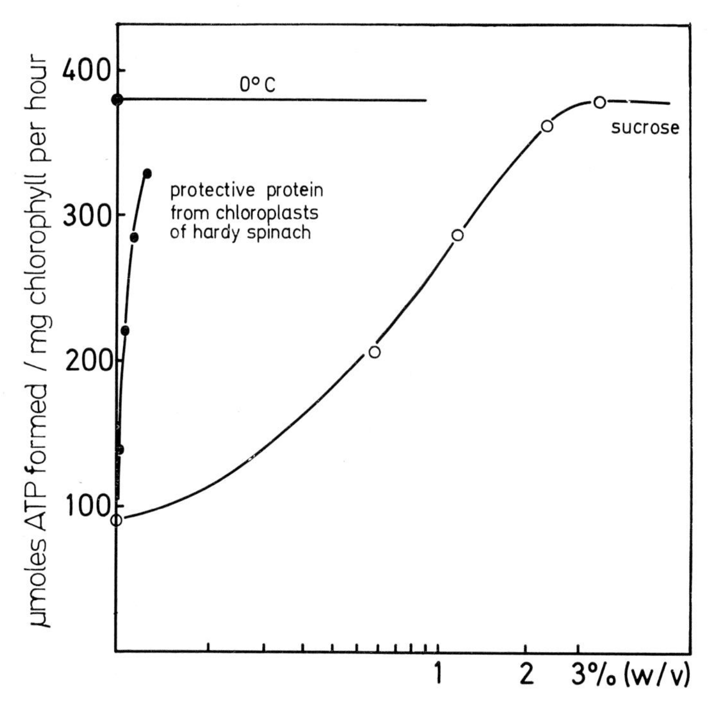

FIG. 1. Protection of thylakoid membranes against freezing by protein factor I and sucrose. After preincubation for 20 min at 0°C, the thylakoids were frozen in the presence of the indicated amounts of protective agent for 3 hr at −25°C. After thawing, light-dependent ATP formation by the membranes was determined as described previously (Heber, 1967; Heber and Santarius, 1964). Controls were kept at 0°C. Note the logarithmic scale on the abscissa.

factors (Fig. 1). From the composition of the protective fractions there is no doubt that the protein part is responsible for the protective effect. In accordance with this, trypsin digestion results in loss of activity.

Salts counteract the protection afforded by the protein factors. The higher the salt concentration the more protein factor is required for protection. Conversely, the higher the concentration of protein factor in the system, the more salt is needed for inactivation by freezing. Principally the situation is very similar to that of sucrose-protected thylakoid membranes (Santarius, 1969), but membranes protected by the protein factors have a higher resistance to the destructive effects of salts on freezing than membranes protected by sucrose. A solution of 0·1 M-NaCl is required to overcome the protection provided by 0·2 per cent protein factor I.

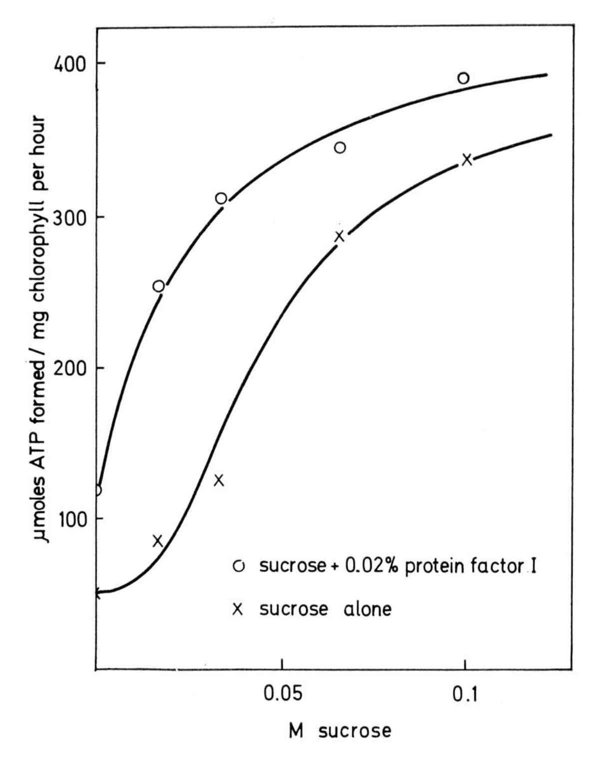

FIG. 2. Protection of thylakoid membranes by sucrose and effect of 0·02 per cent protein factor I on protection. Procedure as in legend to Fig. 1.

In the presence of very low levels of the protein factors the effectiveness of other cryopreservative agents such as sucrose is considerably increased. Fig. 2 demonstrates that in the presence of 0·02 per cent protein factor the concentration of sucrose required for full protection is reduced by about half. Lower concentrations of the protein factors are also effective. Conversely, a concentration of sucrose which by itself is scarcely protective even increases the already high effectiveness of the protein factors, as shown in Fig. 3. This cooperation between protective proteins and sucrose is difficult to explain if sucrose exerts protection on a purely colligative basis, as is often assumed for protection by additives of low molecular weight. Clearly the colligative concept cannot apply for the protein factors as their concentration is much too low and their molecular weight too high for considerable colligative action.

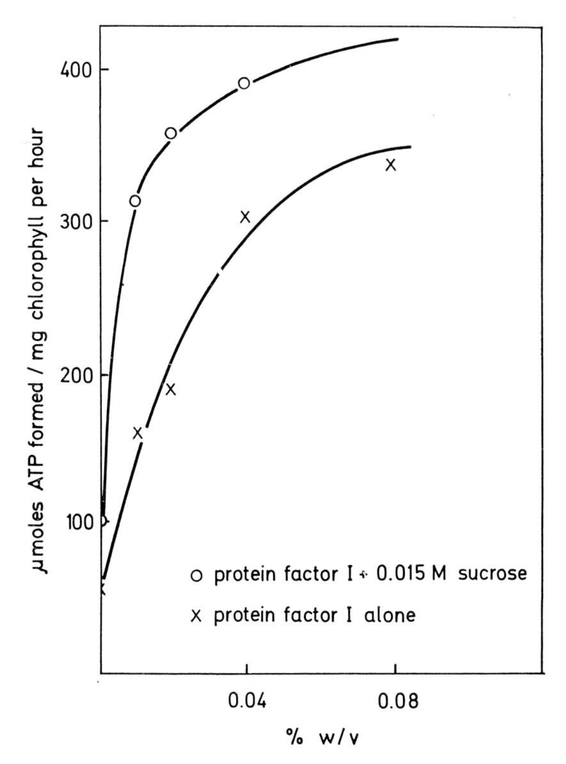

FIG. 3. Protection of thylakoid membranes by protein factor I and effect of 0·015 M-sucrose on protection. Procedure as in legend to Fig. 1.

The significance of the cooperative phenomena is obvious. Even very low intracellular levels of the protein factors can substantially increase the protection afforded by sugars. Levels of sugars quite insufficient to provide protection alone can help to protect cells in the presence of only modest levels of protein factors. Although the frost hardiness of plants often parallels the content of soluble sugars, exceptions have been observed where considerable frost hardiness develops in the absence of significant sugar accumulation. In these cases hardiness may be explained as a consequence of the formation of proteins capable of protection. In fact, accumulation of soluble proteins has often been reported during frost hardening (Siminovitch and Briggs, 1954; Levitt, 1966).

SPECIFICITY

Apart from protein factors I and II only a small third fraction, presumably of a molecular weight somewhat lower than that of protein factor II (cf. p. 177), was found to be highly protective. There was only a little protection of thylakoid membranes by the faster component separating from protein factor I on Sephadex G-75 filtration. Other proteins from hardy leaves were not protective at all. No protection was observed with fractions from non-hardy summer spinach obtained by the procedure employed for the separation of protein factors I and II from winter material. In addition, some preparations of the protein factors, although obtained from winter material, provided much less protection than others. All this indicates a certain specificity of the protein factors. It appears that evolution has led, at least in some plants and possibly also in other organisms exposed to subfreezing temperatures, to the formation of cryoprotective agents of protein or polypeptide nature which are much superior to protective cell constituents of low molecular weight, such as sugars. However, specificity certainly is limited. There is no species specificity, as protein factors from rye or poplar protect thylakoid membranes from spinach. Also, there is some or even considerable protective action of bovine albumin (Heber and Santarius, 1964) and of peptone (Santarius, 1969) in the thylakoid membrane test. Commercial peptone (Merck) subjected to the fractionation procedure used for the separation of protein factor II yields an active fraction which is less effective than good preparations of protein factor II only by a factor of 2 to 4. Thus it appears that molecular size of a polypeptide is one of the factors determining effectiveness.

MECHANISM OF PROTECTION BY THE PROTEIN FACTORS

As would be expected, the protein factors protect not only photophosphorylation of thylakoid membranes against the effects of freezing but also other membrane activities to the same extent, such as proton transport and, under proper conditions, photoinduced shrinkage of the thylakoid vesicles. In this respect they do not differ from the protection afforded by sucrose. However, the large differences in the concentrations required for protection make it likely that different mechanisms underlie protection.

It is commonly accepted that freezing injury is caused by the dehydration accompanying freezing. There appear to be two different ways to protect membranes against freeze-inactivation. The first is protection on a colligative basis by high concentrations of water-soluble, hydrogen-bonding neutral compounds which, by lowering the freezing point, serve to retain water in the system and to dilute salts. The second is a direct interaction between protective agent and membrane. It has already been mentioned that a comparison of the protection provided by sucrose or glycerol and that provided by the protein factors rules out the colligative concept for the protein factors in view of the low concentrations required for protection and their high molecular weight. There remains an interaction with the membranes. Unfortunately now the picture becomes unclear. Only negative evidence can be listed. Direct interaction with the membranes may be permanent or may become possible only during freezing, when the system becomes concentrated and higher concentrations of the protein factors come into contact with the membranes. The first possibility is unlikely, as indicated by the following observation. Addition of thylakoid membranes to a protective concentration of protein factor I (0·1 per cent) and subsequent removal of the membranes by high speed centrifugation did not significantly lower the protective capacity of the supernatant, indicating that little binding of the protein factor to the membranes had occurred. The membranes which had been in contact with the protective protein were again sensitive to freezing after its removal.

Another possible explanation of protection is that interaction of the protein factors with the membranes may cause changes in their configuration rendering them insensitive to freezing. Configurational changes should be accompanied by changes in permeability. There are three highly sensitive indicators, which do not indicate significant changes in permeability, at least in the presence of low concentrations of protective protein. On illumination of intact thylakoid membranes, there is a transfer of protons into the interior of the membrane vesicles. On darkening the protons are released through the membrane into the external medium. The kinetics of the dark release are not influenced by the presence of protective concentrations of the protein factors, indicating that there are no significant changes in permeability.

Similarly, in a suitable buffer system illumination causes thylakoid shrinkage due to osmotic loss of water. On darkening swelling occurs owing to the passive re-entry of solute and water into the vesicles. As shown in Fig. 4, no significant changes in the kinetics of swelling are seen in the presence of protective protein, again indicating unaltered permeability. The same is indicated by essentially unchanged rates of ATP synthesis in the presence of 0·1 per cent protein factor. ATP synthesis requires "intact" membrane vesicles.

There is still the possibility that permeability changes and thereby protection are caused only by the high concentrations of the protein factors resulting from the dehydration of the system by freezing. If this

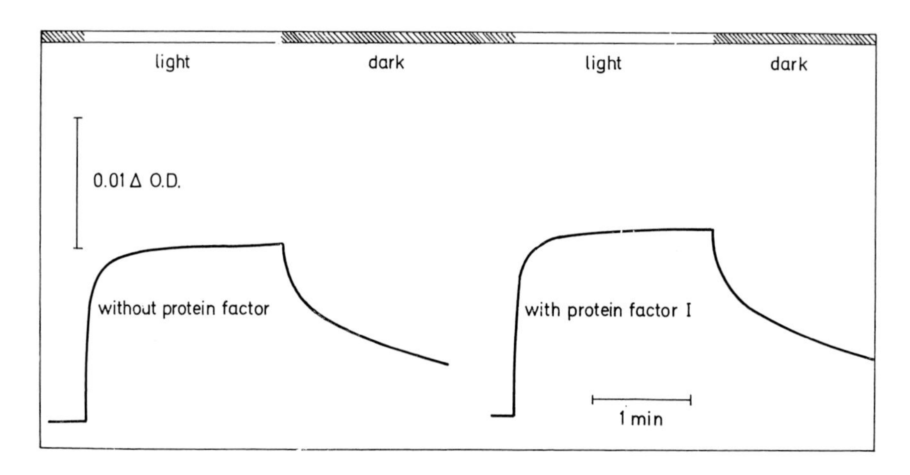

FIG. 4. Light-dependent changes in the optical density (O.D.) at 530 nm of a suspension of thylakoid membranes in the absence and in the presence of protein factor I. Positive changes (increase in light scattering) denote shrinkage of thylakoid vesicles owing to loss of solute and water, negative changes swelling owing to uptake of solute and water. Thylakoid membranes were incubated for 3 hr at 0°C with 0·1 per cent protein factor I or water and then added to 0·025 M-Tricine (*N*-tris (hydroxymethyl) methylglycine, Calbiochem) and $6 \cdot 5 \times 10^{-5}$ M-phenazine methosulphate in 0·1 M-sodium acetate, pH 7·6, to a final concentration of 50 μg chlorophyll/ml. Thylakoid shrinkage was caused by illumination with 60 000 erg cm^{-2} sec^{-1} of red light. Light path was 2·0 mm. The photocurrent produced in a photomultiplier by a measuring beam of 530 nm passing the sample reflected the changes in O.D. and was recorded continuously.

were so, then extremely fast freezing by rapid injection into liquid nitrogen of a thin stream of a thylakoid suspension in a solution of 0·1 per cent of protein factor I should be injurious, while slow prefreezing to −25°C and subsequent transfer to liquid nitrogen should be protective. This expectation is not borne out by experiment. In fact, in the presence of protein factor I fast freezing (injection) in liquid nitrogen appears to be somewhat more protective than slow freezing. While unprotected controls were completely inactivated in liquid nitrogen, 0·09 per cent protein factor I offered more than 50 per cent protection, as shown in Fig. 5. Sucrose was also protective. These experiments also demonstrate that protection is provided even at extremely low temperatures.

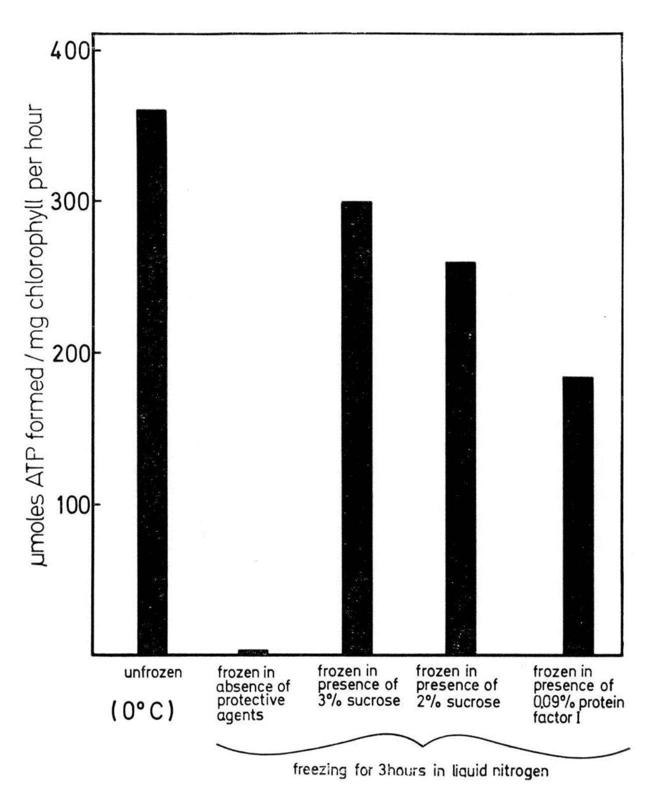

FIG. 5. Protection of thylakoid membranes by sucrose or protein factor I against fast freezing to −196°C.
After preincubation for 20 min at 0°C suspensions of thylakoid membranes in a thin stream were injected under pressure directly into liquid nitrogen. Duration of freezing, 3 hr. After thawing light-dependent ATP formation was discovered.

SUMMARY

Freezing of isolated chloroplast thylakoids results in the irreversible alteration of permeability properties of the membrane systems. As a consequence the ability of the membranes to synthesize ATP from ADP and phosphate in the light is lost. ATP synthesis and other membrane activities are protected against the effects of freezing by the addition, before freezing, of three heat-stable protein fractions isolated from frost-hardy plant material. Other protein fractions were scarcely or not at all protective. On a unit weight basis, the protective proteins were much more effective in protecting chloroplast membranes than cryoprotective compounds of low molecular weight such as sugars, glycerol or dimethylsulphoxide. Very low concentrations of the protective proteins considerably reduced the amount of other cryoprotective agents needed to give full protection of the membranes. Salts decreased the effectiveness of the protein factors. These appear to be involved in the protection of frost-hardy plant tissue against the effects of freezing.

REFERENCES

DOEBBLER, G. F., ROWE, A. W., and RINFRET, A. P. (1966). In *Cryobiology*, pp. 407–450, ed. Meryman, H. T. London & New York: Academic Press.

HEBER, U. (1967). *Pl. Physiol., Lancaster*, **42,** 1343–1350.

HEBER, U. (1968). *Cryobiology*, **5,** 188–201.

HEBER, U., and ERNST, R. (1967). In *Cellular Injury and Resistance in Freezing Organisms*, vol. 2, pp. 63–77, ed. Asahina, E. Sapporo: Institute of Low Temperature Science, Hokkaido University.

HEBER, U., and SANTARIUS, K. A. (1964). *Pl. Physiol., Lancaster*, **39,** 712–719.

LEVITT, J. (1966). In *Cryobiology*, p. 528, ed. Meryman, H. T. London & New York: Academic Press.

SAKAI, A. (1960). *Nature, Lond.*, **185,** 698–699.

SAKAI, A. (1962). *Contr. Inst. low Temp. Sci. Hokkaido Univ., B*, **11,** 1–40.

SANTARIUS, K. A. (1969). Submitted to *Planta*.

SIMINOVITCH, D., and BRIGGS, D. R. (1954). *Pl. Physiol., Lancaster*, **29,** 331–337.

DISCUSSION

Mazur: When you observed protection with 0·05 per cent protein, how much sodium chloride was in the system?

Heber: Its concentration was fairly low, of the order of 0·1 to 0·2 per cent. With higher concentrations of sodium chloride we would have to add more of the protective protein. We haven't investigated the quantitative relationship between salt and protective protein carefully, but very likely the situation is basically similar to that of sucrose protection (Heber and Santarius, 1964). However, in the presence of protective proteins the membranes are more resistant to salt injury than in the presence of sucrose.

Meryman: In R. J. Williams' experiments in our laboratory with your protein, Dr Heber, he attempted to prevent injury during freezing in higher concentrations of salt. He did have to use a higher concentration of the protein, nearly 1 per cent, to protect grana initially suspended in 25 mM-NaCl.

Mazur: It is interesting that Dr Heber's protein protects even when the ratio of sodium chloride to protein is about 2000 to 1.

Heber: For the best preparations of the protective proteins this calculation is about correct. In contrast, when we plot the extent of damage against the sodium chloride concentration in freezing experiments with sucrose as a protective agent we can see how much sucrose is needed to overcome the destructive effect of the salt. Damage is complete when the molar ratio of salt to sugar is higher than 2. At a 1:1 ratio there is still considerable protection by sucrose. These relations hold for a rather broad range of concentrations (Heber and Santarius, 1964). Santarius also found inactivation of the grana at 0°C by high salt concentrations, and protection by sucrose (Santarius, 1969).

Mazur: The curves for protein should be similar.

Heber: We haven't done that yet. Basically the situation is probably similar, but the concentrations of protective protein needed for protection would be much lower.

Levitt: Does it make any difference whether for instance there is a little bit of calcium chloride to balance the sodium chloride?

Heber: Santarius found an influence, but not a very pronounced one (K. A. Santarius, unpublished experiments).

Levitt: Has the effect of calcium been tried with blood cells, Dr Farrant?

Farrant: Not as far as I know.

Does your protein protect in hypotonic conditions?

Heber: Our system is insensitive to hypotonic conditions up to fairly high dilutions, although in distilled water activity is lost. Normally it doesn't need protection against hypotonic stress. We did not check, but I would expect that under extreme conditions (very low salt level) a low concentration of the protein would protect.

Farrant: This is therefore an important difference between your system and the red blood cell.

Levitt: Do the membranes rupture under certain conditions?

Heber: The grana need a few ions around to retain activity. We wash twice in distilled water but we do not remove all the salts. Under these conditions activity goes down from around 1000 μmoles ATP synthesis to around 400 μmoles/mg chlorophyll per hour. That means we inactivate an appreciable part of the grana, presumably due to hypotonic stress, but we are interested in the remaining activity.

Levitt: Physically some of these grana are ruptured. Uribe and Jagendorf (1968) centrifuged down the ruptured ones, and the ones that weren't ruptured remained in the supernatant solution. One can therefore separate those that have been ruptured.

Meryman: One can load the grana with sucrose by exposing them to hypertonic sucrose which leaks in, and then when the grana are returned to distilled water, sucrose leaks out.

REFERENCES

HEBER, U., and SANTARIUS, K. A. (1964). *Pl. Physiol., Lancaster*, **39,** 712–719.
SANTARIUS, K. A. (1969). Submitted to *Planta.*
URIBE, E. G., and JAGENDORF, A. T. (1968). *Archs Biochem. Biophys.*, **128,** 351–359.

THE LOW TEMPERATURE DENATURATION OF CHYMOTRYPSINOGEN IN AQUEOUS SOLUTION AND IN FROZEN AQUEOUS SOLUTION

JOHN F. BRANDTS, JOAN FU and JOHN H. NORDIN*

*Department of Chemistry and *Department of Biochemistry, University of Massachusetts, Amherst, Massachusetts*

THE subject of protein denaturation has long held the attention of many investigators and various theories relating to these strikingly cooperative reactions have been put forward through the years. The picture of denaturation reactions which has now become generally accepted is that fostered by Kauzmann (1954, 1959) in which denaturation, in its simplest form, is regarded as a monomolecular transition from a compactly folded native protein to a highly unfolded denatured protein. Many of the groups along the polypeptide chain, shielded from contact with solvent in the native state, become exposed and more or less fully solvated. For many proteins, at least, the unfolding reaction is readily reversible if solution conditions are adjusted properly. In these cases, the extent of denaturation will be governed by thermodynamics in accordance with the relative free energies of the folded and unfolded states.

Even though such reactions may be intrinsically reversible the reversibility is sometimes impaired, due to aggregation of the denatured protein. For example, the chymotrypsinogen thermal denaturation is completely reversible in the pH range 1·7 to 3·0 (Brandts, 1964*a*) at low ionic strength and low protein concentration, but the formation of an irreversible aggregate proceeds relatively fast under other conditions, so that only limited reversibility to the native state may be achieved. It nevertheless seems likely that the initiating step for the irreversible process is the reversible monomolecular unfolding of the molecule.

Although protein denaturation reactions have been extensively studied at room temperature and above, there is little information available at lower temperatures, particularly at temperatures below the normal freezing point of water. It had long been thought that native proteins in aqueous solution became more stable as the temperature is lowered, and although this is certainly true as one proceeds from very high temperatures down to room temperature, there now seems to be good evidence suggesting that it is not true as the temperature is lowered much below room temperature (Brandts, 1964*a*,*b*, 1967, 1969).

The denaturation of chymotrypsinogen has been studied in our laboratory over a wide range of temperature, from about −80° to +80°C, in order to provide a more comprehensive picture of the denaturation process. Since reversibility can be achieved readily under certain conditions, this reaction may be subjected to thermodynamic analysis. Although the purpose of this paper is to focus attention on the low-temperature stability of chymotrypsinogen, it is important to relate this to earlier findings on the "heat denaturation" since it appears that the thermodynamic properties of this protein are continuous over the entire temperature range and that the low temperature behaviour is in part predictable on the basis of what has been observed previously at high temperatures.

EXPERIMENTAL

The techniques and materials used in this study have for the most part been described elsewhere (Brandts, 1964*a*). The studies of denaturation in frozen solution were conducted using the precipitation technique of Eisenberg and Schwert (1951). Samples of chymotrypsinogen solutions of 1·0 ml, contained in a thin-wall test tube, were immersed directly into a constant temperature bath and allowed to equilibrate for a given period of time. In the samples equilibrated above −10°C, very small flakes of ice were added to prevent supercooling. After equilibration, the samples were thawed as rapidly as possible at 5°C and a 5·0 ml sample of the precipitating buffer (1·2 M-NaCl, 0·2 M-glycine buffer, pH 3·0) was added immediately. Control experiments were done to ascertain whether or not any renaturation occurred during the interval between removal of the sample from the low temperature bath and the addition of the precipitating buffer. Although the renaturation process occurs in thawed solutions, its rate is slow enough for the amount of renaturation during the short interval in question to be insignificant.

After precipitation, the denatured protein was centrifuged off and the amount of native protein in the supernatant was determined spectrophotometrically, using an extinction coefficient of 20·7 at 282 nm.

HEAT DENATURATION

Fig. 1 shows the temperature dependence of the free energy for chymotrypsinogen denaturation, obtained from spectrophotometric data assuming a two-state (Lumry, Biltonen and Brandts, 1966) transition, i.e.

$$N \rightleftharpoons D \qquad \Delta F^{\circ} = -RT\ln\frac{(D)}{(N)}$$

where (N) and (D) represent the concentrations of native and denatured protein and where ΔF° is the standard free energy of denaturation. These curves were compiled from the usual sigmoid-type denaturation curves over the pH range 1 to 3 (Brandts, 1964*a*,*b*,). In a reaction in which the

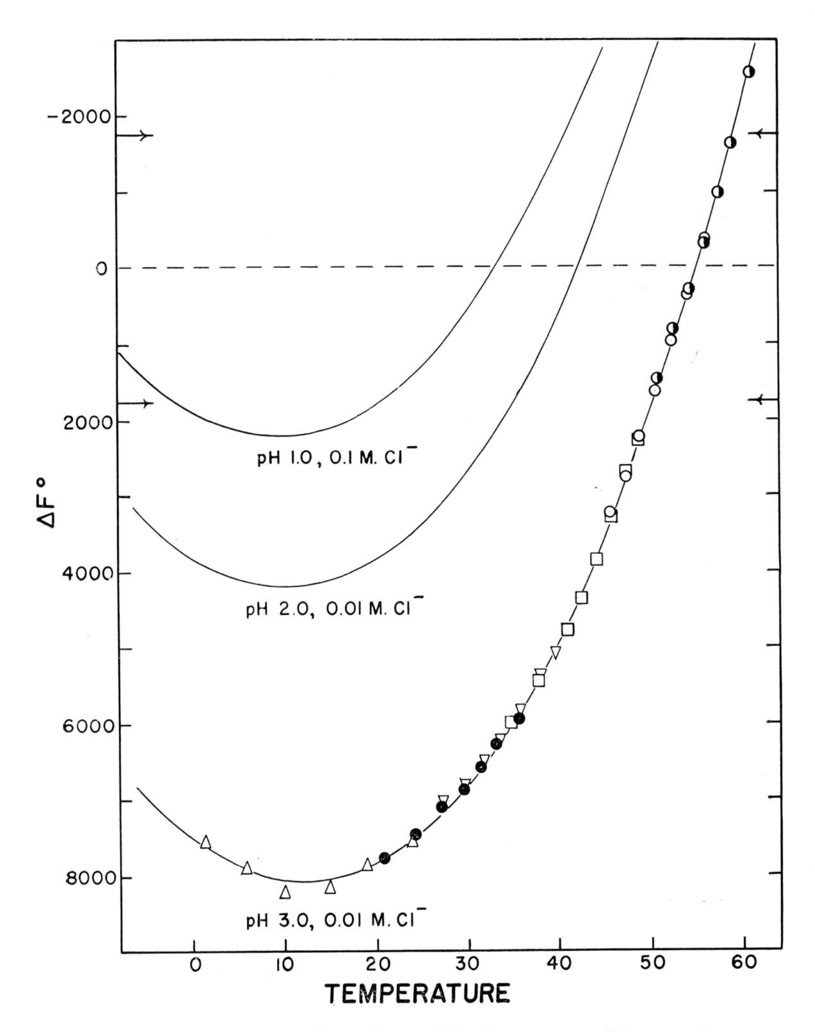

Fig. 1. The temperature dependence of the free energy of denaturation of chymotrypsinogen at different pH values and ionic strengths. From Brandts (1964*a*), with permission of the American Chemical Society.

enthalpy change, ΔH°, and the entropy change, ΔS°, are temperature-independent, ΔF° will vary linearly with temperature. For the chymotrypsinogen denaturation, the definite curvature in the free energy-temperature profiles suggests that the denatured protein has a very much larger heat capacity than does the native protein ($\Delta C_p \cong 3000$ cal [$10^3 \times 12 \cdot 55$ J] mole^{-1} deg^{-1} at 40°C) and this difference in heat capacity imparts a large temperature dependence to ΔH° and ΔS°. For example, ΔH° changes from about 120 kcal ($10^3 \times 502$ J) mole^{-1} at 50°C to a value of zero at 10°C and becomes negative below this temperature.

The implication from these curves is that the free energy profile is nearly parabolic in shape, with the apex located at about 12°C, which corresponds to the temperature of maximum stability (T_{max}) of native chymotrypsinogen. The principal effect of lowering the pH, as shown in Fig. 1, is to transpose the entire curve to more negative ΔF° values and thereby reduce the transition temperature (the temperature at the mid-point of the transition, i.e. where each curve intersects the $\Delta F^\circ = 0$ line).

Since this behaviour was initially observed for chymotrypsinogen, several other protein denaturation reactions have been carefully examined and similar free energy profiles observed (summarized in Brandts, 1969), so that this may be a fairly general feature of denaturation reactions which had been overlooked for many years. The temperature of maximum stability does vary for different proteins, being as low as 0°C for ribonuclease (Brandts and Hunt, 1967).

Any reversible protein denaturation reaction which displays heat capacity effects leading to free energy behaviour of the type shown in Fig. 1 for chymotrypsinogen will *in principle* exhibit two "branches" to its denaturation reaction, corresponding to high temperature denaturation ($\Delta H^\circ > 0$) and low temperature denaturation ($\Delta H^\circ < 0$). This is illustrated in Fig. 2 for a hypothetical situation in which T_{max} is 20°C. The upper solid curve shows the situation for conditions in which the native protein is very unstable ($\Delta F^\circ \leqq 0$ at all temperatures). As the temperature is lowered from 50°C native protein begins to appear. However, ΔH° changes sign at about 20°C when only 50 per cent native protein is reached, and further reduction in temperature leads to increased denaturation.

In conditions which favour increased stabilization of the native protein (more favourable pH, for example), high and low temperature branches of the denaturation tend to separate on the temperature scale, as shown in the dashed curve in Fig. 2. Here there exists an intermediate temperature region over which virtually 100 per cent of the protein is native.

This is the situation which prevails for most of the common proteins under most solution conditions. The high temperature denaturation is readily accessible for study since it occurs in a convenient temperature range. However, the low temperature denaturation will only occur to a very limited extent at temperatures above the phase point. The dashed line below 0° indicates the expected behaviour if the solvent properties were continuous, as they would be for example in supercooled aqueous solutions.

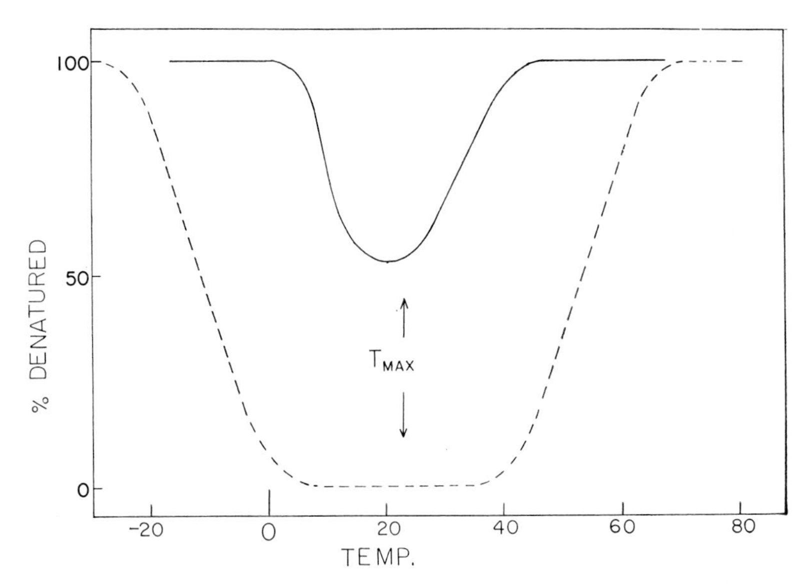

FIG. 2. Schematic illustration of the fraction of denatured protein for a protein with a temperature of maximum stability (T_{max}) of 20°C. The solid curve refers to solution conditions under which the native protein is fairly unstable, while the dashed curve refers to conditions under which the native protein is moderately stable.

A tentative explanation for this rather peculiar temperature dependence of denaturation reactions has been given previously (Brandts, 1964*a*,*b*) and discussed in some detail (Brandts, 1967, 1969; Tanford, 1968). The large heat capacity effects appear to be associated with the exposure of apolar side chains which occurs during the denaturation process. It is known from studies of model compounds that non-polar and slightly polar aliphatic groups have a partial molal heat capacity in water which is about threefold greater (Kresheck and Benjamin, 1964) than the heat capacity of the corresponding groups in organic solvents or in the gas phase. This very large excess heat capacity of hydrophobic solutes in water has been attributed (Frank and Evans, 1945; Kauzmann, 1959) to the additional energy which must be put into the system in order to "melt" the ordered water structures ("icebergs", "flickering clusters", "clathrates", etc.) which are thought to form about such groups in aqueous solution at low temperature. Thus, it is reasonable to expect that denatured proteins, with their greater exposure of hydrophobic side chains to aqueous solvent, would have a larger heat capacity than native proteins and this is probably the primary reason for the very unusual temperature effects associated with denaturation reactions.

LOW TEMPERATURE DENATURATION ABOVE THE FREEZING POINT

The suggestion that chymotrypsinogen has a low temperature denaturation is based largely on the extrapolation of thermodynamic data obtained for high temperature denaturation. This native protein is too stable, even at very acid pH values, to permit one to "see" the small amount of denaturation which presumably occurs from 12° down to 0°C. Alpha-chymotrypsin, the enzymically active form of chymotrypsinogen which differs from the zymogen by the loss of four peptide residues, is a less stable protein and in this case the low temperature branch of the denaturation can be clearly seen (Brandts, 1967; Biltonen and Lumry, 1969). The value of T_{max} at acid pH is about 12°C and thus identical to the value predicted for chymotrypsinogen in similar conditions. A number of other proteins, some 20 in all (see, for example, the discussions in Jarabak, Seeds and Talalay, 1966, or Brandts, 1969), have been found to undergo "cold denaturation" below 25°C and it seems possible that many or all of these situations arise from the type of free energy–temperature relationship alluded to earlier (p. 190), *et seq.* and that this is brought about by the peculiar thermodynamic characteristics of hydrophobic bonding.

In certain conditions, the low temperature denaturation of chymotrypsinogen can be observed without having to rely on extrapolation. For example, the addition of moderate amounts of urea acts to unstabilize the native protein without altering the temperature characteristics of the denaturation reaction to any great extent. In 2·3 M-urea, both branches of the denaturation can be seen and there is an indicated T_{max} value of 11°C (Brandts, 1964*a*). A similar procedure has been used (Pace and Tanford, 1968) to observe the thermodynamics of denaturation of β-lactoglobulin over a wide temperature range and T_{max} for this protein is 32°C in 4·4 M-urea.

The application of hydrostatic pressure acts to depress the freezing point (ice I) of water to a minimum value of $-22°C$ at a pressure of 2000 atmospheres ($10^3 \times 202\,650$ N m^{-2}) (Bridgman, 1912). Pressurizing beyond this point results in the formation of ice III with a density greater than the liquid, so that further increases in pressure raise the freezing point. By working under moderate pressures, it is then possible to increase the temperature range available for the study of proteins in the liquid state. Recently, we have examined chymotrypsinogen solutions in an optical pressure cell, using the difference spectra technique in the wavelength region of the aromatic bands. The results in Fig. 3 show changes in the magnitude of the 293 nm peak at a constant pressure of 1370 atmospheres ($10^3 \times 138\,815{\cdot}25$ N m^{-2}). Under these conditions the high temperature denaturation (not shown) begins to take place at about 30°C, and is accompanied by a decrease in extinction at 293 nm. As the temperature is lowered below 20°C, a second cooperative process is observed to occur although it does not go to completion at temperatures above the freezing point ($-14°C$). The difference spectra produced in the low temperature process are similar to those associated with the heat denaturation in terms of peak positions and relative heights. Reversal of the temperature from $-14°$ to $+20°C$ results in the re-establishment of the native spectrum, so that the process is virtually 100 per cent reversible at this pressure.

From these observations it seems very likely that both the high and low temperature processes correspond to denaturation reactions. The near identity of the difference spectra suggests that aromatic residues are exposed to solvent in the low temperature denaturation as well as in the thermal denaturation (Brandts, 1964*a*), so that the overall conformational changes may be very similar. If this is true, the extent of ordering of the solvent about exposed hydrophobic side chains must be very different in the high and low temperature cases in order to account for the difference in sign of $\Delta H°$ and $\Delta S°$.

The total change in extinction coefficient (293 nm) for denaturation at high temperature is known to be about 1·2 absorbance units. If it is assumed that the low temperature process is characterized by the same change in extinction coefficient, then it is possible to estimate the extent of denaturation at low temperature even though the complete transition cannot be observed at temperatures above the freezing point. This assumption leads to an estimate of about 50 per cent denaturation at $-14°C$, as shown in Fig. 3.

The application of hydrostatic pressure will, in addition to lowering the freezing point, exert effects on the thermodynamics of denaturation. It is known from earlier studies (Oliveira, 1967) of the thermal transition of chymotrypsinogen under pressure that a volume change, ΔV°, is associated with denaturation and that ΔV° exhibits a temperature dependence, being positive at high temperature and negative below about 30°C at pH 2. Thus, high pressures will tend to stabilize native chymotrypsinogen at high temperatures, relative to the situation at one

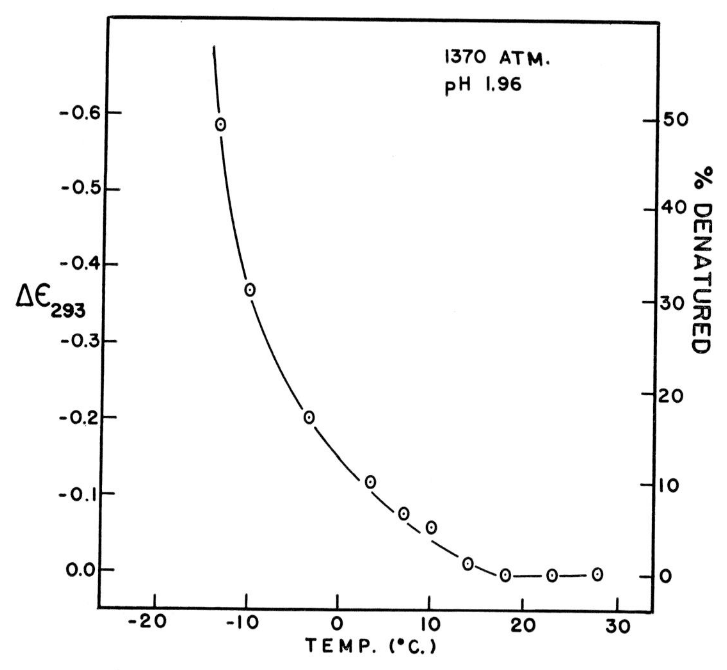

FIG. 3. The change in extinction coefficient (293 nm) of chymotrypsinogen (0·03 wt %) at pH 1·96 and 1370 atmospheres. Data points correspond to equilibrium values at each temperature. The right-hand ordinate shows an estimate of the extent of denaturation (see text).

atmosphere (101 325 N m^{-2}), but act to promote denaturation at low temperatures. This combination of effects will increase the temperature of maximum stability as the pressure is increased. This is in agreement with the fact that the indicated value of T_{max} at 1370 atmospheres is above 20°C as compared to a value of 12°C at one atmosphere.

DENATURATION IN FROZEN SOLUTIONS

The study of the denaturation of proteins at low temperature in frozen solutions, rather than in the liquid state as above, presents many additional problems. The biggest complication is the existence of multiple phases in the frozen solution, which can result in local concentrations of solutes and protein quite different from the bulk concentrations. A second problem is that, due to light-scattering problems, most of the convenient and accurate optical parameters used to study denaturation in the liquid state are difficult to use quantitatively in frozen solutions. Fortunately, a non-optical method exists of following the denaturation of chymotrypsinogen in frozen solutions. Eisenberg and Schwert (1951) noted that thermally denatured chymotrypsinogen could be selectively precipitated

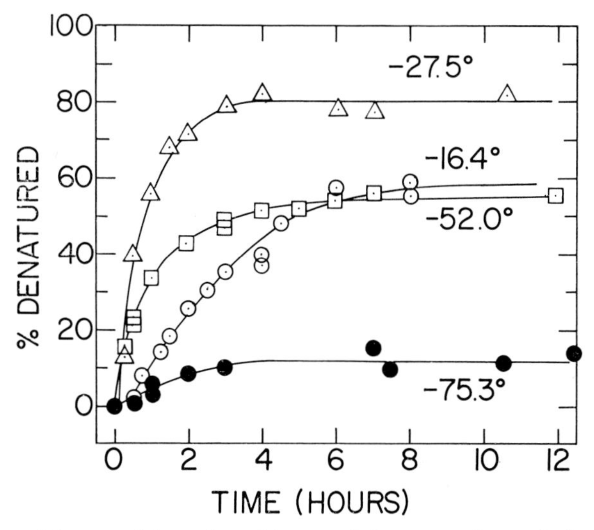

FIG. 4. The rate of formation of denatured chymotrypsinogen in frozen solutions at four different temperatures. Solutions contain only protein (0·25 wt %) and HCl (pH 1·78 at 25°C).

from an equilibrium mixture by the addition of a high salt buffer (1·2 M-NaCl, glycine-HCl, pH 3·0), the native protein remaining in solution. A parallel situation is found in working with frozen solutions. A solution of chymotrypsinogen, frozen at a given temperature for a period of time, is found to be free of precipitate when rapidly thawed at 5°C. The addition of the high salt buffer immediately after thawing, however, leads to the formation of a precipitate in certain cases. For purposes of this study, the precipitate is defined as denatured protein, in analogy with the high temperature case, and that remaining in solution as native protein.

The amount of denaturation observed is a function of the temperature at which the sample is equilibrated in the frozen state, the pH of the solution, the concentration of protein, and the presence of other solutes such as salts, urea, etc. It is also a function of the time spent in the frozen state at any given temperature. This is illustrated in Fig. 4, where the kinetics of the denaturation reaction are shown for a sample of pH 1·78 (HCl only) and protein concentration of 0·25 per cent. These data show that the half-times for the denaturation reaction are of the order of two hours at −16·4°C and 0·5 hours at −52·0°C, showing that the rate constant for denaturation is larger at lower temperatures. At all temperatures examined, the extent of denaturation reaches a constant value after several hours and remains constant for times in excess of 24 hours. This implies that one of two situations must exist in these frozen solutions, according to whether thermodynamic or kinetic factors are dominant:

(1) An equilibrium exists between native and denatured protein and the fact that steady-state values of less than 100 per cent denatured protein are achieved reflects the fact that the equilibrium constant is not too different from unity.
(2) The denaturation reaction goes essentially to completion for all protein molecules "capable" of being denatured in these frozen solutions. This would imply that the protein must exist in more than a single "phase" and that the denaturation reaction cannot take place in at least one of these phases. The amount of protein which exists in each of the separate phases may then be a simple or complex function of the temperature and composition of the frozen solutions.

Later it will be shown that the first alternative cannot be the correct one, so that one must seek an explanation in terms of kinetic arguments.

Fig. 5 illustrates the extent of denaturation achieved in the time-independent region. All the points below 0°C correspond to frozen solutions, since supercooling was avoided by seeding. At pH 1·78, denaturation becomes appreciable at temperatures below −10°C, increases to a maximum of 80 per cent at −20°C, and becomes progressively smaller as the temperature is lowered to −75°C. Thus the variation in the extent of denaturation with the temperature is complex, the presence of a sharp maximum suggesting that at least two separate effects are operative.

At pH values slightly higher than 1·78 no denaturation is observed, as shown for the pH 2·55 case in Fig. 5. We have found that this situation prevails through the neutral pH range as well and it is not until a pH in excess of 10 (NaOH) is reached in the alkaline region that significant denaturation is again observed. Thus both an acid- and a base-induced

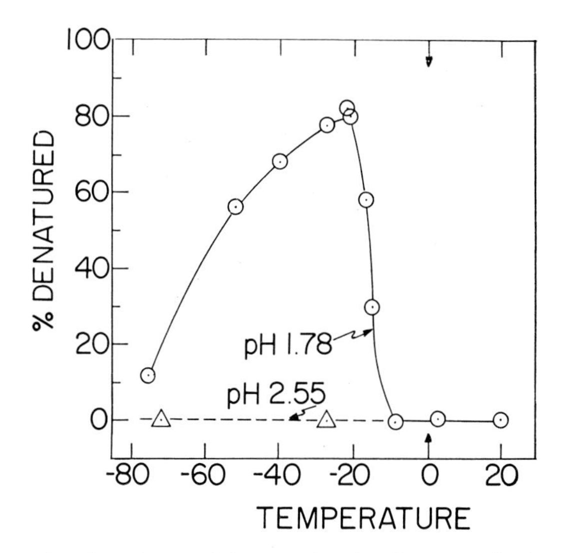

FIG. 5. The dependence of denaturation (steady-state values) on the temperature of equilibration of the frozen samples. Protein concentration is 0·25 per cent and pH values (25°C) are given. The two data points above 0°C are for unfrozen samples, using the same method for estimation of denaturation.

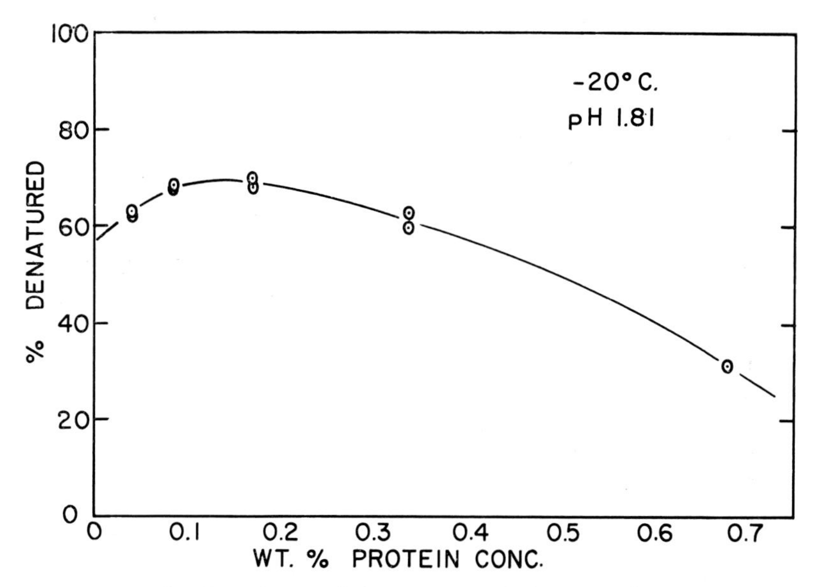

FIG. 6. The dependence of denaturation of chymotrypsinogen on protein concentration at −20°C, pH 1·81.

denaturation appear to exist in frozen solutions of chymotrypsinogen, just as in liquid solutions.

As might be anticipated, the bulk concentration of protein is an important variable in frozen solutions, as shown in Fig. 6. Over a 16-fold concentration range, the extent of denaturation varies by nearly a factor of two and reaches a maximum at a concentration of 0·15 per cent. Over most of the range studied, increasing protein concentration exerts a "protective" effect against denaturation. These results parallel those of

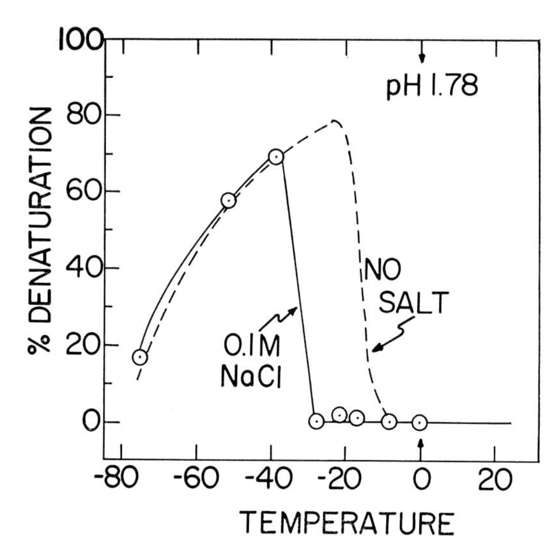

FIG. 7. The effect of 0·1 M-NaCl on the denaturation of chymotrypsinogen in frozen solutions. For comparison, the dashed curve shows the behaviour under identical conditions (0·25 per cent protein, pH 1·78), but with no NaCl present.

Chilson, Costello and Kaplan (1965), who found that the freezing inactivation of lactic dehydrogenase was suppressed by increasing the concentration of the protein, or by the addition of serum albumin to solutions of lactic dehydrogenase.

It is interesting to examine the effects of different small solutes on the denaturation reaction. In Fig. 7, the effect of 0·1 M-NaCl is shown. This electrolyte completely prevents denaturation at all temperatures down to -27°C. A sharp break in the curve occurs near this temperature so that, below -38°C, sodium chloride exerts no protective effect whatsoever.

The discontinuity in the effect of NaCl on chymotrypsinogen is probably caused by the precipitation of the salt. In the two-component system, NaCl–H_2O, a eutectic occurs at $-21°C$. In the four-component system under discussion here, the precise phase relationships are not known but it seems likely that NaCl will begin to precipitate somewhere near its normal eutectic temperature since, in terms of molar concentration, it is by far the dominant solute. Once precipitation of NaCl begins the HCl concentration in the unfrozen liquid will increase markedly, due to volume reduction, so the common ion effect will tend to reduce the solubility of NaCl in the remaining liquid. This will be accentuated with

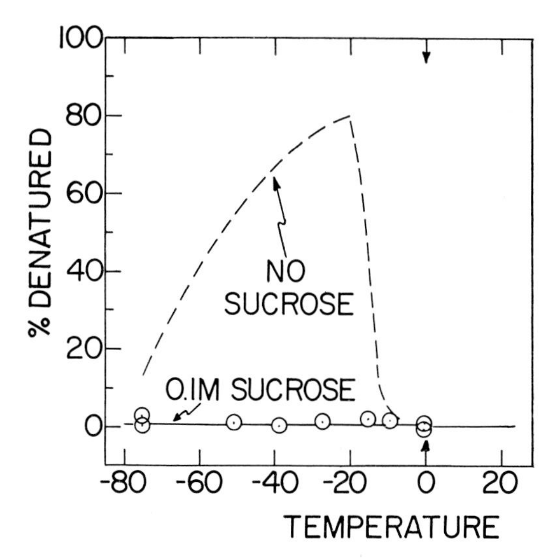

FIG. 8. The denaturation of chymotrypsinogen in the presence of 0·1 M-sucrose. Reference curve included for comparison. Conditions are those of Fig. 5, with the exception of the added sucrose.

further reduction in temperature, as more water is removed in the form of ice, so that the composition of the unfrozen liquid will ultimately approach the same composition as exists in the absence of NaCl.

Sucrose, a good protective agent for most biological materials at low temperature, completely prevents the denaturation of chymotrypsinogen over the entire temperature range examined. This is shown in Fig. 8 for a sucrose concentration of 0·1 M. At this low concentration sucrose would exert only a very small effect on denaturation reactions in liquid solutions, so clearly the effectiveness of solutes is amplified on denaturation in frozen solutions. Whether sucrose acts to prevent denaturation by a direct effect on the stability of the protein or by an indirect effect, such as preventing excessively high HCl concentration in the unfrozen liquid, cannot be ascertained from these observations. It has been suggested (Brandts, 1969) that organic solutes similar to sucrose act to increase the stability of native proteins at low temperatures.

In the presence of 0·1 M-urea both the rate and extent of denaturation increase markedly, as seen in Fig. 9. For all temperatures examined, the rate of denaturation was too fast to be measured accurately with the techniques available in this study. The steady-state values were achieved in less than 15 minutes, and this corresponds to a considerably larger rate

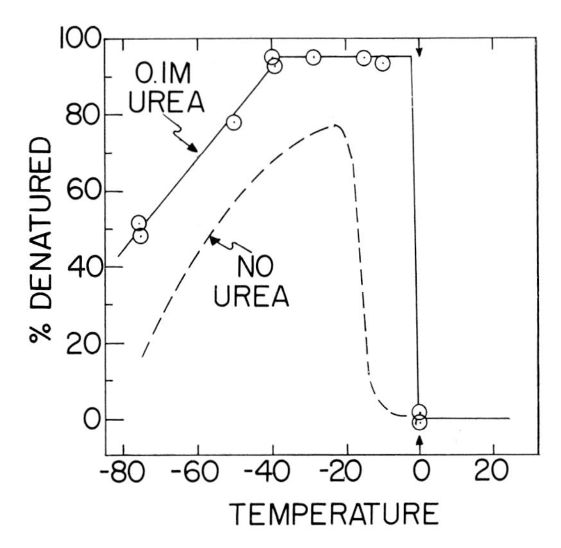

FIG. 9. The effect of urea on the denaturation of chymotrypsinogen. Conditions are identical to those of Fig. 5 except for the addition of 0·1 M-urea. The points at 0°C are for solutions that were not frozen. Otherwise, all samples were frozen.

constant than in the absence of urea but otherwise under identical conditions. Above −40°C, the amount of denatured protein was consistently about 95 per cent. Below this temperature the extent of denaturation is reduced, although the values are always higher than those achieved in the absence of urea.

Of the solutes examined, urea is the only one which increases the amount of denaturation. This is in keeping with its reputation as a strong unstabilizing agent for native proteins. However, it is interesting that, even in the presence of urea, there is always at least a small amount of protein which seems to be incapable of being denatured, and this becomes a more sizeable fraction of the total as the temperature is reduced.

REVERSIBILITY OF DENATURATION IN FROZEN SOLUTIONS

The reversibility pattern in frozen solutions is such as to eliminate any simple thermodynamic interpretation of these data. As shown in Table I, the denaturation which occurs in frozen solutions is almost completely reversible when the solutions are thawed and allowed to remain at 5°C for sufficient time before the addition of the precipitating buffer. Sephadex chromatography shows that there are no changes in the molecular weight of the protein after thawing, using the unfrozen protein as a control. We are therefore led to the conclusion that the denaturation reaction which occurs in frozen solutions is intrinsically reversible, so that renaturation is possible assuming that no additional constraints are imposed on the protein in the frozen solutions.

TABLE I

RENATURATION OF CHYMOTRYPSINOGEN AFTER DENATURATION IN FROZEN SOLUTIONS AND THAWING

T (°C)	*% denatured after more than 5 hr at T °C*	*% denatured after samples thawed at 5°C for 2 hr*	*% denatured after samples thawed at 5°C for 24 hr*
−15·0	29·5	3·0	—
−22·0	81·5	15·5	—
−27·5	76·5	15·5	5·3
−28·0	72·3	24·7	6·0
−39·2	67·1	—	0·0
−75·3	11·5	—	0·4
(0·1 M-NaCl)			
−39·2	68·6	—	3·2
−51·8	57·1	—	0·0
−75·3	81·1	—	0·0
(0·1 M-urea)			
−28·0	93·9	20·5	1·8
−75·3	49·8	—	2·2

All samples contained HCl (pH 1·78) and 0·25 per cent chymotrypsinogen. The presence of other solutes is indicated in parentheses.

However, the situation is actually not so simple as this might suggest, as can be readily demonstrated by the examination of reversibility in the absence of thawing. This is illustrated in Fig. 10 for solutions containing only protein and HCl. A sample initially equilibrated at −76°C, for example, contains 10 per cent denatured protein. If the temperature of the sample is then changed to −26°C and sufficient time for equilibration allowed, the amount of denatured protein is 83 per cent, which is nearly identical to what is observed if the sample is taken directly to −26°C without prior equilibration at −76°C. However, if a sample is initially equilibrated at -26°C near the maximum in the temperature profile, and then re-equilibrated at either a higher (-9°C) or lower (-76°C) temperature, as shown in Fig. 10, the amount of denatured protein observed is equal to that at the maximum and considerably different from what would have been observed had the sample not been initially equilibrated at -26°C. The extent of denaturation observed at any single temperature

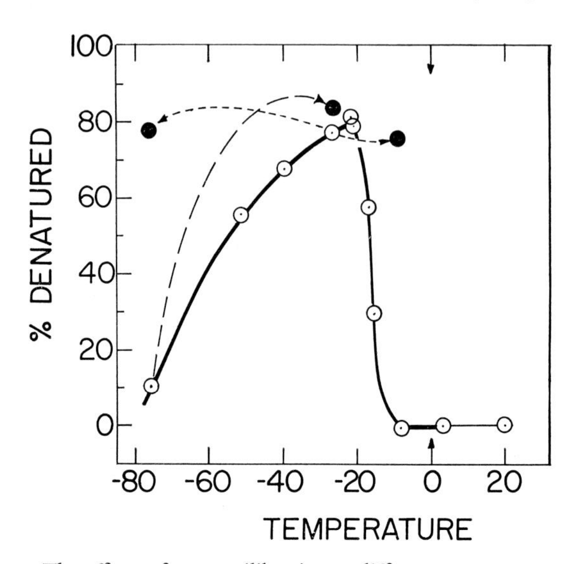

FIG. 10. The effects of pre-equilibration at different temperatures. The open circles are for samples taken immediately to the temperature indicated (Fig. 5). The filled circles are for samples initially equilibrated at a different temperature (indicated by the dashed line) before equilibration at the final temperature indicated by the placement of the filled circles on the temperature axis.

therefore depends upon prior exposure of the sample to other temperatures (assuming no intermediate thawing) so that the systems exhibit no reversibility and *always display the maximum amount of denaturation attainable at any of the temperatures to which it was exposed.* In other words, the process of renaturation does not occur in these frozen solutions, even though it occurs readily upon thawing.

DISCUSSION

Although it is possible to be reasonably confident about the effect of low temperature on protein stability above the freezing point of the solution, it does not necessarily follow that the principal cause of denaturation in frozen solutions is the exposure to low temperature. In fact, this would be an oversimplification of the actual situation. When protein solutions are frozen, the protein molecules may reside in any of at least three locations:

(1) Part of the protein must certainly be dissolved in the liquid "pools" which remain after freezing. These pools will contain, in addition to highly concentrated protein, all or nearly all of the HCl, which has no eutectic above −86°C in these dilute solutions.
(2) Since the protein is greatly concentrated in the liquid regions after freezing, its solubility may be exceeded in certain cases, so a separate crystalline protein phase may also exist in these frozen solutions.
(3) Since freezing in these experiments was accomplished fairly rapidly by simple immersion of the sample in a bath of the desired temperature, part of the protein may be mechanically "trapped" in some way in the crystallizing ice structure and therefore it could not be regarded as either dissolved or crystallized.

It seems reasonable to suppose that proteins in a crystalline phase or in a trapped state will be unable to denature. We are inclined to feel that the protein which is capable of undergoing denaturation is exclusively that which resides in the pools in which the freezing point has been depressed by concentration of electrolytes and other solutes. The large effects observed on addition of low concentrations of solutes such as urea and sucrose confirm this, implying that the local concentrations of these additives in the vicinity of the protein molecules which denature are much greater than the bulk concentration.

Even if this supposition is accepted, it is difficult to provide a simple and reasonable interpretation of all the data. It is tempting to try to fit our results to existing data and interpretations in the literature on other systems in frozen solution. For example, Curti, Massey and Smudka (1968) have observed that L-amino acid oxidase undergoes inactivation when frozen from 0·2 M-tris-HCl buffer, pH 7·8. The temperature inactivation profile is extremely similar to that shown for chymotrypsinogen in Fig. 5 with a maximum in inactivation at about −30°C. The reversibility pattern is also similar. They conclude that "this conformational change may be induced by the ordered packing of water molecules in certain of the ice structures, with subsequent disruption of the water shell of the protein." They propose a model for the transition to the inactive form (I) in which the enzyme is in equilibrium with a metastable state (E*) which is produced as a result of ice formation, i.e.

$$E \rightleftharpoons E^{\star} \rightarrow I$$

wherein the equilibrium process has a large negative ΔH° and the kinetic step has a small positive activation energy. The maximum in the temperature profile is then attributed to a combination of kinetic and thermodynamic effects, with temperature being the primary variable of importance. This mechanism taken in its simplest form is not consistent with the data on chymotrypsinogen. A model of this sort necessitates that, given enough time in the frozen state, one would eventually observe 100 per cent denaturation (inactivation) since the equilibrium between E and E* would eventually be pulled completely to the right by the formation of I. This is contrary to our findings, so that the model is too simple for the chymotrypsinogen case. Since Curti, Massey and Smudka analysed for inactivation after a given time interval and did not measure the time-progress of the reaction, it is impossible to say whether or not the model is totally consistent with the L-amino acid oxidase system.

It has also been suggested (Chilson, Costello and Kaplan, 1965) that denaturation in frozen solution is brought about principally by the exposure of proteins to high local concentrations of various small solutes. It seems possible that this is an important factor. It is known for example that many native proteins are unstabilized by increasing acid concentration in the pH range 5 to 1. The principal reason for the unstabilization of proteins by acid is the fact that, in general, denatured proteins have a greater affinity for protons than do native proteins, owing to the larger protonic association constants of the carboxyl groups in the unfolded form (Laskowski and Scheraga, 1954; Kauzmann, 1954; Brandts, 1964*b*; Tanford, 1968). However, all the carboxyl groups in both native and denatured protein should be titrated by the time pH 1 is reached, so further reduction in the pH will be ineffective with this particular mechanism. This has been demonstrated for chymotrypsinogen in liquid solution above 0°C, where it was observed that increasing the HCl concentration above 0·1 M produced little additional unstabilization (Brandts, 1964*a*).

In the frozen HCl solutions (assuming the osmotic effect of the protein is small relative to the HCl) the local concentration of HCl will change from about 2 molal at −10°C to 4 molal at −20°C, the temperature span over which the amount of denaturation for chymotrypsinogen changes from 0 to 80 per cent. If it is principally the change in HCl concentration which is responsible for causing denaturation, then the mechanism of action cannot be preferential binding of protons to the carboxyl groups of the denatured form since all these groups would be titrated at HCl concentrations considerably lower than 2 molal. It is possible that these very high HCl concentrations exert a general solvent effect of the type discussed by von Hippel (von Hippel and Wong, 1965; von Hippel and Schleich, 1969) and that this leads to instability of the native protein. It is also conceivable that the amide groups in the polypeptide backbone begin to protonate at these very high acid concentrations, as has been observed to occur in strong organic acids such as trifluoroacetic acid (Hanlon and Klotz, 1965; Stake and Klotz, 1966). The latter could lead to an unfavourable electrostatic situation which could be relieved by unfolding.

Perhaps also pertinent is the fact that many systems containing only small solutes, with no macromolecules present, exhibit an unusual temperature dependence in frozen solution which in some cases bears a strong resemblance to results obtained in this study. To cite one example, Grant, Clark and Alburn (1961) observed that the base-catalysed hydrolysis of penicillin occurs many times faster in frozen solutions. The rate of this reaction approaches a maximum near $-20°C$ and tails off at lower temperatures. The presence of various solutes (glycerol, ethanol) stopped the reaction in the frozen samples. There does not seem to be universal agreement on the factors responsible for these increased rates of reaction of small solutes in frozen solutions. Grant and co-workers (Grant, Clark and Alburn, 1961, 1966; Grant and Alburn, 1965, 1967) argue that, at least in some cases, the ice structure itself plays an active role. Pincock (1969), on the other hand, suggests that the dominant effect is simply the increase in concentration of reactants as the volume of unfrozen solution is reduced when the temperature is lowered. According to the latter explanation, the maximum which is frequently seen in the rate of bimolecular reactions is the result of two compensating effects acting on the overall velocity: the increase in the local concentrations of reactants as the temperature decreases, and the normal decrease in rate constant resulting from a positive activation energy. The first of these effects is dominant on the high temperature side of the maximum, while the second is more important on the lower temperature side. Tong and Pincock (1969) have also studied the denaturation of invertase in frozen solutions containing HCl and other small solutes at $-5°C$, and they conclude that the primary factor leading to denaturation is the high concentration of small solutes, particularly HCl.

Although it would be nice to fit the data on chymotrypsinogen into a simple picture, it is clearly impossible. Our findings indicate that the protein, in frozen solutions, exists in more than a single phase and that the denaturation process is unable to occur in at least one of these phases. It seems likely that the distribution of protein between phases is a function of temperature and concentrations of small solutes, and perhaps depends also on protein concentration. We feel that the denaturation which occurs in frozen solutions is basically similar (i.e. an unfolding reaction) to that which occurs above the normal freezing point, and that it occurs in the unfrozen pools. The exposure to low temperature itself may play some role in the denaturation in the sense that it leads to a general instability of native proteins and makes them susceptible to the action of other factors. However, the complex temperature behaviour must result from a combination of factors including, in addition to pure temperature effects, changes in the concentration of small solutes, changes in protein concentration, separation of protein into multiple phases and, perhaps, specific effects resulting from the "dehydration" of protein molecules which must occur as the vapour pressure of ice is lowered with reduction in temperature. The difficult job of sorting out the importance of these variables requires more extensive experimental studies.

SUMMARY

Chymotrypsinogen is one of many proteins known to exhibit low temperature instability, resulting from the fact that the denatured form has a much higher heat capacity than the native form. Under normal solution conditions it is difficult to demonstrate this fact directly, since the amount of denaturation which occurs at low temperature is very small, due to the relatively high stability of chymotrypsinogen at all temperatures. However, by working under moderately high pressures it is possible to observe two separate denaturation processes: one occurring at high temperature with a positive ΔH° and the other at low temperature with a negative ΔH°. Changes in the absorption properties suggest that the high temperature and low temperature denatured forms may be similar in structure.

Denaturation of chymotrypsinogen also occurs at low temperature in frozen solution. In this case, the temperature dependence of denaturation is complex and is not susceptible of a simple interpretation. The addition of small amounts of solutes such as HCl, NaCl, sucrose, and urea have profound effects on the denaturation. The analysis of the data suggests that the protein which is denatured in frozen solutions is located in the unfrozen pools, although it also appears that a sizeable fraction of the protein, not susceptible to denaturation, is located elsewhere.

Acknowledgements

The authors are indebted to the National Aeronautics and Space Administration (NGL-22-010-029) and to the U.S. Public Health Service (GM-11071) for direct financial support for this work. One of us (J.B.) would also like to acknowledge fellowship support from the Alfred P. Sloan Foundation during the course of these studies.

REFERENCES

BILTONEN, R., and LUMRY, R. (1969). *J. Am. chem. Soc.*, **91**, 4256–4263.
BRANDTS, J. F. (1964*a*). *J. Am. chem. Soc.*, **86**, 4291–4301.
BRANDTS, J. F. (1964*b*). *J. Am. chem. Soc.*, **86**, 4302–4314.
BRANDTS, J. F. (1967). In *Thermobiology*, pp. 25–71, ed. Rose, A. H. London: Academic Press.
BRANDTS, J. F. (1969). In *Structure and Stability of Biological Macromolecules*, pp. 213–290, ed. Timascheff, S. N., and Fasman, G. D. New York: Dekker.
BRANDTS, J. F., and HUNT, L. (1967). *J. Am. chem. Soc.*, **89**, 4826–4838.
BRIDGMAN, P. W. (1912). *Proc. Am. Acad. Arts Sci.*, **47**, 439–558.
CHILSON, O. P., COSTELLO, L. A., and KAPLAN, N. O. (1965). *Fedn Proc. Fedn Am. Socs exp. Biol.*, **24**, suppl. 15, S55–S65.
CURTI, B., MASSEY, V., and SMUDKA, M. (1968). *J. biol. Chem.*, **243**, 2306–2314.
EISENBERG, M. A., and SCHWERT, G. W. (1951). *J. gen. Physiol.*, **34**, 583–606.
FRANK, H. S., and EVANS, M. W. (1945). *J. chem. Phys.*, **13**, 507–532.
GRANT, N. H., and ALBURN, H. E. (1965). *J. Am. chem. Soc.*, **87**, 4174–4177.
GRANT, N. H., and ALBURN, H. E. (1967). *Archs Biochem. Biophys.*, **118**, 292–296.
GRANT, N. H., CLARK, D. E., and ALBURN, H. E. (1961). *J. Am. chem. Soc.*, **83**, 4476–4477.
GRANT, N. H., CLARK, D. E., and ALBURN, H. E. (1966). *J. Am. chem. Soc.*, **88**, 4071–4074.
HANLON, S., and KLOTZ, I. M. (1965). *Biochemistry*, **4**, 37–48.
HIPPEL, P. H. VON, and SCHLEICH, T. (1969). In *Structure and Stability of Biological Macromolecules*, pp. 417–574, ed. Timascheff, S. N., and Fasman, G. D. New York: Dekker.
HIPPEL, P. H. VON, and WONG, K. (1965). *J. biol. Chem.*, **240**, 3909–3923.
JARABAK, J., SEEDS, A. E., Jr., and TALALAY, P. (1966). *Biochemistry*, **5**, 1269–1279.
KAUZMANN, W. (1954). In *The Mechanism of Enzyme Action*, pp. 70–110, ed. McElroy, W. D., and Glass, B. Baltimore: Johns Hopkins.
KAUZMANN, W. (1959). *Adv. Protein Chem.*, **14**, 1–63.
KRESHECK, G., and BENJAMIN, L. (1964). *J. phys. Chem., Ithaca*, **68**, 2476–2486.
LASKOWSKI, M., Jr., and SCHERAGA, H. A. (1954). *J. Am. chem. Soc.*, **76**, 6305–6319.
LUMRY, R., BILTONEN, R., and BRANDTS, J. F. (1966). *Biopolymers*, **4**, 917–944.
OLIVEIRA, R. (1967). M.Sc. Dissertation, University of Massachusetts, Amherst.
PACE, N. C., and TANFORD, C. (1968). *Biochemistry*, **7**, 198–208.
PINCOCK, R. E. (1969). *Accounts chem. Res.*, **2**, 97–103.
STAKE, M. A., and KLOTZ, I. M. (1966). *Biochemistry*, **5**, 1726–1729.
TANFORD, C. (1968). *Adv. Protein Chem.*, **23**, 121–282.
TONG, M., and PINCOCK, R. E. (1969). *Biochemistry*, **8**, 908–913.

DISCUSSION

Farrant: When you added 0·1 M-sodium chloride you delayed the denaturation at low temperatures to lower temperatures. You said that the increase in denaturation took place at a temperature below the eutectic point of sodium chloride. Because protein is present as well as salt, the eutectic temperature of sodium chloride and water (−21°C) will perhaps not be relevant. The temperature at which salt begins to come out of solution may in fact be in the temperature region that you begin to get the increase in denaturation.

Mazur: When the eutectic point of, say, a simple sodium chloride solution is measured by conductometry, the resistance does not become high until the temperature has dropped to −28°C or −29°C, about eight degrees below the eutectic point. Presumably the explanation is that the sodium chloride solution supersaturates about 10°C. However, the electrical resistance shows hysteresis. When the completely frozen solution is warmed, the resistance doesn't drop until the true eutectic point of −20·1°C is reached (Mazur, 1961).

Farrant: Can you move that temperature by changing the amount of salt?

Brandts: We have not tried this.

Meryman: If you get denaturation at the eutectic temperature of the salt, this could result from the removal of all solute that is retaining water colligatively.

Brandts: This was pH 1·78 HCl, and 0·1 M-sodium chloride. So there would still be a small amount of concentrated HCl solution which would be unfrozen and saturated with sodium chloride and with protein.

Meryman: But then the protective sucrose will also retain some additional water, colligatively.

Brandts: Certainly. However, sucrose also has a high eutectic point of about −13°C, which did not show up as a discontinuity in these studies.

Meryman: Without sucrose, after the sodium chloride has precipitated out as denaturation starts, perhaps the denatured protein becomes more effective colligatively. It may retain water which then permits some protein to remain native. The denaturation may unfold the protein and increase its osmotic coefficient, so that it retains more water colligatively.

Heber: To what extent does your model reflect the situation in the cell? Two features, reversibility and protection by salts, are quite inconsistent with what is generally observed in cells. Salts are destructive.

Brandts: It is difficult to generalize our findings to the conditions which exist in frozen cells. First of all, to my knowledge, no one has looked for *reversible* denaturation in cells which have been frozen. Secondly, it is likely that the many different proteins in a cell would show different reversibility patterns and perhaps might respond in different ways to the presence of salts. Even a single protein may respond in different ways to the addition of salt, depending on other variables. Chymotrypsinogen, for example, is known to be protected against heat denaturation by addition of KCl if the pH is below 2, whereas addition of KCl at higher pH facilitates both reversible and irreversible denaturation (Brandts, 1964). In keeping with this, there is evidence that salts may promote denaturation of other proteins in frozen solutions when conditions are less acid than those used in this study (Chilson, Costello and Kaplan, 1965; Tong and Pincock, 1969).

Reversibility of denaturation after thawing will also probably depend upon a number of variables including pH, salt, protein concentration and, most important, the particular protein being examined. This is known to be the case for heat denaturation, and it seems a good guess to assume it will be true for low temperature denaturation in frozen solutions.

Nei: One of my colleagues, Hanafusa, has worked on the denaturation of protein by freezing and freeze-drying. With optical rotatory dispersion and ultracentrifugation he found conformational changes in fibrous proteins such as myosin, but no conformational change in globular protein such as catalase on freezing. The unfolding of myosin and the dissociation of catalase were found in freeze-drying. In recent work he also confirmed the protective action of several additives against denaturation of protein in freezing and freeze-drying (Hanafusa, 1969).

Brandts: Had we denatured chymotrypsinogen at high temperature under the apparent conditions of salt and acid that presumably exist in unfrozen regions in the liquid state, it would have aggregated extensively, but it didn't do this in frozen solution. I don't know why. There were no molecular weight changes after thawing.

Meryman: Could the aggregation have been reversible?

Brandts: Possibly.

Greaves: In the very early days of freeze-drying I worked with the Adairs (Adair, Adair and Greaves, 1940) on serum and crystalline egg albumin. We particularly chose something that had an irreversible denaturation, and we were never able to discover any denaturation after freeze drying. Why was this?

Brandts: Presumably because the conditions were not those which produced instability of the native protein. Most proteins, including chymotrypsinogen, do not denature when frozen in solutions of neutral pH. Only the most unstable proteins will be susceptible to low temperature denaturation in neutral solution. L-Amino acid oxidase is an example of a protein in the latter category, since it is inactivated when frozen from neutral solutions (Curti, Massey and Smudka, 1968).

Leibo: About five years ago we studied the effects of freezing a chromoprotein from a unicellular alga (Leibo and Jones, 1964). In its gross physical properties, such as the sedimentation coefficient and behaviour in electrophoresis, there were no changes. But its optical properties, such as fluorescence and absorbancy, were a more sensitive measure of change. There were in fact drastic changes in these properties after freezing.

Levitt: Could the reversibility of denaturation be related to the size of the molecule? Is irreversibility due to aggregation more likely with a large molecule than with a small molecule?

Brandts: Most of the proteins highly reversible to denaturation are small.

Mazur: Most microorganisms, and plant and animal cells are not damaged by cooling in the absence of freezing, although thermal shock in red cells appears to be an exception.

Levitt: That is also true for higher plants, except for those that are injured by chilling. We have never found that those that are injured by freezing are injured by supercooling, and we have taken some to −20°C.

Mazur: We have supercooled yeast to −21°C without injury. Of course, it is still possible that *reversible* denaturation occurs at these temperatures.

Levitt: Insects too are never injured in the unfrozen state but they can be killed by freezing.

REFERENCES

Adair, G. S., Adair, M., and Greaves, R. I. N. (1940). *J. Hyg., Camb.*, **40,** 548–554.

Brandts, J. F. (1964). *J. Am. chem. Soc.*, **86,** 4291–4301.

Chilson, O. P., Costello, L. A., and Kaplan, N. O. (1965). *Fedn Proc. Fedn Am. Socs exp. Biol.*, **24,** suppl. 15, S55–S65.

Curti, B., Massey, V., and Smudka, M. (1968). *J. biol. Chem.*, **243,** 2306–2314.

Hanafusa, N. (1969). In *Freezing and Drying of Micro-organisms*, pp. 117–129, ed. Nei, T. Tokyo: University of Tokyo Press; Baltimore, Md., and Manchester, England: University Park Press.

Leibo, S. P., and Jones, R. F. (1964). *Archs Biochem. Biophys.*, **106,** 78–88.

Mazur, P. (1961). *Biophys., J.*, **1,** 247.

Tong, M., and Pincock, R. E. (1969). *Biochemistry*, **8,** 908–913.

THE ROLE OF PEPTIDES IN PREVENTING FREEZE-THAW INJURY

J. D. Davies

Department of Pathology, University of Cambridge

The combination of successfully used suspending media, namely glucose, broth and serum, by Fry and Greaves (1951) gave rise to the freeze-drying medium *Mist. desiccans* which proved to be extremely valuable for preserving microorganisms in the freeze-dried state. It was suggested that glucose helped to "buffer" the residual moisture and so prevent over-drying, and that serum acted as a protective colloid and as a "scaffolding" to form a solid dried cake. The necessity for broth could not be explained, but other workers reported the beneficial action of digest broth or peptone solutions when used for preserving freeze-dried viruses (Collier, 1955) and dried bacteria (Annear, 1956). Scott (1960) concluded that the instability of dried cultures at high temperatures was due to the presence of carbonyl groups, and it was then suggested that the amino groups present in broth neutralized carbonyl groups and that the efficacy of such a medium could be improved by replacing glucose with sucrose to prevent the addition of extra carbonyl groups (Scott, 1960; Greaves, 1962). Meanwhile, Meryman and Kafig (1955) and Rinfret (1960) had shown that extracellular additives such as sugars and polymers could afford protection against freeze–thaw injury. Many of these extracellular additives were also known to be essential constituents of successful freeze-drying media and the problem arose of whether the mechanism of protection was the same in both circumstances. We therefore began to study the mechanism of protection against both types of injury, using four moderately frost-sensitive organisms, *Saccharomyces cerevisiae*, *Strigomonas oncopelti*, *Pseudomonas* sp. (1 OH) and the coliphage T4. Our conclusions were that freezing injury played a major role in the overall injury of freeze-drying. Even though there was an additional drying injury, provided that freezing and subsequent drying took place under optimal conditions the survival level of such organisms after freeze-drying could be greatly improved (Greaves and Davies, 1965).

In all this work peptone proved to be an outstanding protective agent and although it had been previously reported as a most valuable freeze-drying additive we were unable to find any evidence that it had been used as a freeze–thaw protective additive, which we found it undoubtedly was (Greaves, Davies and Steele, 1967).

The results of a preliminary investigation into the effects of some additives on the survival of *Pseudomonas* sp. (1 OH) after cooling at 1°C/min to various temperatures followed by a rapid thaw, or alternatively drying from such temperatures, are shown in Table I.

TABLE I

SURVIVAL OF *Pseudomonas* SP. (1 OH) SUSPENDED IN VARIOUS MEDIA AFTER COOLING AT 1°C/MIN TO VARIOUS TEMPERATURES AND FREEZE-DRYING OR THAWING RAPIDLY

	Percentage survival					
	−12·5°C		−20°C		−35°C	
Suspending medium	*F/T*	*F.D.*	*F/T*	*F.D.*	*F/T*	*F.D.*
Distilled water	26		30	3	26	
Sucrose, 12·5%	58		51		31	
Rehydrated in:						
broth		38		33		12
25% sucrose				10		
moist atmosphere				1·2		
Sucrose, 12·5% + PVP, 10%					7	4
Peptone, 20%					47	43

F/T = Frozen/thawed.
F.D. = Freeze-dried.

In addition to the damage to cells suspended in distilled water while cooling at 1°C/min to −20°C, there was further damage during the drying process. Survival was slightly improved by the presence of 12·5 per cent sucrose at all the temperatures studied, but again there was a further drop (of about 20 per cent) in survival on drying in each case. Twenty per cent peptone was the only additive tested which showed much success, in that it was able to maintain survival at 47 per cent on freezing to −35°C and this was only slightly reduced (43 per cent) on subsequent drying at this temperature. It therefore seemed that whereas sucrose was able to protect to some extent against freezing but not nearly so well against drying, peptone was able to protect the cells to a greater extent against freezing damage and, perhaps more significantly, was able to prevent any further damage during the drying stage.

Freeze–thaw damage was further investigated by studying the effects of varying the cooling rate on the survival of *Pseudomonas* suspended in 0·1 ml quantities of various media and thawed rapidly (Fig. 1). The survival of cells suspended in distilled water increased as the rate of cooling was increased from 1°C/min to approximately 60°C/min. Above this rate of cooling the survival began to drop until a minimum was reached at approximately 180°C/min. Survival then increased progressively again up to 900°C/min, the fastest cooling rate used in this study. This survival was reduced by about a half after a slow thaw. These results therefore seem to support the proposition that there are two possible mechanisms of damage which can occur during the freezing of cells, depending upon the

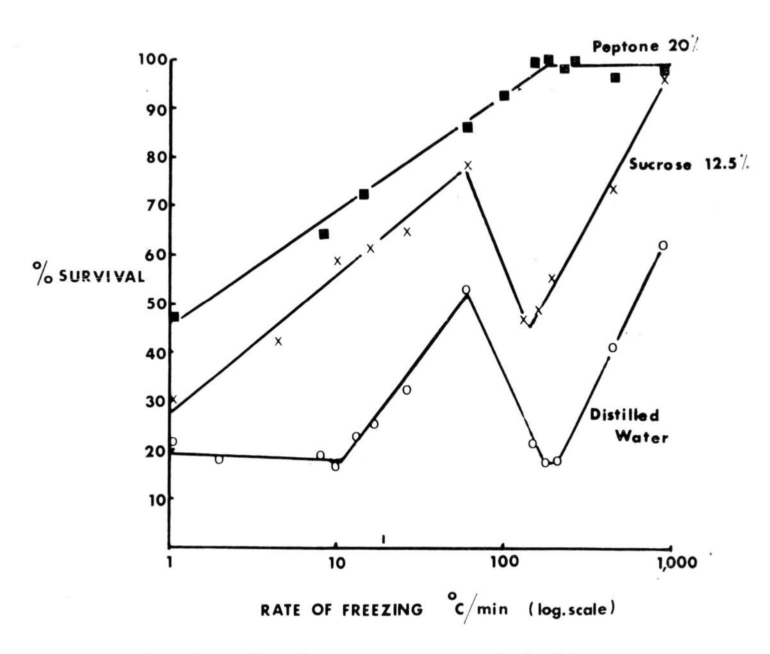

FIG. 1. The effect of cooling rate on the survival of *Pseudomonas* sp. (1 OH) suspended in various media.

rate at which they are cooled (Mazur, 1965). Sucrose (12·5 per cent) was found to confer approximately the same degree of protection over the whole range of cooling rates studied.

Fig. 1 shows that not only did 20 per cent peptone protect better than sucrose and distilled water at the slower rates of cooling but it also increased survival at faster rates of cooling until, at approximately 150°C/min, 100 per cent survival was achieved. In contrast to the curves obtained with distilled water and sucrose there was no further reduction in survival with the medium rates of cooling, and maximum survival was retained over the remainder of the range.

Thermal analysis and electrical conductivity measurements have clearly shown that, in the concentrations used, both sucrose and peptone are capable of forming glass-like structures known to be relatively stable at low temperatures and thought to occupy the interstitial spaces in which cells congregate (Davies, 1966*a*). It has been suggested that by preventing the complete removal of water in the form of crystalline ice such structures reduce the possible effects associated with an increase in salt concentration or intracellular ice. The ability of such compounds to prevent the complete removal of water and in fact to promote structured water during freezing seems to agree well with the numerous reports suggesting that for the preservation of viability at low temperatures the retention of at least some water, particularly that involved with macromolecular structure, in an unfrozen state is essential (Davies, 1968). Both sucrose and peptone may act to some extent in this way but the results of freeze-drying and freezing *Pseudomonas* at high cooling rates suggest that peptone has an additional mechanism of action. The possibility that damage on drying and that associated with fast rates of cooling are similar has not yet been supported by any further experimental evidence.

If peptone has a more specific mode of action than sucrose in preventing such injury one also has to consider the fact that peptone is present in the growth medium of microorganisms and perhaps may account to a greater or lesser degree for the inherent resistance found in different microorganisms. We therefore investigated the effects of varying the growth media on the subsequent sensitivity of cells to freeze–thaw damage. Fig. 2 shows the comparative survival of *Pseudomonas* cells, taken at different stages of growth from media both possessing and lacking organic nitrogenous compounds, when suspended in distilled water and cooled at varying rates. Although the growth or morphological characteristics of the cells did not seem to alter when grown in either type of medium, both age and growth medium cause considerable variation in susceptibility to freeze–thaw injury. Cells grown in the logarithmic stage of growth appear to be more susceptible than cells in the stationary phase at certain rates of cooling; the reverse is true at other rates. The central peak is assumed to correspond to the balance between increasing salt damage and intracellular ice formation (Mazur, 1965), and the fact that this occurs at higher cooling rates in logarithmic phase cells suggests that at least one of the factors thought to affect the permeability of the membrane to water varies with age. As these experiments were originally designed to investigate the role of peptone in increasing the resistance of cells to freeze–thaw injury, it was somewhat surprising that on comparing cells at the same stage of growth in the two media, more cells survived in the simple salt medium than in broth over the whole range of cooling rates studied.

On adding each salt present in Koser's citrate medium separately to broth and so varying the culture media, and by using the salts in suspending media, it was possible to show that any increase in survival brought about by the addition of a salt must be brought about during growth rather than in the suspending medium (Table II). Although magnesium sulphate, ammonium dihydrogen phosphate and dipotassium hydrogen

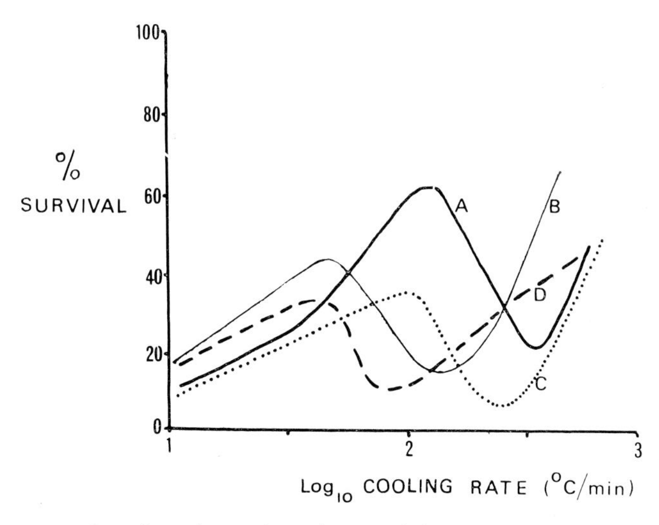

FIG. 2. The effect of growth medium and the age of the cell on the survival of *Pseudomonas* sp. (1 OH) cooled at different rates when suspended in distilled water.

A: Logarithmic phase cells in Koser's citrate medium.
B: Stationary phase cells in Koser's citrate medium.
C: Logarithmic phase cells in broth.
D: Stationary phase cells in broth.

phosphate all had effects at the cooling rates which had been chosen to typify both kinds of possible damage, the potassium salt was particularly effective at high rates of cooling. Similarly, increasing the concentration of sodium chloride in the suspending medium was effective in increasing survival at the higher cooling rates, whereas at the lower rate concentrations above 0·1 per cent seemed to be deleterious. The latter results may reflect a certain amount of osmotic dehydration, thus limiting the likelihood of intracellular ice formation. For cells grown in the citrate media, however, sodium chloride had a much smaller effect, though in this particular case magnesium sulphate gave good survival at both cooling

TABLE II

THE INFLUENCE OF GROWTH AND SUSPENDING MEDIA ON THE PERCENTAGE SURVIVAL OF LOGARITHMIC PHASE *Pseudomonas* sp. (1 OH) AFTER COOLING AT 100°C/MIN AND 500°C/MIN FOLLOWED BY A RAPID THAW

Additives	*Cells grown in broth with appropriate salt and suspended in distilled water*		*Cells grown in broth with appropriate salt in suspending medium*		*Cells grown in Koser's citrate medium and suspended in appropriate salt*		*Cells grown in Koser's citrate media deficient in appropriate salt and suspended in distilled water*	
	100°C/min	500°C/min	100°C/min	500°C/min	100°C/min	500°C/min	100°C/min	500°C/min
Water	—	—	(35)	(35)	(63)	(36)	—	—
Digest broth	(35)	(35)	—	—	—	—	—	—
Koser's citrate	—	—	42	43	38	65	—	—
0·1% K_2HPO_4	58	78	13	16	52	65	60	73
0·1% $NH_4H_2PO_4$	52	56	4	27	36	50	—	—
0·02% $MgSO_4 \cdot 7H_2O$	48	52	12	25	65	67	62	84
0·2% Na-citrate	33	46	30	35	—	—	—	—
0·1% NaCl	—	—	34	66	—	—	—	—
0·5% NaCl	35	77	4	63	18	42	55	65
1·0% NaCl	—	—	3	49	—	—	—	—
2·0% NaCl	—	—	1	73	—	—	—	—

Parentheses signify control samples

rates (Table II). By omitting from Koser's citrate medium those salts which did not seem to be essential for the growth of the cells, the greater effect on survival was found to occur at the higher rate of cooling. This series of experiments emphasizes the importance and interdependence of both the growth medium and suspending medium on the survival of cells subjected to freeze–thaw injury. The fact that the greatest effect of salts was seen at high cooling rates again suggests that intracellular ice formation may be modified in some way. Certainly in considering possible changes in permeability of cell membranes it is interesting to reflect on the work of Winslow and Haywood (1931). They associated an increase in viability of the bacterial cell with an increase in permeability of the cell which was evident when cells were suspended in low cation concentrations, and the opposite with high concentrations. Bivalent ions were eight to ten times more potent than univalent ions in this protection. Some bacterial cells also leak compounds which protect against freeze–thaw injury. Major, McDougal and Harrison (1955) working with *Pseudomonas aeruginosa* were able to correlate the survival of these cells during freezing and thawing with an increase in the amount of substances liberated and hence with an increase in the initial cell concentration.

Many workers have analysed these effects further. Moss and Speck (1966) showed that *Escherichia coli* cells frozen and stored in phosphate buffer were more resistant to freezing damage than those frozen and stored in distilled water. Similarly they showed that greater amounts of what they considered to be protective material leaked from the cells suspended in the phosphate buffer than from those suspended in distilled water. On analysis the liberated substances were found to contain peptides of relatively low molecular weight, which on addition to suspensions of other cells showed protective activity and also allowed the recovery of metabolically injured cells. We are now trying to determine whether the action of salts in the suspending medium or in the growth medium in bringing about the changes in *Pseudomonas* sp. (1 OH) described in Table II can be correlated with such leakage of peptide material. However, as broth cultures are not nearly so resistant as those grown in the citrate medium, any leakage may merely signify an alteration in the permeability of the membrane which at certain rates of cooling may prove to be more significant than the presence of peptides in the external medium.

In contrast to the results described with *Pseudomonas* cells, the yeast cells *Saccharomyces cerevisiae* depended greatly on organic nitrogenous material being present in the growth medium for their resistance to freeze–thaw damage. When *Saccharomyces cerevisiae* was grown in Nickerson's medium and in the same medium excluding the yeast extract, though there was no obvious change in growth and morphological characteristics, the susceptibility to freeze–thaw injury varied significantly. Although such injury could be avoided by adding 0·01 per cent yeast extract to the basic medium, greater concentrations of yeast extract in the growth medium did not further increase survival (Table III). The presence of

TABLE III

THE EFFECTS OF GROWTH AND SUSPENDING MEDIA ON THE SURVIVAL OF *Saccharomyces cerevisiae* AFTER COOLING AT VARIOUS RATES FOLLOWED BY A RAPID THAW

		% *Survival* Cooling *rates* (°C/min)			
Growth medium	*Suspending medium*	1	10	50	450
Nickerson medium					
(0·01% yeast extract)	Water	72	68	50	80
+ 0·1% yeast extract	Water	—	78	62	85
+ 1·0% yeast extract	Water	—	72	62	80
+ 1·0% yeast extract	5·0% yeast extract	—	75	64	93
+ 10·0% yeast extract	Water	—	74	52	83
+ 10·0% yeast extract	5·0% yeast extract	—	90	95	90
Modified Nickerson medium	Water	30	30	35	23
	0·05% yeast extract	—	56	42	80
	0·5% yeast extract	—	56	—	80
	5·0% yeast extract	58	—	40	90
	5·0% peptone	—	—	45	94
+ 5·0% peptone	Water	70	—	58	88
+ 5·0% peptone	5·0% peptone	70	—	40	90
Broth	Water	40	38	30	60
Modified Nickerson medium	Broth (half-strength)	27	30	30	20

yeast extract in the suspending medium also helped to restore the survival of cells, especially at the higher rates of cooling. However, at the lower rates protection was less than that obtained by growing the cells in the presence of yeast extract. Peptone solutions gave similar results to those obtained with yeast extract (Table III). Digest broth, though rich in peptides, did not protect cells when used in low concentration as a suspending medium. Nevertheless, when used as a growth medium broth increased the survival over a similar range of cooling rates as did the yeast extract.

In low concentrations peptone and yeast extract seem to be more effective during the growth of the cell and to increase its resistance, particularly at the higher rates of cooling. In higher concentrations the effects found during growth and the formation of a glass-like structure seem to be synergistic.

When they added peptone to give final concentrations of between 1 and 10 per cent in T4 phage suspensions, Steele, Davies and Greaves (1969*a*) showed that the survival of the phage could be progressively increased after cooling slowly and thawing rapidly, and that the temperature range of inactivation could be lowered by about 2–5°C. However, increasing the peptone concentration was much more effective in increasing the survival of such organisms after a slow thaw and this effect seemed to be related to a complete alteration in both the rate and temperature range of inactivation. A similar increase in survival of slowly thawed T4 phage, obtained by increasing the initial phage titre, could be greatly diminished by washing the phage with distilled water, and could even be prevented by preparing the T4 phage in cells grown in a salt medium lacking organic nitrogenous compounds. The concentration effect may therefore have been caused by peptides or amino acids adsorbed on to the phage from the original broth lysate.

On Sephadex fractionation of the actual components of peptone certain fractions gave good protection to slowly thawed phage and the maximum survival was in fact increased by the fractionation. The latter effect was probably due to the removal of inhibitory components from the crude peptone solution. The fractions, however, had no effect on the viability of rapidly thawed phage. Further fractionation enabled the protective activity to be correlated with those fractions containing relatively small peptides, while amino acids had no significant protective effect (Steele, Davies and Greaves, 1969*a*). Both oxytocin and vasopressin were highly effective in protecting T4 phage but they had little activity in protecting *Pseudomonas* sp. (1 OH), which was, however, protected by corticotropin (Davies, Greaves and Steele, 1967).

As peptone seemed to protect against slow thaw damage and to some extent against rapid thaw damage, while the peptide fractions only protected against slow thaw damage, it was necessary to distinguish the possible mechanisms of such injury. It has since been shown that the damage caused by thawing rapidly can be correlated with osmotic shock (Anderson, 1953; Leibo and Mazur, 1966; Steele, Davies and Greaves, 1969*c*). Since peptone solutions in fairly high concentrations are capable of reducing this damage it may well be that they act as an osmotic buffer and so prevent the sudden influx of water back into the cells during thawing. However, as described earlier, even in fairly low concentrations peptone and peptides can also protect cells against a second type of damage which occurs even in the absence of osmotic shock. This damage, which is manifest after thawing slowly, is directly related to the eutectic temperature of the suspending medium: the higher the eutectic temperature the greater the damage (Steele, Davies and Greaves, 1969*a*). These results seem to support the idea that the injury may be caused by the removal of the final traces of water at the eutectic point.

Of particular interest, especially after having considered the importance of dipotassium hydrogen phosphate in the growth medium of *Pseudomonas* sp. (1 OH), was the observation that in the absence of phosphate buffer there was an additional inactivation of the T4 phage above the eutectic temperature of the suspending medium. Furthermore, the degree of this inactivation was dependent on the species of ions in the suspending medium, whereas below the eutectic temperature the inactivation was independent of the type of ions present (Steele, Davies and Greaves, 1969*b*). The relative effect of various ions in causing inactivation above the eutectic temperature was found to be in the order $NH_4 < Na^+, K^+, < Li^+ < Mg^{++}$; $CH_3COO^- < Cl^- < Br^- < I^-$, which seems to agree well with the series for the destabilizing effects of such ions on the conformational stability of macromolecules (von Hippel and Wong, 1964). The protection afforded by phosphates, especially the dipotassium salt, suggests that the phosphate ion is able to exert a dominant effect against such destabilization. Although increasing the rate of cooling increased the survival against such damage, after equilibration at a particular temperature it was found that the survival was in fact independent of the cooling rate and also of the initial concentration of salt.

In spite of the fact that the T4 phage was capable of withstanding solutions of salts up to 3 M concentration at room temperature, at similar concentrations reached during freezing at about $-12°C$ there was a marked inactivation of the phage. More remarkable, perhaps, was the observation that the degree of inactivation was related to the salt concentration even in the absence of ice formation, and was dependent upon the temperature and the particular species of ion present (Steele, Davies and Greaves, 1969*b*).

Cooling tends to weaken hydrophobic bonds and the combination of cooling and the ionic destabilization of macromolecules described earlier may magnify such damage, so explaining the results obtained on cooling and storage at temperatures above the eutectic temperature in the absence of ice. In fact the presence of ice seemed to protect cells against such ionic damage and this may be related to the structuring effect which is thought to occur in narrow channels (Bangham and Bangham, 1968; Willis *et al.*, 1969). Certainly the evidence on the effect of hydrophobic bonds in such inactivation is partly supported by the increased survival obtained when D_2O was used in place of H_2O (Steele, Davies and Greaves, 1969*b*). Furthermore, glycerol prevented the ionic damage which occurred in the absence of ice, and again this effect may depend upon glycerol promoting a structural effect on the water (Davies, 1968).

When the effect of peptides on such ionic damage was investigated a tryptic digest of globin was used in place of peptone, as the amino acid sequence of the component peptides had already been determined (Braunitzer *et al.*, 1966; von Ehrenstein, 1966; Hunt, Hunter and Munro, 1968), making any correlation between effect and structure easier to determine. Globin tryptic digest (10 mg/ml) gave complete protection against slow freeze–slow thaw inactivation of the T4 phage both above and below the eutectic temperature, though the latter protection was considerably inhibited by the addition of amino acids: the longer the side chain the greater the inhibition (Table IV). On fractionation of the

TABLE IV

THE INHIBITORY EFFECT OF AMINO ACIDS WHEN ADDED TO T4 PHAGE FROZEN IN 5 MG/ML GLOBIN TRYPTIC DIGEST (GTD). SAMPLES WERE SUSPENDED IN 0·13 M BUFFER AND COOLED AT 1°C/MIN TO −45°C AND THAWED SLOWLY

Added amino acid (0·01 M)	*Percentage survival*	
	Control	*Test*
None	12	90
Glycine	8	51
Alanine	11	29
Valine	8	20
Leucine	4	18
Isoleucine	6	18
Serine	11	49
Threonine	13	27
Methionine	6	18
Phenylalanine	0·6	18
Tryptophan	10	35
Histidine	11	37
Arginine	14	33
Lysine	24	36
Aspartic Acid	12	35
Glutamic Acid	12	35
Proline	—	19

globin digest, as was found with peptone, protective ability could be correlated with the smaller peptides, and in particular the di- and tripeptides (Figs. 3*a*, *b*). However, at temperatures above the eutectic temperature protection was obtained with most fractions and seemed to depend upon peptide concentration rather than peptide size.

The specific protection below the eutectic point possibly depends upon a direct substitution within the membrane, while the non-specific effect

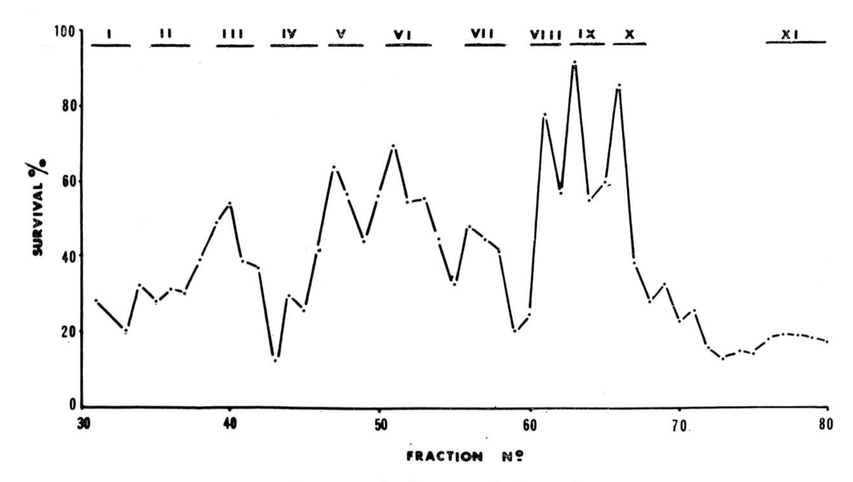

FIG. 3*a*. Protection by peptide fractions below the eutectic temperature. Samples of T4 phage suspended in phosphate buffer and appropriate fraction were cooled at 1°C/min to −45°C and thawed slowly. Roman numerals refer to the fractions used in the determination of the constituent peptides (Hunt, Hunter and Munro, 1968).

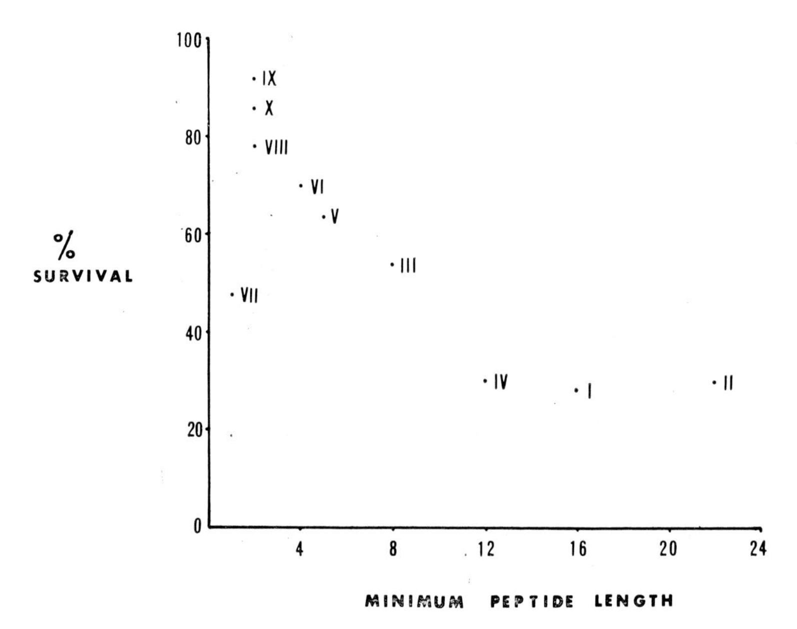

FIG. 3*b*. A possible correlation between peptide length and protection. The figure shows the peaks of survival of Fig. 3*a* as a function of the shortest peptide (amino acid units) in the respective fraction.

above the eutectic temperature is possibly a stabilizing effect upon the water structure and hence on hydrophobic bonds.

If, as postulated, changes in water structure play a significant part in the denaturation of macromolecules during freezing, thawing and drying (Davies, 1968), it is conceivable that the major portion of damage to the cell is localized at the delicately balanced lipoprotein complex and hence in particular to the membrane structures. Although the protein coat of the phage has been of tremendous value in studying the mechanisms of damage during freezing, verification of damage to the membrane structure necessitates a study of such effects on a complete lipoprotein membrane. Although the results with *Pseudomonas* sp. and *Saccharomyces cerevisiae* indicate a possible change in permeability of the membrane with the use of peptides, any deleterious effects caused by the increased permeability resulting in swelling and lysis of the cells during thawing may be masked by the cell being surrounded by a rigid cell wall. The use of human red blood cells as an indicator system has therefore proved to be of particular advantage in relating the importance of some of the factors described in previous sections to the freezing and thawing of mammalian cells.

Seeman (1966) had demonstrated that a range of pharmacologically active compounds such as phenothiazines, steroids, local anaesthetics and some surface-active antibiotics were capable in low concentrations of stabilizing red cell membranes against hypotonic stress. It seemed to us that such compounds were either stabilizing the membrane by a direct combination, or alternatively they altered the permeability of the membrane so as to allow equilibration to occur more readily. If such a change in permeability was the factor involved then it might also be effective in preventing the osmotic stress which occurs after the cell has attained a minimum cell volume under conditions of hypertonic stress (Meryman, 1968). If these experiments proved positive, then such compounds might prevent the hypertonic stress which is thought to occur during slow cooling and the hypotonic stress which might be expected during rapid thawing. In addition, the fact that many short-chain peptides themselves possess a pharmacological action as well as their cryoprotective ability supported the possibility that peptone and peptides may act against hypotonic stress in a similar (stabilizing) way to chlorpromazine and lignocaine, the compounds chosen for this investigation.

The effect of these compounds on the release of haemoglobin from red cells under hypotonic stress was studied and, just as with chlorpromazine (5×10^{-5} M) and lignocaine (6×10^{-2} M), there was a concentration of

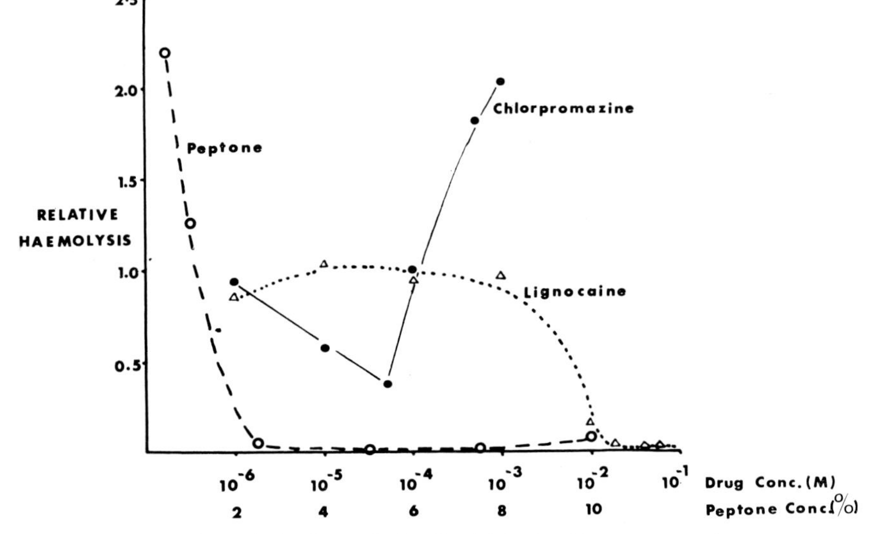

FIG. 4. The relative haemolysis of red blood cells exposed to hypotonic stress in the presence of various compounds.

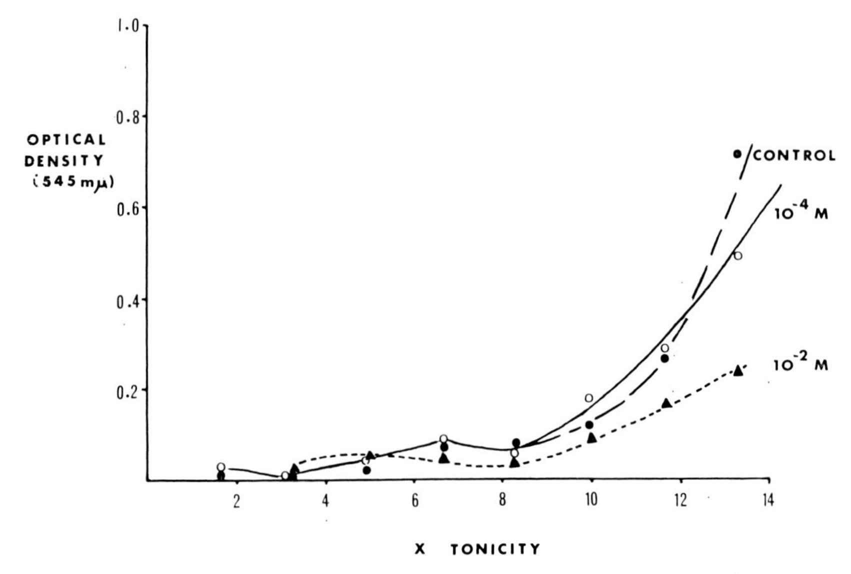

FIG. 5. The stabilization of red blood cells by lignocaine against the leakage of haemoglobin in hypertonic solutions.

peptone (7·5 per cent) which gave a maximum stabilizing effect before going on to labilize the membrane (Fig. 4). During hypertonic stress, although chlorpromazine showed hardly any effect, 10^{-2} M-lignocaine and concentrations between 1 and 5 per cent peptone all tended to stabilize the membrane to a small extent against release of haemoglobin over a range of hypertonic saline solutions greater than 1·5 M (Fig. 5).

In spite of such a stabilization of the membrane against hypotonic and hypertonic stress by lignocaine and to a lesser degree by chlorpromazine, neither drug led to any survival of red cells after cooling at rates varying

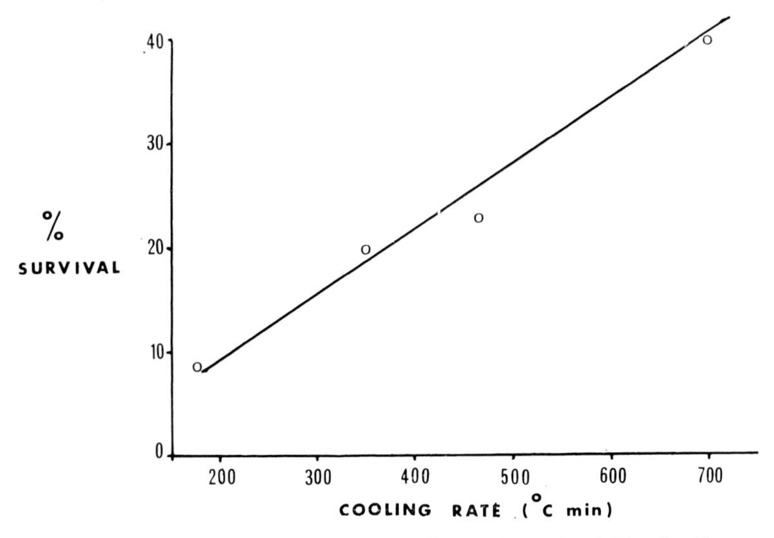

FIG. 6. The effect of cooling rate on the survival of red blood cells suspended in 2·5 per cent peptone and thawed rapidly.

between 5 and 2300°C/min followed by a rapid thaw. In fact the effect on the membrane seemed to be such that the damage was increased as compared to the survival obtained with a control suspension cooled at the optimum rate of 2300°C/min (Rapatz, Sullivan and Luyet, 1968). Peptone on the other hand produced a marked increase in the survival of red cells, particularly at high rates of cooling (Fig. 6). It was thought that, as was suspected with the *Pseudomonas*, this increase might be a result of osmotic dehydration. Although increasing the salt concentration to twice isotonic considerably increased survival at a cooling rate of 2300°C/min, such a dehydration did not give as significant an effect as that obtained in peptone solutions giving equivalent cell shrinkage. Hence, although dehydration seems to be involved to some extent it is not in itself sufficient to explain the increase in survival obtained with peptone

TABLE V

EFFECT OF PEPTONE ON SURVIVAL OF RED BLOOD CELLS AFTER COOLING AT VARIOUS RATES FOLLOWED BY A RAPID THAW

Cooling rate	*% Survival*		
(0°C/min)	*5% Peptone*	*2·5% Peptone*	*Isotonic saline*
5	—	—	—
10	—	—	—
140	—	—	—
180	—	8·5	—
230	—		—
350	—	20	—
460	—	23	—
700		40	—
1100			—
1250	84·5		—
1750	81·5		50
2300	92	70	45
3500	46		35
4500	15·4		—

— indicates zero survival

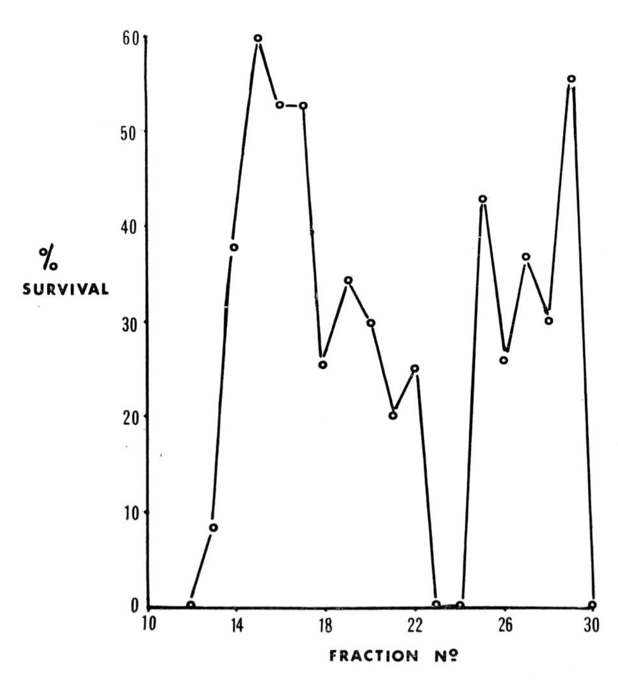

FIG. 7. The effect of Sephadex (G25) fractions of peptone on the survival of red blood cells after cooling at 2300°C/min and thawing rapidly.

over the range of cooling rates studied (Table V). On fractionation of the peptone certain fractions were again found to increase the survival of the red cells (Fig. 7). But when the two most effective fractions were combined, the survival after cooling at 2300°C/min was less than with either fraction alone. The concentration of the fractions used was equivalent to that which would have been present in 5 per cent peptone solution, and it was interesting to note that at rates between 1250 and 2300°C/min survival was less than that obtained in the crude 5 per cent peptone but was greater at rates above 2300°C/min. Work is now continuing in an effort to determine the type of peptides involved in such protection and the mechanism whereby they protect against rapid freeze–rapid thaw injury.

Taking into account the recent publication of Meryman (1968) it is not surprising that peptone and peptides are unable to prevent the osmotic stress within the cell after it has reached its minimum volume during a slow freeze. Peptone will in itself exert an additional osmotic

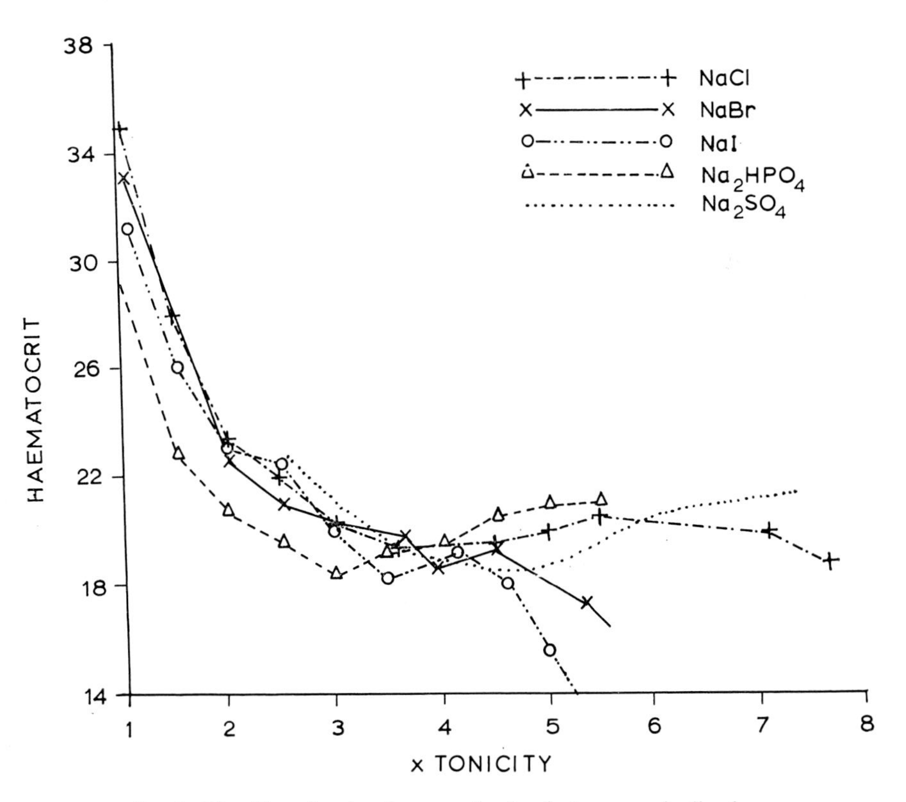

FIG. 8. The effect of various hypertonic salt solutions on red cell volume.

pressure to that caused by the increasing salt concentration, and the stabilizing action on the membrane is only significant in the higher ranges of hypertonicity, well beyond the critical 0·8 M concentration at which such irreversible osmotic stress occurs. If cell shrinkage could be avoided by any means there may perhaps be additional factors which would influence the maintenance of the cell as an organized unit. With the coliphage T4 ionic damage was a possible cause of damage during cooling, so the possibility of such damage was examined in red blood cells. The effects of osmotic shrinkage in hypertonic solutions of different salts showed that although the minimum cell volume was obtained at approximately the same tonicity in each salt, the tonicity at which one obtained direct leakage of haemoglobin from the cell varied in a somewhat similar relationship to the series of ions referred to earlier for their destabilization of macromolecules (von Hippel and Wong, 1964) (Fig. 8). The liberation of haemoglobin during storage in various salt solutions in the absence of ice at $-10°C$ was greater in every case than that in samples maintained at room temperatures (Table VI). The significance of these results and any possible correlation with damage at high rates of cooling have yet to be determined.

TABLE VI

THE EFFECTS OF COOLING AND STORING FOR 30 MINUTES AT $-10°C$ IN THE PRESENCE OF VARIOUS SALTS ON THE LIBERATION OF HAEMOGLOBIN FROM RED BLOOD CELLS

Suspending medium	*Haemoglobin absorption* (545 *nm*)	
	Control at room temp.	*Test*
1·5 M-NH_4Br	0·75	1·01
1·5 M-NH_4Cl	0·64	0·66
1·5 M-NH_4I	1·1	1·4
Saline	0·019	0·025
Water	1·2	1·4

Although it is not possible to extrapolate results obtained with micro-organisms to mammalian cells, certain salient features regarding the fundamental effects of cooling and freezing on lipoprotein membranes are beginning to emerge.

Damage which may occur in a cell during freezing and subsequent thawing can now be classified into (*i*) thermal shock, (*ii*) ionic damage, (*iii*) osmotic stress, (*iv*) eutectic damage, (*v*) intracellular ice formation, (*vi*) osmotic shock. Thermal shock and ionic damage may well be inter-dependent in that thermal shock can be influenced by the ions present, while the ionic damage can be amplified by cooling. Both effects may in fact be manifestations of the weakening of hydrophobic bonds by the respective conditions, and they may result in the disruption, which may prove in some cases to be reversible, of the macromolecular configuration. The actual nature of the damage brought about by osmotic stress is not yet known, though the disruption in the membrane may be brought about either by a direct physical strain or possibly by interaction of active groups brought into proximity by the shrinkage of the membrane. Intracellular ice may be injurious by its mere physical presence though it could equally result in the removal of structured water. Similarly eutectic damage which is thought to coincide with the removal of the final traces of water—possibly of the "bound" water (Davies, 1968)—may involve a general breakdown in the structural arrangement of the macromolecules and allow interaction of previously separated active groups. On rehydration these newly formed bonds may remain and the altered permeability may allow increased swelling, with subsequent lysis of the cell as found in osmotic shock damage, particularly in cases where the membrane lacks a certain degree of elasticity (Davies, 1966*b*). Slight changes in configuration which are not in themselves lethal may occasionally be detected as alterations in the antigenic or biochemical characteristics of the cell.

The role of peptides in preventing these various forms of damage to some extent depends upon the concentrations used. High concentrations of peptone are able to form glass-like structures which in turn increase the stabilization of structured water during slow cooling and also act as osmotic buffers to prevent swelling of the cell on rehydration. In lower concentrations peptone and even peptides seem to confer a more specific effect. It is suggested, though it is by no means certain at this stage, that they may act as stabilizers for hydrophobic bonds. There is also increasing evidence which suggests that peptides are able to influence the permeability of cell membranes to various ions (Podleski and Changeux, 1969; Haynes, Kowalsky and Pressman, 1969). The precise mechanism of inducing such changes in permeability is uncertain, though they may act as a bridge across the membrane, their water of hydration helping to form a continuous aqueous channel; alternatively they may act as a specific carrier molecule across the membrane. More simply a direct combination of such peptides with the protein surface of the cell, with the resulting change in configuration, may well influence the intimate bonding between the protein and lipid components and so alter the so-called "pores" across the membrane in some way. Certainly the inhibition of such protective effects by amino acids and other peptides suggests a competitive combination with the receptor site. Alternatively neutralization of the active group on the peptide by such compounds would prevent adequate bonding with the membrane, or the increase in size may decrease its efficiency as a substitute for structured water between active groups. In the freezing of complex tissues and organs, peptides may well prove to be effective cryoprotective agents in their own right, enabling a range of suitable cooling rates to be used, thus allowing an effective range of cooling rates to act across a thick piece of tissue. The possible alteration in permeability brought about by peptides may also allow easier penetration by the more commonly used cryoprotective agents. At a more fundamental level perhaps their greatest contribution will be in allowing a greater understanding of the basic principles involved in membrane structure and function.

SUMMARY

Experiments on the effects of cooling and thawing on a range of micro-organisms and human red cells indicate that there are certain fundamental consequences of cooling and freezing on lipoprotein membranes which are common to all cells.

Ionic damage, with the possible weakening of hydrophobic bonds and the resulting destabilization of macromolecular configuration, is thought to be a primary cause of damage on cooling to temperatures above the eutectic temperature of the medium; such damage can be prevented by compounds which are thought to promote structure formation within water, e.g. phosphates, acetates, glycerol and peptides. Below the eutectic temperature peptide protection seems to be more specific; it may act either by altering the permeability of the cell membrane, so preventing the formation of intracellular ice, or by a direct substitution within the membranes, so preventing cross-linkage between active groups which could lead to the destruction of the cell on thawing.

Acknowledgements

I am grateful for the assistance of N. S. Davies, R. M. du Bois and R. G. Jezzard in the collection of some of the experimental data presented in this article and to P. R. M. Steele for permission to reproduce some of his results obtained with the T4 phage.

This project has been supported in part by the Office of Naval Research under contract F61052-68C-0041.

REFERENCES

ANDERSON, T. F. (1953). *Cold Spring Harb. Symp. quant. Biol.*, **18**, 197–206.

ANNEAR, D. I. (1956). *J. Hyg., Camb.*, **54**, 487–508.

BANGHAM, A. D., and BANGHAM, D. R. (1968). *Nature, Lond.*, **219**, 1151–1152.

BRAUNITZER, G., BEST, J. S., FLAMM, V., and SCHRANK, B. (1966). *Hoppe-Seyler's Z. physiol. Chem.*, **347**, 207–211.

COLLIER, L. H. (1955). *J. Hyg., Camb.*, **53**, 76–101.

DAVIES, J. D. (1966*a*). In *Advances in Freeze-Drying*, pp. 9–20, ed. Rey, L. Paris: Hermann.

DAVIES, J. D. (1966*b*). Ph.D. thesis, University of Cambridge.

DAVIES, J. D. (1968). In *Low Temperature Biology of Foodstuffs*, pp. 177–204, ed. Hawthorn, J., and Rolfe, E. J. Oxford: Pergamon.

DAVIES, J. D., GREAVES, R. I. N., and STEELE, P. R. M. (1967). *Cryobiology*, **3**, 371 (abstr.).

EHRENSTEIN, G. VON (1966). *Cold Spring Harb. Symp. quant. Biol.*, **31**, 705–714.

FRY, R. M., and GREAVES, R. I. N. (1951). *J. Hyg., Camb.*, **49**, 220–246.

GREAVES, R. I. N. (1962). *J. Pharm. Pharmac.*, **14**, 621–640.

GREAVES, R. I. N., and DAVIES, J. D. (1965). *Ann. N.Y. Acad. Sci.*, **125**, 548–558.

GREAVES, R. I. N., DAVIES, J. D., and STEELE, P. R. M. (1967). *Cryobiology*, **3**, 283–287.

HAYNES, D. H., KOWALSKY, A., and PRESSMAN, B. C. (1969). *J. biol. Chem.*, **244**, 502–505.

HIPPEL, P. H. VON, and WONG, K. Y. (1964). *Science*, **145**, 577–580.

HUNT, T., HUNTER, T., and MUNRO, A. J. (1968). *J. molec. Biol.*, **36**, 31–45.

LEIBO, S. P., and MAZUR, P. (1966). *Biophys. J.*, **6**, 747–777.

MAJOR, C. P., McDOUGAL, J. D., and HARRISON, A. P. (1955). *J. Bact.*, **69**, 244–249.

MAZUR, P. (1965). *Ann. N.Y. Acad. Sci.*, **125**, 658–676.

MERYMAN, H. T. (1968). *Nature, Lond.*, **218**, 333–336.

MERYMAN, H. T., and KAFIG, E. (1955). *Proc. Soc. exp. Biol. Med.*, **90**, 587–589.

MOSS, C. W., and SPECK, M. L. (1966). *J. Bact.*, **91**, 1105–1111.

PODLESKI, T., and CHANGEUX, J. P. (1969). *Nature, Lond.*, **221**, 541–545.

RAPATZ, G., SULLIVAN, J. J., and LUYET, B. J. (1968). *Cryobiology*, **5**, 18–25.

RINFRET, A. P. (1960). *Ann. N.Y. Acad. Sci.*, **85**, 576–594.

SCOTT, W. J. (1960). In *Recent Research in Freezing and Drying*, pp. 188–202, ed. Parkes, A. S., and Smith, A. U. Oxford: Blackwell.

SEEMAN, P. M. (1966). *Int. Rev. Neurobiol.*, **9**, 145–221.

STEELE, P. R. M., DAVIES, J. D., and GREAVES, R. I. N. (1969*a*). *J. Hyg., Camb.*, **67**, 107–114.

STEELE, P. R. M., DAVIES, J. D., and GREAVES, R. I. N. (1969*b*). *J. Hyg., Camb.*, in press.

STEELE, P. R. M., DAVIES, J. D., and GREAVES, R. I. N. (1969*c*). *J. Hyg., Camb.*, in press.

WILLIS, E., RENNIE, G. K., SMART, C., and PETHICA, B. A. (1969). *Nature, Lond.*, **222**, 159–161.

WINSLOW, C. E. A., and HAYWOOD, E. T. (1931). *J. Bact.*, **22**, 49–69.

[For discussion of this paper see pp. 246–250.]

MECHANISMS OF FREEZING DAMAGE IN BACTERIOPHAGE T4*

STANLEY P. LEIBO and PETER MAZUR

Biology Division, Oak Ridge National Laboratory, Oak Ridge, Tennessee

BACTERIOPHAGES—because they are well understood genetically, chemically, and physically—were used in a continuing series of experiments designed to determine the mechanisms by which biological systems may be altered by freezing and thawing. Initially, we studied the inactivation of wild-type bacteriophage T4 suspended in several simple salt solutions and exposed to subzero temperatures between 0° and −80°C. We compared these results with those obtained using a related phage mutant that has somewhat different biological and physical properties. On the basis of the results we hypothesize that three factors are primarily responsible for the inactivation of frozen–thawed T4 phage. (1) When phage frozen in dilute solutions of salts such as NaCl, KCl, and KNO_3 are thawed rapidly, they are exposed to a rapid dilution of salts that concentrated during freezing. This "osmotic shock" (Anderson, Rappaport and Muscatine, 1953; Leibo and Mazur, 1966) destroys phage infectivity. (2) When phage are frozen in more concentrated solutions (> 0·1 molal) of these salts, however, inactivation results from the complete crystallization of water and the precipitation of all salts. (3) When phage are frozen in solutions of other salts, such as NaBr and $MgCl_2$, they are inactivated simply by exposure to toxic levels of these salts that result from the removal of water as ice.

FREEZING OF PHAGE IN CONCENTRATED SOLUTIONS

Unlike most cells, bacteriophages are stable at +25°C in concentrated (and even saturated) solutions of many salts—such as NaCl, KCl, CsCl, NH_4Cl, and $MgSO_4$. For convenience, we shall refer to these as Class I salts. There are, however, other salts (Class II) in which the phage are

* Research sponsored by the U.S. Atomic Energy Commission under contract with the Union Carbide Corporation.

completely stable at +25°C in one concentration, but are completely inactivated in somewhat higher concentrations (Fig. 1). Similar inactivation occurs in $BaCl_2$, NaI, and $NaClO_4$. Although the mechanism of this inactivation is not known, it may be related to the fact that salts which inactivate phage at +25°C (Class II) are sufficiently soluble to

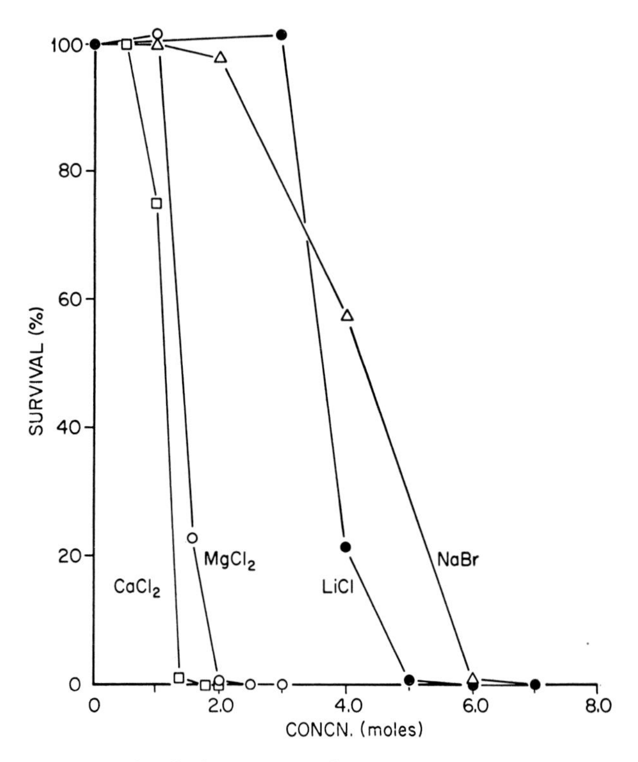

FIG. 1. Survival of phage T4B after suspension in various initial concentrations of the indicated salts at +25°C. The phage were diluted slowly from these solutions to avoid osmotic shock. The experiments were repeated with phage T4B*os*, and gave essentially identical results, regardless of whether the dilutions were carried out rapidly or slowly.

reduce the water activity of the solution to below 0·7. (This is not true for any of the Class I salts.) Phage inactivation in Class II salts is not an osmotic shock effect: it occurs as well in T4B*os*, a phage mutant that is resistant to large, rapid changes in the osmotic pressure of its environment, as in T4B, the parent strain that is sensitive to such changes. When phage are frozen in solutions of Class II salts, inactivation is not correlated with the eutectic point, but occurs when ice formation raises the concentration to the lethal level. Support for this conclusion is derived by plotting phage survival both as a function of salt concentration at +25°C, and as a function of the salt concentration produced by freezing to various subzero temperatures (Figs. 2 and 3). The curves almost coincide.

When phage are frozen in concentrated solutions of the non-toxic Class I salts, on the other hand, several lines of evidence indicate that

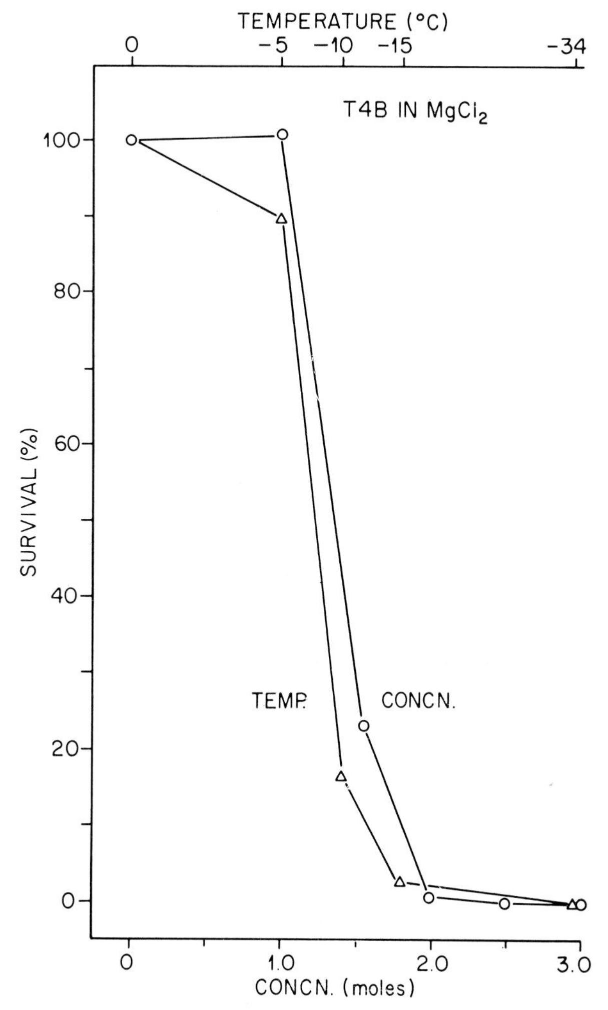

FIG. 2. Survival of phage T4B as a function of $MgCl_2$ concentration at +25°C (labelled "Concn.") and as a function of the subzero temperature at which ice and the solution at those concentrations are in equilibrium (labelled "Temp."). In the latter case, the phage were suspended in 1·0 molal $MgCl_2$, cooled to various temperatures above the eutectic point, seeded with ice, held 10 min, and then thawed slowly. Since the equilibrium concentration is a curvilinear function of subzero temperature, the temperature scale has been adjusted appropriately.

inactivation results only upon complete crystallization of water and the precipitation of all salts. We have found that cooling alone is innocuous. One hundred per cent of the phages T4B and T4B*os* suspended in the eutectic concentration of NaCl survived cooling to temperatures as low as $-40°C$ when the solution remained supercooled (Leibo and Mazur, 1969). But when the solution froze, almost 100 per cent of the phage were inactivated. The inactivation was a step-function of temperature,

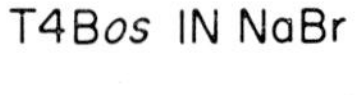

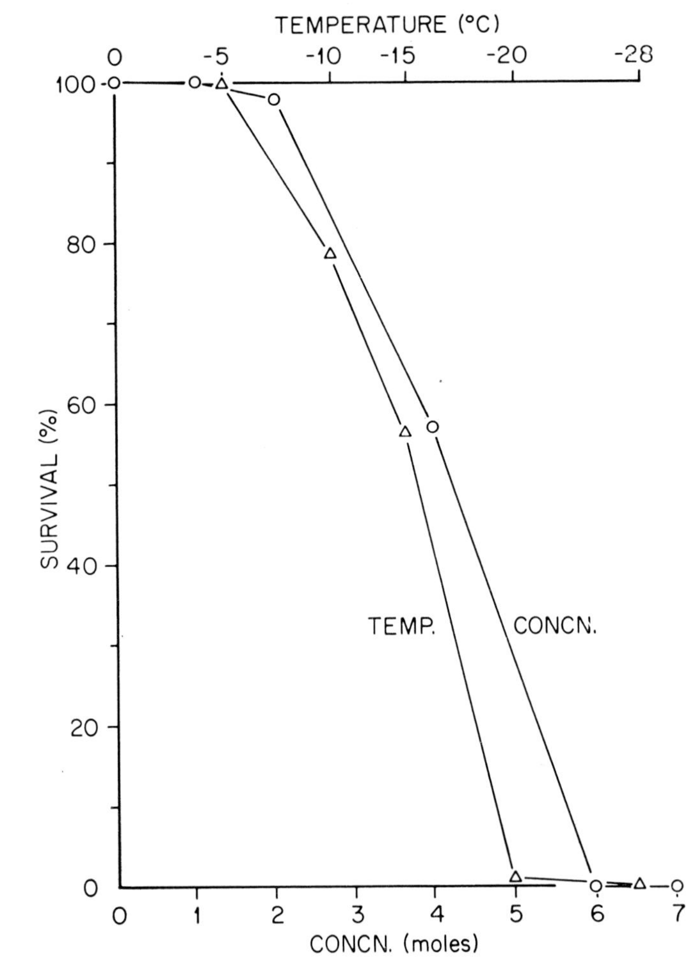

FIG. 3. Survival of phage T4B*os* as a function of NaBr concentration at $+25°C$ and as a function of corresponding subzero temperatures. The treatment was the same as that described in Fig. 2 except that the phage to be subjected to freezing were initially suspended in 1·35 molal NaBr, and the frozen suspensions were thawed rapidly.

and occurred about 10°C below the eutectic point of the solution with both T4B and T4B*os*, regardless of the warming rate. It occurred in both phages even though T4B is sensitive to rapid osmotic pressure changes and T4B*os* is not (Leibo and Mazur, 1969).

Phage frozen in eutectic concentrations of other Class I salts are also abruptly inactivated about 10°C below the eutectic point of the solution (Fig. 4). That is, inactivation occurred below −30°C in 5·2 molal NaCl (eutectic point: −21·1°), below −20°C in 3·3 molal KCl (eutectic point: −10·7°), and below −10°C in 1·15 molal KNO_3 (eutectic point: −2·8°). The addition of as little as 0·0005 molal NaCl to the KNO_3 solution depressed the temperature for almost total inactivation to −30°C.

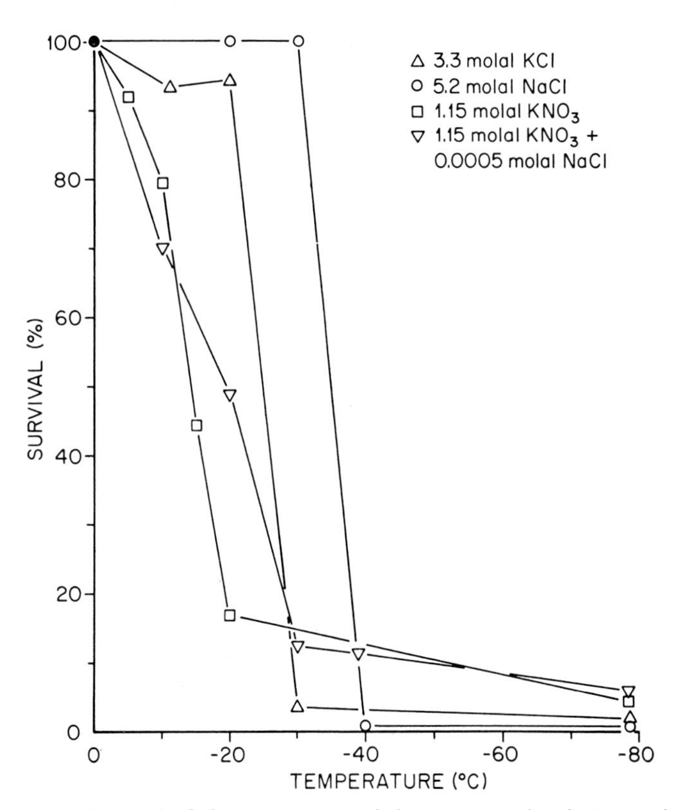

FIG. 4. Survival of phage T4B suspended in various salt solutions and cooled to various subzero temperatures. The suspensions were cooled directly to the appropriate temperatures, seeded with ice, held 10 min, and thawed slowly. They were then slowly diluted out of the suspending media, and assayed for infectivity.

That inactivation required cooling to 10°C below the eutectic point may reflect the fact that concentrated solutions supersaturate to that extent. To test this possibility, we cooled phage T4B*os* in 5·2 molal NaCl to several subzero temperatures, seeding these suspensions with salt crystals rather than with ice as in the previous experiments. When this was done, inactivation occurred *just* below the eutectic point of the solution rather than 10°C below the eutectic point (Leibo and Mazur, 1969).

Thus, inactivation apparently occurs when both salt and ice precipitate, and the solution consequently undergoes complete solidification.

FREEZING OF PHAGE IN DILUTE SOLUTIONS

The inactivation of phage frozen in *dilute* solutions of Class I salts has a different explanation: osmotic shock. Osmotic shock is the phenomenon responsible for the inactivation of T-even phages at room temperature when they are abruptly transferred from concentrated to dilute salt solutions (Anderson, Rappaport and Muscatine, 1953). They are not inactivated if the transfer is carried out slowly; and there are phage mutants, designated *os*, that are refractory to both rapid and slow transfer (Leibo and Mazur, 1966).

Phage frozen and thawed in dilute solutions are exposed to an analogous sequence: a salt solution concentrates during freezing as water is removed in the form of ice, but is diluted to its initial concentration during thawing as the ice melts. Therefore, one might expect phage inactivation from osmotic shock to occur during freezing and thawing, and a variety of data indicate that it does.

The results in Fig. 5 demonstrate that the survival of the shock-sensitive phage T4B differed from that of the shock-resistant phage T4B*os* when both were frozen in dilute NaCl solutions. When T4B was frozen and rapidly thawed, its survival decreased to a minimum at about −20°C, but then increased with decreasing temperature (Fig. 5*a*). When T4B*os* was similarly treated, however, its survival simply decreased progressively with decreasing temperature (Fig. 5*c*). The survival pattern of slowly thawed T4B mimicked that of rapidly thawed T4B*os* (compare Figs. 5*b* and 5*c*). All these results are explicable in terms of osmotic shock. The minimum survival of T4B at −20°C reflects the fact that an NaCl solution reaches its maximum concentration of 5·2 molal at −21·1°C, the eutectic point. Phage frozen to −20°C are therefore exposed to a highly concentrated liquid solution. Rapid thawing of such a suspension dilutes the solution rapidly, the sequence that produces osmotic shock at room temperature. Phage frozen to higher temperatures are exposed to lower salt concentrations, and the osmotic shock is less severe. Phage frozen to lower temperatures are exposed only very briefly to a concentrated liquid solution before the entire suspension freezes. Another consequence is that permeation of salts is reduced, and, consequently, the osmotic shock is decreased. Slow warming of frozen T4B is analogous to a slow dilution of phage suspended in a concentrated solution at room

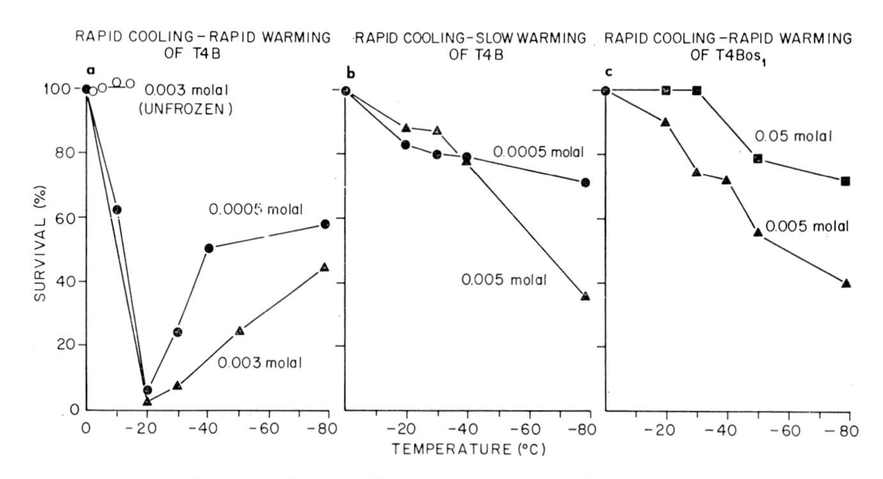

FIG. 5. Survival of phages T4B and T4B*os* exposed to subzero temperatures in dilute NaCl solutions. The phages were suspended in the indicated NaCl solutions and cooled to various subzero temperatures. The open circles in (*a*) indicate that these suspensions contained no ice. All other suspensions froze. (Modified from Fig. 1, Leibo and Mazur, 1969.)

temperature. In both cases, the phage survives. Phage T4B*os* is neither inactivated by osmotic shock at room temperature nor by rapid thawing of a frozen suspension.

To test the hypothesis that osmotic shock is primarily responsible for the inactivation of rapidly thawed T4B, we utilized the fact that shock-resistant phages may be isolated, by repeated cycles of osmotic shock at room temperature, from wild-type phage stocks treated with mutagens (Leibo and Mazur, 1969). If our explanation is correct, it should also be possible to isolate a shock-resistant phage from a wild-type stock treated with mutagens by repeated cycles of freezing and rapid thawing. But since a genetic analysis would be necessary to demonstrate the identity of two mutants isolated by different methods, phages T4D and T4D*os* were used for these experiments. These phages are similar to T4B and T4B*os* both in osmotic shock and freezing sensitivity (Leibo and Mazur, 1966; 1969), but their genetic composition is known in much greater detail.

Since severe osmotic shock ought to occur when a phage is frozen in an NaCl solution to −20°C and then thawed rapidly, this treatment was used as the selection procedure. Five phage stocks, treated with the mutagen 5-bromodeoxyuridine, were suspended in 0·05 molal NaCl, frozen to −20°C, and rapidly thawed. The survivors were regrown to a high titre, and the freezing treatment repeated. After four cycles of freezing and regrowth, we were able to isolate phage (designated *op* mutants) that exhibited 100 per cent survival both after freezing and thawing from −20°C, and after osmotic shock at room temperature.

By means of mixed infections of a non-permissive host with phage T4D*op* and four different *amber* mutants, we were then able to demonstrate that the *op* mutation is located in the same gene as the *os* mutation that confers osmotic shock resistance (Leibo and Mazur, 1969). This fact argues that T4D*op*, isolated by freezing and thawing, is identical to T4D*os*, isolated by osmotic shock; and it strongly supports the contention that most of the inactivation of wild-type T-even phage after freezing in dilute salt solutions is caused by osmotic shock.

SIGNIFICANCE OF THE RESULTS

Osmotic phenomena may be involved in the freezing inactivation of biological entities other than phage. Meryman (1968) has suggested that cells are inactivated when osmotic withdrawal of water during freezing causes damaging shrinkage of the cells. Lovelock (1953) accounted for the haemolysis of erythrocytes caused by freezing in terms quite similar to those described here for phage T4. He concluded that as red cells are exposed to NaCl solutions more concentrated than 0·8 M during freezing, they become leaky to cations. Sodium flows into the cells, and they haemolyse during thawing because of what Lovelock called an "excessive internal osmotic pressure due to their burden of sodium ions." As with phage, the chief support for Lovelock's hypothesis came from his demonstration that freezing damage can be mimicked by osmotic shock experiments carried out on unfrozen suspensions. That is, red cells suspended in concentrated NaCl solutions undergo extensive haemolysis when—and only when—they are transferred to isotonic saline. In this respect, phage T4 and erythrocytes are remarkably similar: neither is destroyed when simply exposed to concentrated NaCl solutions, but both are destroyed when rapidly diluted from such solutions (Fig. 6*a*). If one plots Lovelock's data for erythrocyte recovery (simply the opposite of haemolysis) both as a function of NaCl concentration at 20°C, and as a function of the subzero temperature at which ice and the solution at that concentration are in equilibrium, the curves are approximately coincident (Fig. 6*b*). Lovelock's hypothesis rests mainly on this coincidence. The

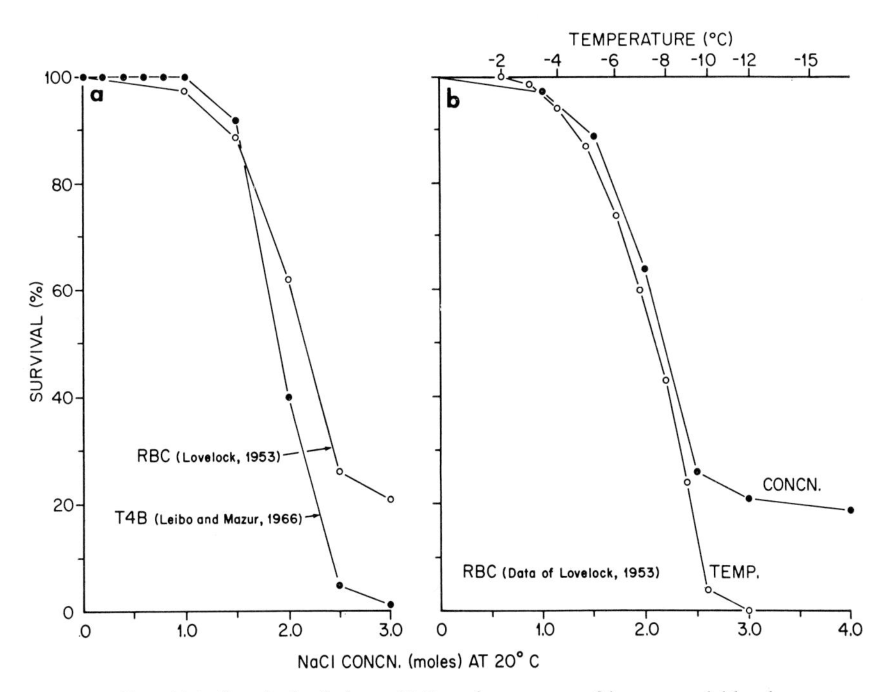

FIG. 6 (*a*). Survival of phage T4B and recovery of human red blood cells (RBC) as a function of the initial NaCl concentration at +20°C. The samples were abruptly diluted either with nutrient broth (for the phage) or with 0·15 M-NaCl (for the red blood cells). Survival was assayed appropriately. The data for the red cells are from Table II of Lovelock (1953).

(*b*) Recovery of human red blood cells both as a function of NaCl concentration at +20°C, and as a function of the equilibrium subzero temperature corresponding to those concentrations. The data are from Table IV of Lovelock (1953).

inactivation of phage in solutions of NaBr and $MgCl_2$ appears similar to the haemolysis of red cells in solutions of NaCl in this respect (compare Figs. 2 and 3 with Fig. 6*b*).

There are further similarities between phage and red cells in their response to similar freezing treatments. For example, the recovery of bovine red blood cells (100% minus % haemolysis) as a function of subzero temperatures (Rapatz and Luyet, 1965) is similar to the survival of phage T4B (Leibo and Mazur, 1969). For both systems, survival decreases to a minimum at about −20°C, but then increases with decreasing temperature (Fig. 7). There are, of course, differences—such as the fact that the cooling rates used by Rapatz and Luyet for erythrocytes were many-fold greater than those we used for phage. But even if one compares the recovery of phage and red blood cells as a function only of

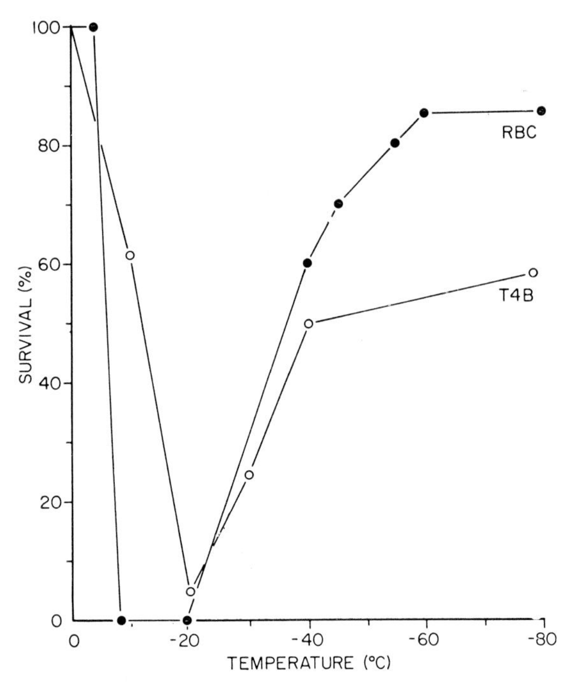

FIG. 7. Survival of phage T4B and recovery of bovine erythrocytes as a function of subzero temperature. The phage or cells were suspended in dilute salt solutions, cooled directly to various temperatures, held briefly, and then warmed rapidly. The data for the red blood cells are redrawn from Fig. 1 of Rapatz and Luyet (1965).

the rate at which the samples were cooled to −100°C, the similarity still holds (Fig. 8). In both these cases, the data are consistent with osmotic shock. The minimum survival at about −10° to −20°C could reflect the fact that the salt concentration in the unfrozen portion of the suspension becomes sufficiently high to sensitize the phage or cell. Rapid thawing then produces rapid dilution, which, at room temperature, haemolyses the red cell and inactivates the phage. Increasing recovery with increasing cooling rate may reflect the fact that the time of exposure to highly concentrated liquid solution decreases with increasing cooling rate, and the extent of haemolysis or inactivation would therefore be reduced.

These data are consistent with osmotic shock being the cause of the inactivation observed both during dilution and during thawing. But the mere fact that a biological entity is destroyed when exposed to concentrated salts and then diluted does not prove osmotic shcck. It is demonstrated, instead, by an effect of dilution rate, slow dilution being less damaging than rapid. For example, phage T4B suspended in concentrated NaCl solutions at room temperature is destroyed by rapid, but not slow, dilution (Leibo and Mazur, 1966). In an analogous way, T4B frozen

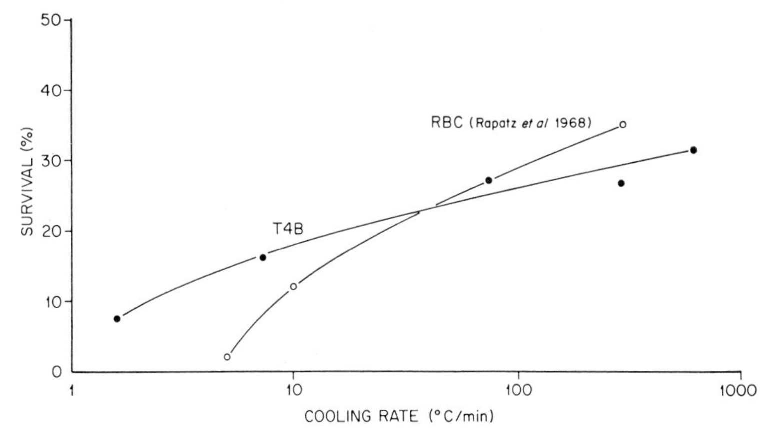

FIG. 8. Survival of phage T4B and recovery of human erythrocytes as a function of the rate at which the samples were cooled to about -100°C. After being cooled to -196°C, the samples were held briefly, and then thawed rapidly. The data for the red blood cells are redrawn from Fig. 4 of Rapatz, Sullivan and Luyet (1968).

in dilute NaCl solutions is destroyed by rapid, but not slow, thawing (Leibo and Mazur, 1969). This is not the case for phage in $MgCl_2$ and NaBr, however. The shock-sensitive T4B is destroyed even with slow dilution from these solutions, and the shock-resistant T4B*os* is equally sensitive.

Like phage, red cells are "inactivated" (haemolysed) by the sequence of exposure to concentrated NaCl and dilution. But unlike phage, there is no published information on whether this inactivation is dependent on the dilution rate, or, like phage T4 in $MgCl_2$, occurs regardless of dilution rate. There is information on the effect of thawing rate on haemolysis (Lovelock, 1953), but this information argues against osmotic shock being the primary factor responsible for haemolysis of frozen–thawed red cells. Lovelock found that rapid thawing was much less deleterious to slowly frozen red cells than was slow thawing, despite the fact that slow thawing ought to have minimized the osmotic shock. This raises the question of whether the haemolysis of slowly frozen red cells is primarily a reflection of osmotic shock, as Lovelock and we define it.

SUMMARY

Because their physical, chemical, and genetic properties are well understood, bacteriophages are a convenient model for studying some of the mechanisms by which freezing and thawing affect biological systems. Three factors appear to be responsible for the inactivation of frozen–thawed T4 phage. (1) When phage frozen in dilute solutions of many salts such as NaCl, KCl, and KNO_3 are thawed rapidly, they are inactivated by osmotic shock. (2) When phage are frozen in solutions of these salts more concentrated than 0·1 molal, however, they are inactivated by the complete crystallization of water and the precipitation of all salts. (3) When phage are frozen in solutions of other salts, such as NaBr and $MgCl_2$, they are inactivated simply by exposure to toxic levels of these salts that result from the removal of water as ice. These conclusions were based on observations of the freezing sensitivities of phage T4B, which is sensitive to large rapid changes in the osmotic pressure of its environment, and phage T4B*os*, which is resistant to such changes.

Phage are not cells, yet their response to concentrated salt solutions and to various freezing treatments mimics that of erythrocytes in many respects. Nevertheless, there are differences that lead us to question whether osmotic shock is sufficient to explain the haemolysis of frozen–thawed red blood cells.

REFERENCES

ANDERSON, T. F., RAPPAPORT, C., and MUSCATINE, N. A. (1953). *Annls Inst. Pasteur, Paris*, **84**, 5–14.
LEIBO, S. P., and MAZUR, P. (1966). *Biophys. J.*, **6**, 747–772.
LEIBO, S. P., and MAZUR, P. (1969). *Virology*, **38**, 558–566.
LOVELOCK, J. E. (1953). *Biochim. biophys. acta*, **10**, 414–426.
MERYMAN, H. T. (1968). *Nature, Lond.*, **218**, 333–336.
RAPATZ, G., and LUYET, B. (1965). *Biodynamica*, **9**, 333–350.
RAPATZ, G., SULLIVAN, J. J., and LUYET, B. (1968). *Cryobiology*, **5**, 18–25.

DISCUSSION

Luyet: Dr Davies, you observed that peptones exert their protective action above the eutectic point of the salts present. Is it implied that protection by peptones—and perhaps by other cryoprotective agents—

is limited to temperatures above the eutectic of some of the solutes in the medium? If so, the experiments in which the salt present was sodium bromide would not be conclusive, since the eutectic point of sodium bromide is −27°C and the temperatures reported did not extend below −20°C.

Davies: We chose sodium bromide because the bromide ions are fairly damaging and any protection obtained against such damage would be quite obvious. We were studying only ionic damage above the eutectic point and the possible effects on what we believe might be hydrophobic bonding. Below the eutectic point completely different effects may be involved. The fact that glycerol can be used to protect cells during storage at −196°C has shown that glycerol obviously protects at temperatures below eutectic points, and in fact by differential thermal analysis and electrical conductivity measurements we have shown that glycerol completely masks any eutectic formation. The point of the particular experiment in question was to show that glycerol is capable of preventing the ionic damage which may occur even in supercooled suspensions.

Farrant: Were the concentrations of the different substances equal, in the experiment where you showed percentage survival as a function of the minimum peptide length and got increased survival at the shorter lengths?

Davies: The concentration of peptides in each fraction was not equal and in fact the concentration by weight in the most protective fractions was considerably lower than in some of the less effective fractions. On a mole basis, however, they were virtually equivalent.

Mazur: The question is do the short-chain peptides protect better on a unit weight or molar basis, or do they protect better because the mass or molar concentration is higher in the active fractions?

Davies: There are four fractions which contain dipeptides. Three are very protective, the dipeptides in order of activity being Tyr-His, Tyr-Arg and Leu-Arg. The fourth contains Leu-Arg together with Val-Lys. Although this might be expected to be more effective than the others on a mole basis this peak also contains free lysine which we now know to have a considerable inhibitory effect against the protection afforded by such peptides.

Lee: Dr Heber, what is the chemical nature of the peptone fraction which provides protection in your system?

Heber: I don't know much about the chemical and physical properties of the fraction isolated from commercial peptone. As for the protective fractions from spinach, there are about 50 to 150 amino acid residues per molecule of protective protein. These molecules are large if compared with the small peptides used by Dr Davies and Professor Greaves.

Levitt: Have you done any amino acid analyses?

Heber: I have hydrolysed the fractions in strong acid and compared the amino acid pattern with that of hydrolysed albumin. Qualitatively it didn't look very different.

Levitt: Is it essentially a normal protein in terms of the amino acids?

Heber: Our fractions are not homogeneous. One should do a quantitative analysis with the components after they have been isolated.

Leibo: Dr Davies, how do you prepare phage in the absence of organic nitrogenous compounds?

Davies: We grow the host cells in a chemically defined medium without any peptides or amino acids present. It would appear that the presence of such compounds in the external medium rather than in the host cell is the critical factor.

Heber: Could you say more about the mechanism of the protective action of glycerol and hydrophobic bonding?

Davies: I have no particular evidence as to how this works; possibly glycerol is promoting a structuring effect within the water, which in turn helps to stabilize the macromolecules.

Elford: Do your peptide fractions only protect T4 phage against damage from a slow freeze–slow thaw procedure?

Davies: They are not protective against a fast thaw; in other words they are unable to prevent osmotic shock.

Elford: You postulate either that the peptide from your fractionation procedure is associated with the membrane, giving it greater strength, or that it may change membrane permeability. You also said that vasopressin and oxytocin produce protective effects similar to those of your peptide fractions. One can be more specific about the effects of vasopressin (ADH) on permeability as it is known to cause a marked increase in the osmotic water permeability of toad bladder (Hays and Leaf, 1962) and muscle fibres (Zadunaisky, Parisi and Montoreano, 1963). Yet the increase in the permeability to ions is much less (Leaf and Hays, 1962). As you quite rightly interpret, during slow thawing in the presence of ADH the osmotic stress set up across the membrane by salt concentration gradients will probably produce a smaller transient swelling of phage than in the absence of ADH. However, during rapid thawing a high concentration gradient of salts across the phage membrane is produced very quickly and an increased permeability to water in the presence of

ADH is not sufficient in this case to prevent critical volumes from being reached.

Davies: ACTH, which we found to be effective with *Pseudomonas*, is a much larger peptide than vasopressin (ADH) or oxytocin, which were effective with the T4 phage. I think the specific effect of peptides in this type of damage is going to vary in different cells. Furthermore I doubt whether a permeability change by itself can account for the protection found below the eutectic temperature. Dr Heber's protein which protects the membrane of the chloroplast grana is also very large and it has a highly specific effect. I think we have to have more evidence before we can directly correlate results obtained with the phage protein membrane and those of other workers with lipoprotein membranes.

Klotz: You said dipeptides were the most highly protective of the peptides examined. Synthetic dipeptides are readily available and might therefore be more worth investigating than digests which might or might not be pure and are more difficult to define chemically.

Davies: We intend to do that, and also to see whether the particular amino acids in a dipeptide are important. We are also going to look at the inhibition of the effects below the eutectic point.

Klotz: Synthetic oligopeptides of various lengths are now also available, so one could tell whether it is a question of length or of composition.

Davies: Certainly, but at the moment it is a matter of cost.

Klotz: In connexion with osmotic shock, are the halophilic bacteria particularly stable to freezing, Dr Leibo? These bacteria are often cultured at concentrations of 4 or 5 M-salt at room temperature and this would be an interesting test of some of the osmotic shock propositions.

Leibo: People surveying enormous numbers of different organisms have looked for recovery, but the results are not quantitative, and have been expressed only as "growth" or "no growth".

Klotz: Among the microorganisms thermophiles are very stable at the upper temperature ranges. Has anybody frozen any thermophiles?

Leibo: A purified enzyme system of one thermophile, *Cyanidium caldarium*, has been shown to be no more active at very high temperatures than that of mesophiles (Ascione, Southwick and Fresco, 1966).

Klotz: Even if the enzyme is sensitive when pure, clearly in its natural environment something is protecting it because it can survive high temperature.

Levitt: Higher plants show both effects. You can get a correlation between thermophilic resistance and freezing resistance, but you can also get a type of thermophilic resistance in the absence of freezing resistance.

REFERENCES

Ascione, R., Southwick, W., and Fresco, J. R. (1966). *Science*, **153**, 752–755.
Hays, R. M., and Leaf, A. (1962). *J. gen. Physiol.*, **45**, 905.
Leaf, A., and Hays, R. M. (1962). *J. gen. Physiol.*, **45**, 921.
Zadunaisky, J. A., Parisi, M. N., and Montoreano, R. (1963). *Nature, Lond.*, **200**, 365.

THE EFFECT OF FREEZING AND REWARMING ON FELINE BRAIN TISSUE: AN ELECTRON MICROSCOPE STUDY

H. A. D. Walder*

Department of Neurosurgery, Catholic University, Nijmegen, Holland

In an attempt to gain some impression of the submicroscopic changes in brain tissue exposed for some time to low temperatures, we studied such tissue (non-medullated nerve fibres and cells from the brain of the anaesthetized cat) with the electron microscope. Experiments of two types were set up for this purpose.

(1) A temperature of −60°C was applied to the brain tissue for five minutes with a freezing-probe. The tissue was immediately rewarmed and removed for electron-microscopic examination. The control was a corresponding tissue fragment from the contralateral hemisphere.

(2) A temperature of −60°C was applied to brain tissue for five minutes with a freezing-probe which was then brought to room temperature (in about 30 seconds) and removed, the tissue being left *in situ*. After 30 minutes pre-marked tissue fragments were removed and prepared for electron-microscopic examination. The test animals remained under anaesthesia during the 30-minute period. As controls, tissue fragments were obtained from the untreated hemisphere both before and after cold application. In the latter case it was ensured that only such tissue was removed as had in fact been frozen.

MATERIAL AND TECHNIQUES

Immediately after cold application the tissue fragments were transferred to a 6·25 per cent glutaraldehyde solution buffered with 0·1 M-sodium cacodylate, pH 7·4 (Sabatini, Bensch and Barnett, 1963). The tissue was cut with a razor into 1 mm cubes and fixed for two hours. It was then

* This study was carried out in cooperation with G. F. J. M. Vrensen, Electron Microscopy Centre (Head: A. M. Stadhouders), Catholic University, Nijmegen.

washed for one to twelve hours in 0·1 M-sodium cacodylate buffer, pH 7·4, to which 0.25 M-sucrose had been added. After washing, the tissue was post-fixed for 90 minutes in a 1 per cent solution of osmium tetroxide buffered at pH 7·4 (Palade, 1952). The tissue fragments were dehydrated in ethyl alcohol and embedded in Epon 812.

Ultra-thin sections were collected on copper grids covered with a carbon-reinforced collodion film. They were then stained with a lead citrate solution (Reynolds, 1963). Electron micrographs were taken with either a Philips EM 200 or a Philips EM 100b electron microscope.

Illustrations

Figs. 1–6 show control tissue.

Figs. 7–11 show tissue frozen for five minutes before removal from the brain.

Figs. 12–14 show tissue frozen for five minutes and left *in situ* for 30 minutes before removal from the brain.

The following abbreviations are used on the figures:

a-f = astroglial filament
as-f = astroglial process
As = astrocyte
b-m = basal membrane
Cap = capillary
chr = chromatin material
d-b = dense body
End = endothelial cell
Ery = erythrocyte
g = Golgi field
gl = glycogen
g-v = dense Golgi vacuoles
i-s = intercellular space
ly = lysosome
max = myelinated axon
mi = mitochondrion
n = nucleus
n-e = nuclear membrane
Neu = neuron
Oli = oligodendrocyte
p-g = pigment granule
po-s = postsynaptic element
pr-s = presynaptic element
rer = rough endoplasmic reticulum
ri = ribosomes
s-m = synaptic membrane swelling
s-v = synaptic vacuoles
syn = synapse
tu = tubule

RESULTS

In order to provide an adequate basis for comparison a brief description of control tissue follows. Our observations were confined to the grey matter.

(A) *Neurons*

We intend to discuss only the nerve cell bodies (perikaryons).

Nuclei. The nuclei are oval-shaped and usually show an undulating surface (Figs. 1 and 2). The perinuclear space between the membranes

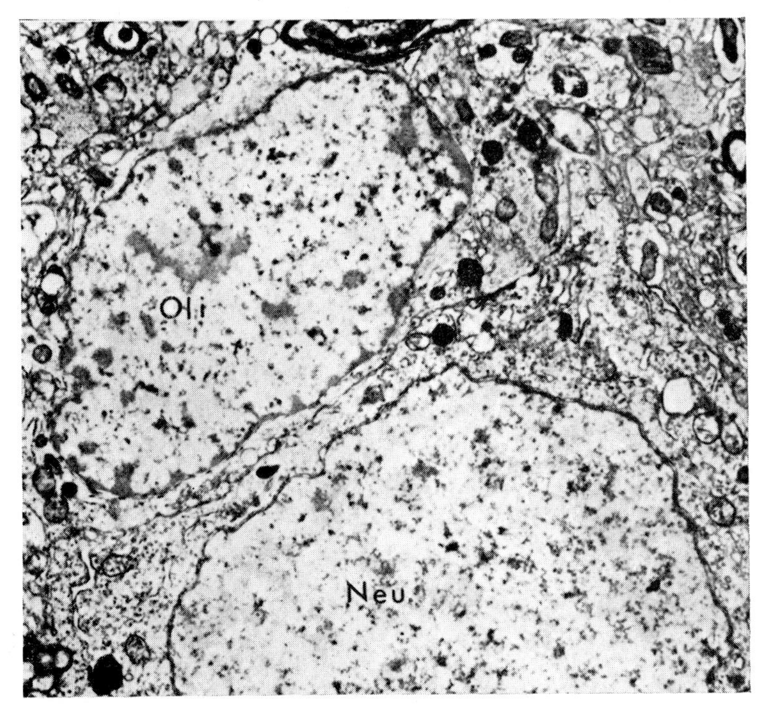

FIG. 1. Neuron with an oligodendrocyte (satellite cell), surrounded by a continuous neuropil (control; × 6400).

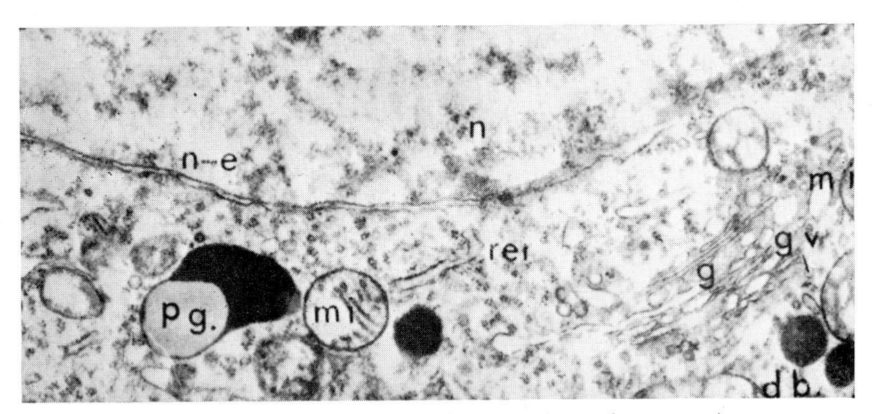

FIG. 2. Detail of the perikaryon of a neuron (control; × 12 000).

varies. Nuclear pores are in evidence. The nucleoplasm is homogeneous without chromatin clumpings. There is a prominent nucleolus.

Mitochondria. The mitochondria are spherical to rod-shaped with a relatively clear matrix (Fig. 2).

Endoplasmic reticulum and ribosomes. The cytoplasm of neurons contains numerous cisternae of the endoplasmic reticulum, which carry ribosomes. Apart from these membrane-bound ribosomes there are free ribosomes. This so-called "rough" endoplasmic reticulum, and the free ribosomes, were identified by Palay and Palade (1955) as Nissl substance.

Golgi zones (lamellar-vacuolar zones). A number of parallel cisternae with optically clear contents are seen (Fig. 2). At their periphery one observes clear vacuoles. In addition one finds smaller vacuoles with optically dense contents chiefly in the vicinity of the cisternal system.

Dense bodies. In the neurons we have often observed spherical granules enveloped by a single membrane, which is separated from the optically dense contents by a clear zone (Fig. 2). These structures have been described as "dense bodies" (Palay and Palade, 1955).

Pigment granules (Fig. 2) are also seen.

(B) *Glia Tissue*

Astroglia, protoplasmic astrocytes (Figs. 3, 4). The most conspicuous feature of this cell type is its exceedingly clear ground cytoplasm and small number of organelles. The nucleoplasm is optically clear with peripheral accumulations of chromatin. The cytoplasm includes some cisternae of rough endoplasmic reticulum and small groups of ribosomes. The ground cytoplasm in addition contains a few very small optically dense granules which we later observed even more clearly in the astroglial process. On the basis of their stainability with fluids of high lead concentration, these granules can be identified as glycogen particles (Fig. 3).

Oligodendroglia, oligodendrocytes (Fig. 1). This cell type is easily confused with smaller, optically dense neurons. The oligodendrocytes differ from the neurons in showing the following features:

(1) Accumulation of chromatin material against the nuclear membrane.

(2) An exceedingly small perikaryon.

(3) A larger number of pigment granules with an eccentric clear substructure.

(4) Smaller mitochondria, with a dense matrix.

(5) A smaller ribosome population.

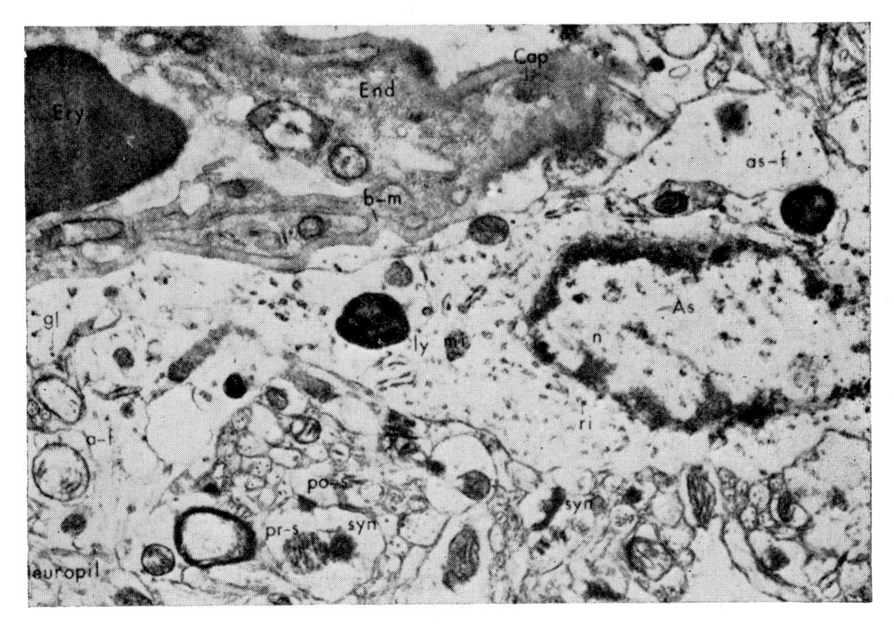

FIG. 3. Ultrastructure of an astrocyte cell body and its relation to capillary and neuropil (control; × 10 620).

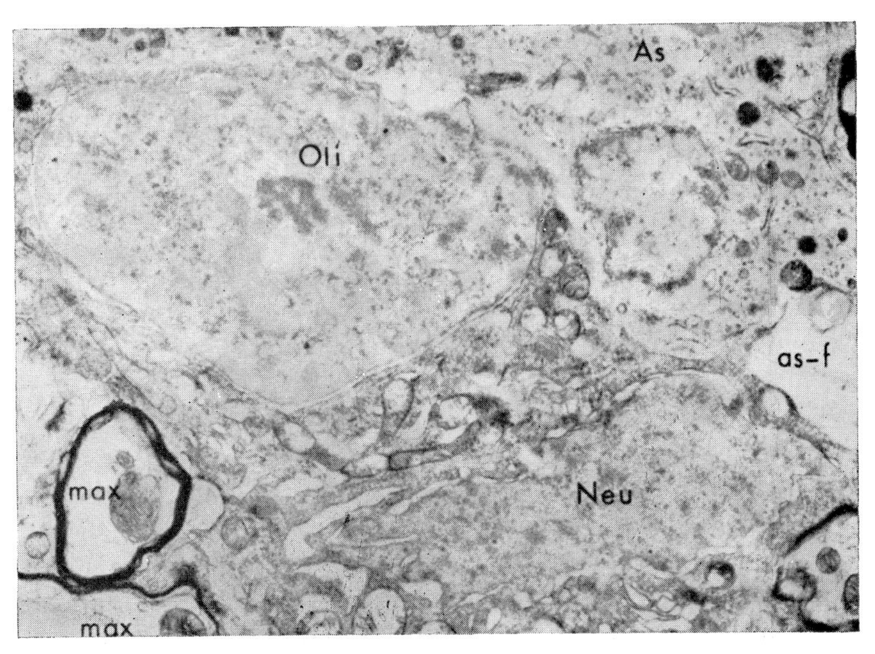

FIG. 4. Neuron, oligodendrocyte and astrocyte in relation to one another and to the neuropil (control; × 6300).

(C) *Neuropil*

Brain tissue consists for the most part of the so-called neuropil. This is characterized by a closely woven network of processes of both neurons and glia cells, visible by electron microscopy as a continuous arrangement of polymorphous profiles. The profiles of the various processes are closely packed so that the extracellular space is limited to intercellular crevices ranging in size from 100 to 200 Å (10 to 20 nm). The profiles visible in the sections studied were identified as follows.

Myelinated axons. These neuropil profiles are highly conspicuous in that they are enveloped by a highly osmiophilic sheath of concentric membranes. These profiles are found in large numbers in both longitudinal and transverse sections. Apart from these clearly recognizable myelinated axons the neuropil consists of a large number of unmyelinated processes of neurons, astroglia and oligodendroglia. These processes show a singularly delicate ramification and interlocking, so that very small profiles of various types are present.

Astroglial processes (Figs. 3, 6). The astrocyte cell bodies and their direct processes around the capillaries, which we identified with certainty, show a characteristic which is very useful in the identification of other astrocyte processes: they contain glycogen particles. These particles occur in the neuropil also, in profiles which are otherwise optically clear and possess no other shaped structures.

While a certain identification of the neuropil profiles is thus possible, interpretation—of smaller profiles in particular—is exceedingly difficult, if not impossible.

Synapse. The last of the structures commonly observed in the neuropil is the synapse (Fig. 5).

(D) *Capillaries*

Blood capillaries were a fairly frequent feature in the sections we examined. Fig. 6 shows a normal capillary surrounded by astroglial processes, which are screened from the surrounding neuropil by a membrane.

THE DIRECT EFFECT OF FREEZING AND REWARMING

(A) *Disturbances of Cellular Components*

Two phenomena stand out when the changes in neurons and oligodendrocytes are examined. The ultrastructure of the mitochondria is changed appreciably. Although the mitochondria of the neurons show some swelling in the control tissue too, this is considerably more striking

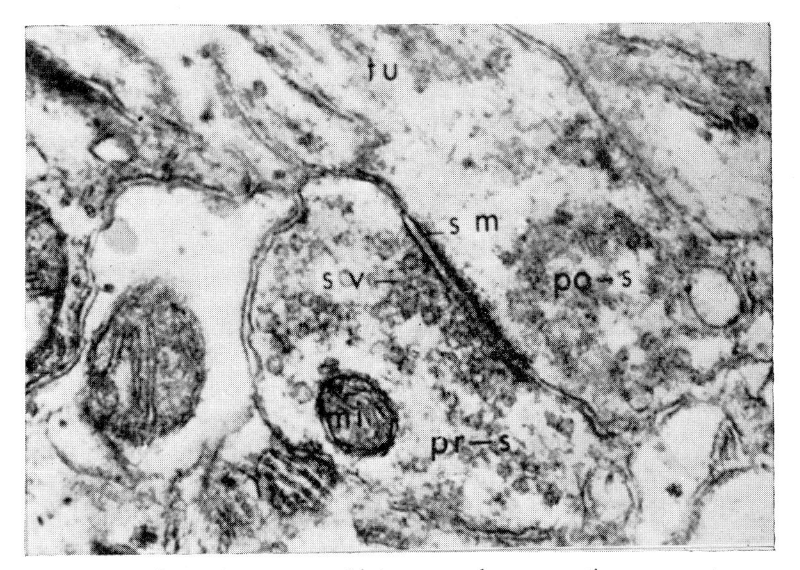

FIG. 5. Synaptic structure with its pre- and postsynaptic components and membrane swelling. The presynaptic structure encompasses synaptic vacuoles and a mitochondrion. The postsynaptic structure shows dendrite tubules (control; × 36 000).

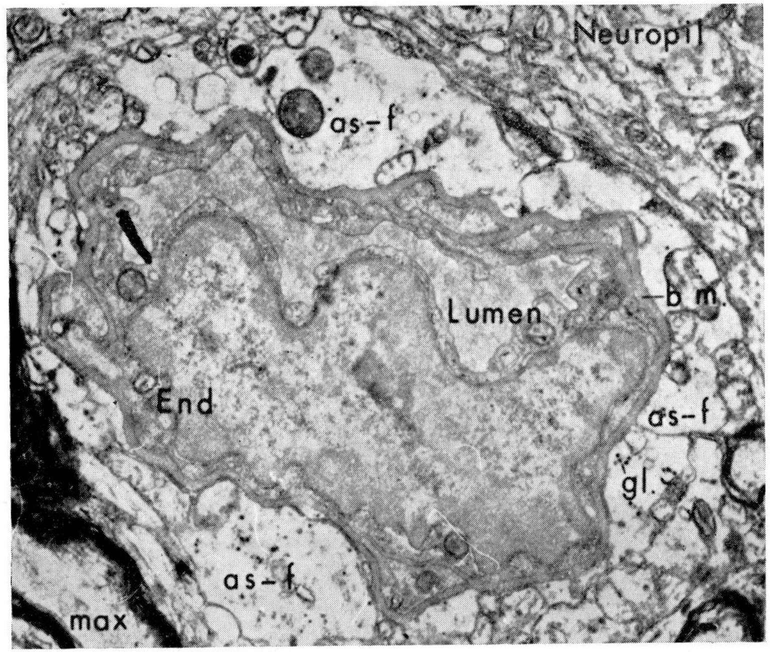

FIG. 6. Capillary with endothelial lining and basal membrane, surrounded by astrocyte processes (end-plates) (control; × 11 200).

in frozen tissue. The difference is particularly distinct in the oligodendrocytes, whose mitochondria possess a dense matrix in the control tissue. The swelling is distinct also, and more conspicuous, in the neuropil. The ultrastructure of the neurons otherwise shows few changes. There is no dilatation of the endoplasmic reticulum cisternae; neither the ribosomes on these cisternae nor the ribosome rosettes show any changes.

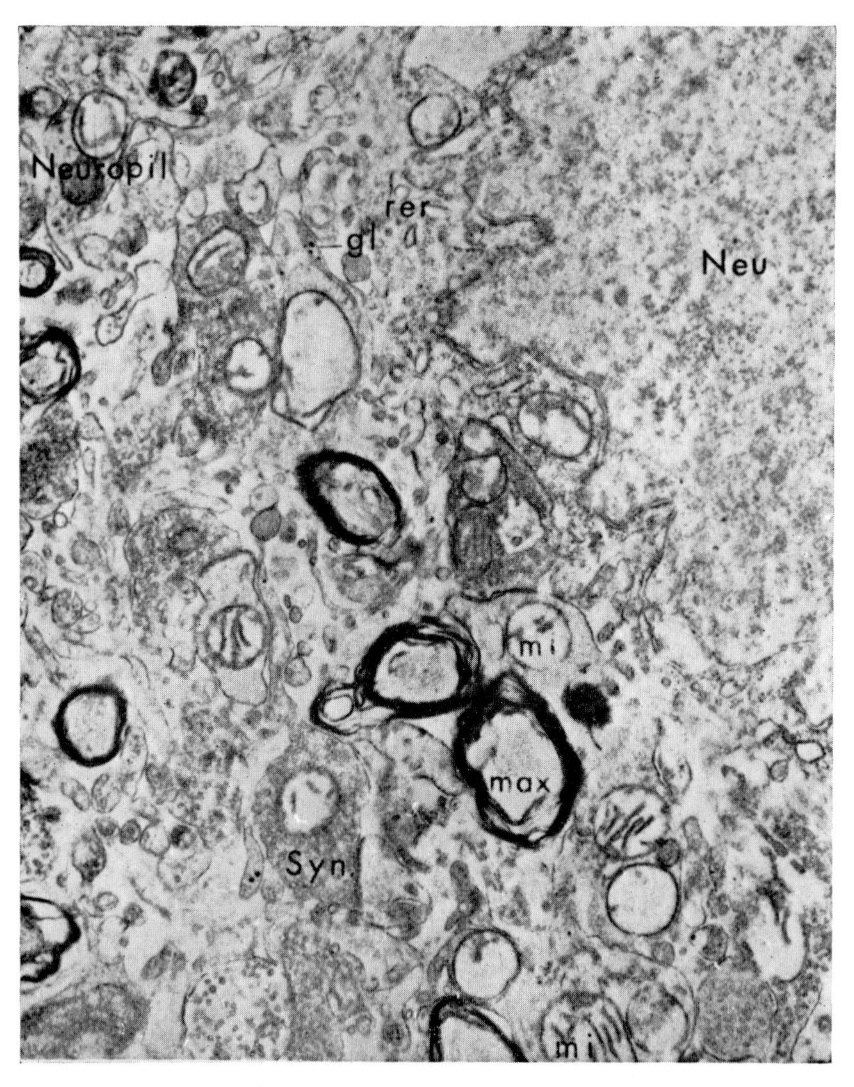

FIG. 7. Tissue frozen for 5 min. Part of neuron surrounded by neuropil, with a disturbed interrelation. Astrocyte processes can no longer be identified as such. Mitochondria are swollen (× 16 500).

The disturbance in the cell membrane of the neuron is much more conspicuous. In the control tissue the neurons are unmistakably enveloped by a continuous cell membrane, and either astroglial processes or other neuropil profiles lie against this membrane or even form synaptic swellings. The neuron shown in Fig. 7 demonstrates that this situation is greatly disturbed in the treated tissue. We are unable to distinguish the perikaryon from the surrounding neuropil; only the ribosome distribution gives some indication of the extent of the cytoplasm. In the treated specimens the ribosome-rich cytoplasm of the neurons is often surrounded by clear areas, not enclosed by a membrane. The neuropil profiles are localized in the same areas and surrounding membrane-enclosed astroglial processes have become rare. The cytoplasm is as a rule surrounded by a poorly defined clear zone. A similar situation is seen in the astrocyte (Fig. 8). The astrocyte is no longer clearly enveloped by a membrane, and the neuropil profiles surrounding the astrocytes are rounded off and likewise localized in a poorly defined clear zone. The astrocyte cytoplasm seems slightly denser than in control tissue, and the mitochondria also appear slightly swollen.

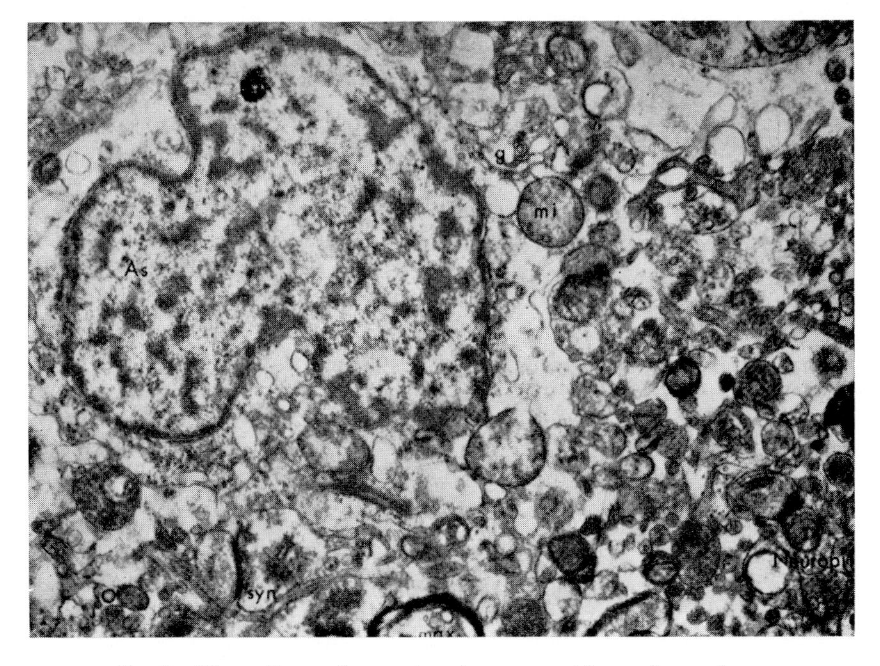

FIG. 8. Tissue frozen for 5 min. Astrocyte with a cell membrane lacking continuity. The surrounding neuropil is disorganized (× 7275).

(B) *Capillaries*

The capillaries show features reminiscent of the changes observed in neurons and astrocytes. The ultrastructure of the endothelial lining shows that the mitochondria appear to be somewhat swollen, and the endoplasmic reticulum seems to be slightly dilated (Fig. 9). A striking feature is that the pattern of surrounding astroglial processes is again disturbed (Fig. 9). The capillary is enclosed by a zone of slightly higher density than

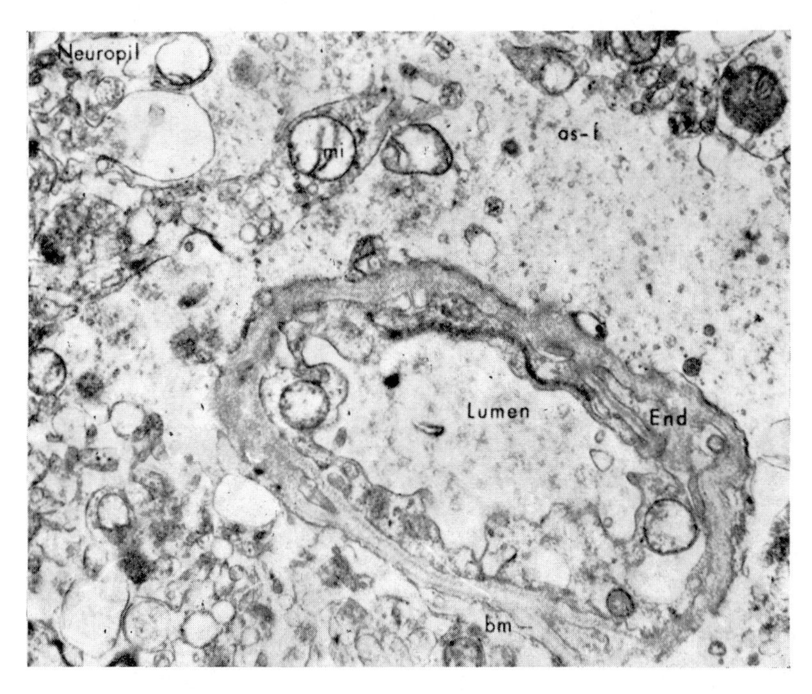

FIG. 9. Tissue frozen for 5 min. Capillary with swollen endothelial lining and divided basal membrane. The surrounding astrocyte process is not membrane-enclosed and the neuropil profiles are disconnected (× 4720).

the normal astroglial processes. This zone is not separated from the neuropil profiles by a membrane, as is normally the case. The profiles are scattered in this clear zone and moreover are rounded off. A study of the treated tissue discloses that this packing of profiles is disturbed, and that the profiles have assumed more rounded shapes. Fig. 10 shows that the profiles are scattered free in a clear zone. The profile membranes are not disturbed, and the synaptic junctions are intact. This situation does not always take such an extreme form, as Fig. 11 shows.

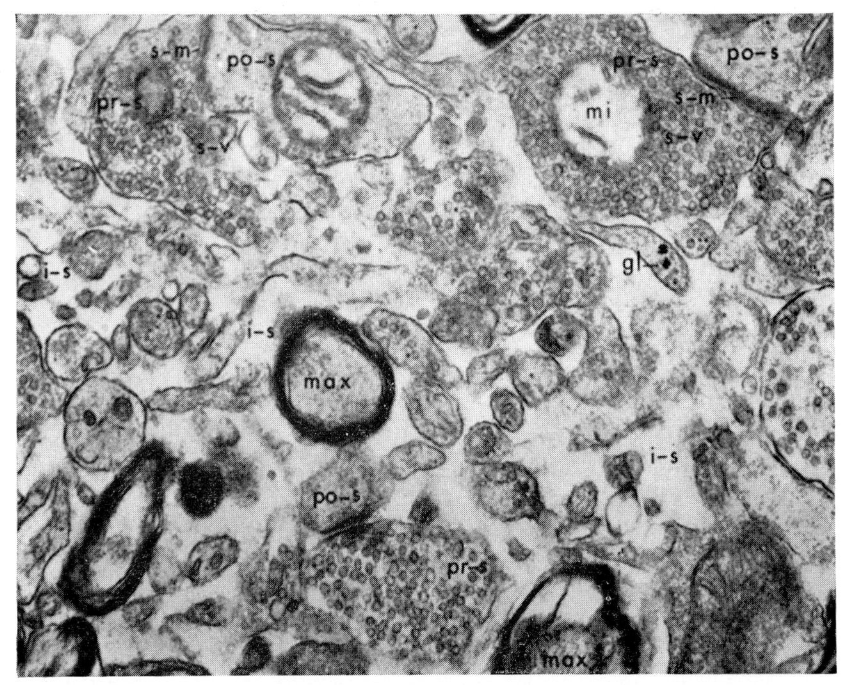

FIG. 10. Tissue frozen for 5 min. Detail of neuropil. Note disorganization and enlargement of the intercellular spaces. Profiles and synapses have retained their normal structure. Mitochondria are swollen (× 17 200).

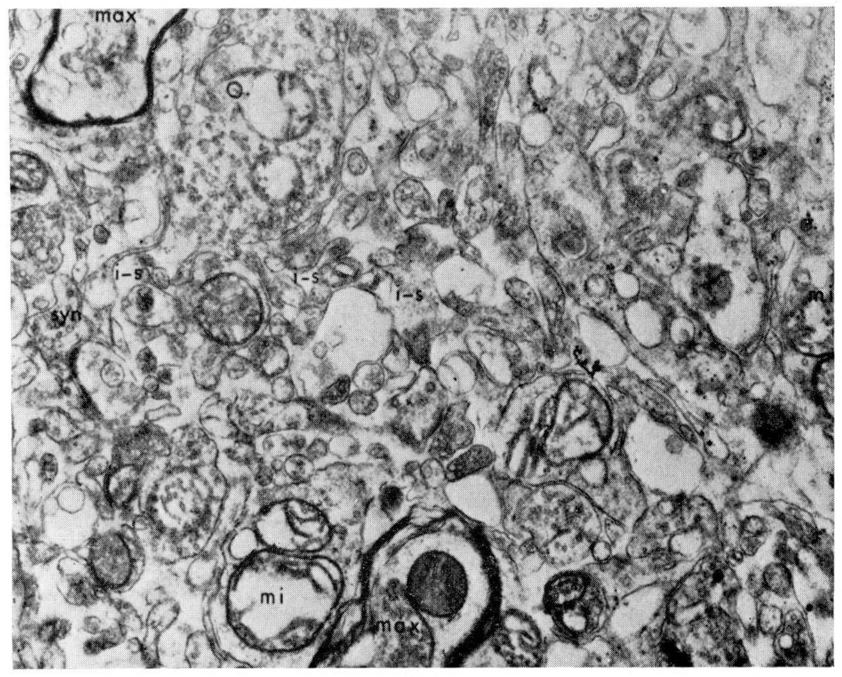

FIG. 11. The same as Fig. 10, but changes are less conspicuous (× 11 200).

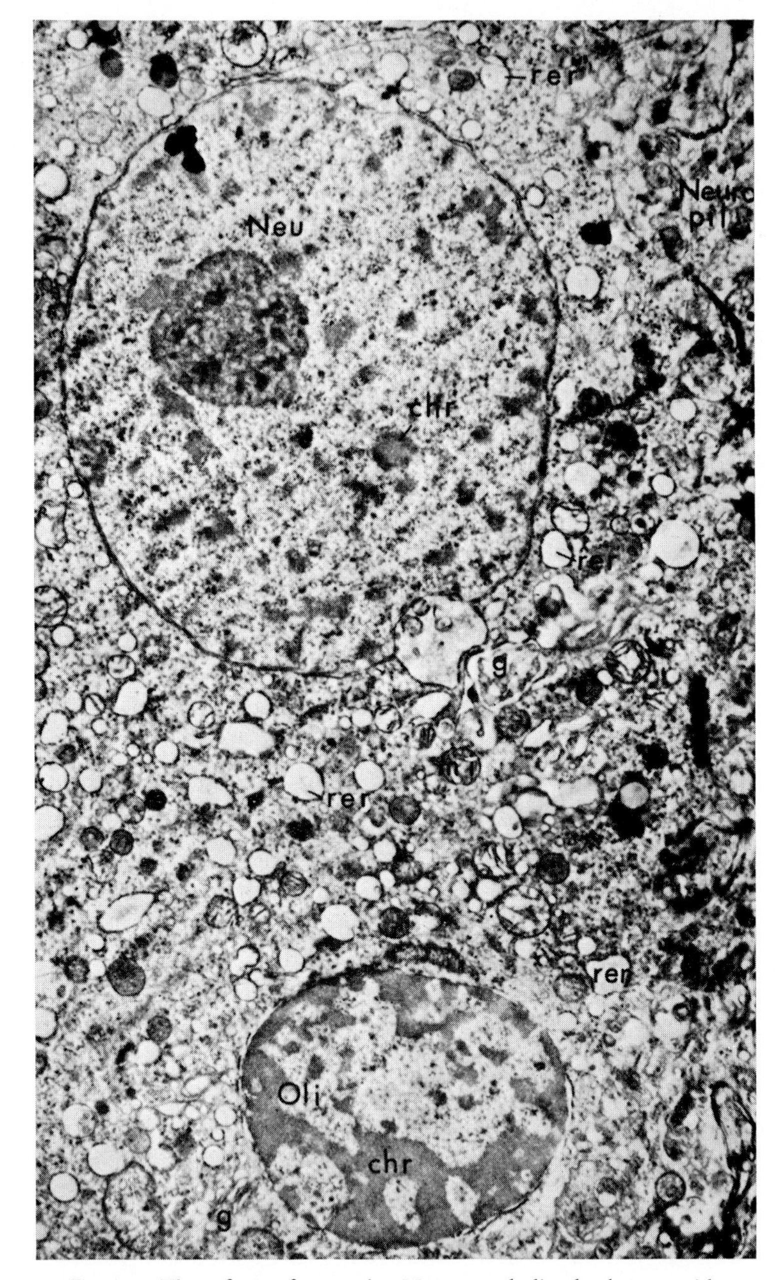

FIG. 12. Tissue frozen for 30 min. Neuron and oligodendrocyte with dilated endoplasmic reticulum (vacuolation) and clumping of chromatin material. Note that mitochondria are generally normal (× 8700).

In summary, we can conclude that the direct consequence of freezing and immediate rewarming is a disturbance of the astroglia. The disturbances are indeed most conspicuous around the capillaries and neurons, where in the control tissue the astroglial processes are most distinct. The observation that the neuropil profiles, although rounded off, are undisturbed, indicates that the astroglial processes are more sensitive to freezing and rewarming.

EFFECT OF FREEZING AND REWARMING ON TISSUE LEFT *in situ* FOR 30 MINUTES

(A) *Disturbances of Cellular Components*

Neurons and oligodendrocytes (Fig. 12) at this stage already show unmistakable ultrastructural changes. The nuclei of the neurons show chromatin clumpings, sometimes accumulating against the nuclear membrane.

The cytoplasm shows a few conspicuous disturbances. The endoplasmic reticulum has assumed a rounded shape and consists mainly of round vacuoles of varying size. The Golgi zone shows dilatation.

The mitochondria, which immediately after freezing show unmistakable swelling, have returned to normal after 30 minutes, although the matrix is sometimes a little denser. However, a few swollen mitochondria are still present. The lack of astroglial processes around the neurons is as conspicuous as in specimens removed immediately after the freezing period.

The oligodendrocytes show changes equivalent to those of the neurons.

The neuropil surrounding the two cell types, although disturbed, still shows unequivocal continuity.

Fig. 14 shows the features of an astrocyte which occurs in this disturbed tissue. The nucleus shows no marked difference in chromatin densification from the control tissue. Most mitochondria appear normal.

A striking feature is the almost total lack of glycogen particles.

(B) *Intracapillary and Pericapillary Changes*

An abnormal endothelial lining is a conspicuous feature of the capillaries (Fig. 13). The endothelial cells are greatly swollen and occupy a good part of the capillary lumen. The cytoplasm is singularly clear, with few organelles. The mitochondria are swollen and the endoplasmic reticulum is dilated.

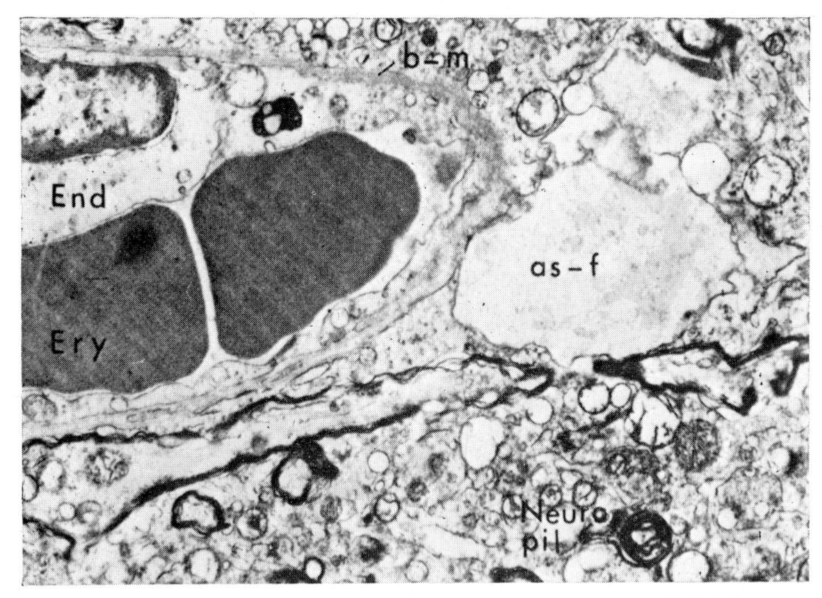

FIG. 13. Tissue frozen for 30 min. Capillary with swollen endothelial lining and swollen astrocyte process. Lumen filled with red cells (× 5360).

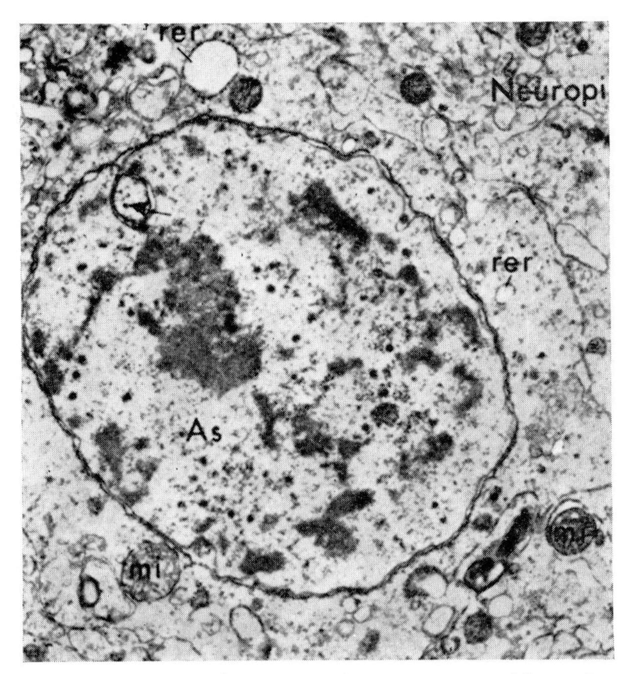

FIG. 14. Tissue frozen for 30 min. Intact astrocyte with continuous cell membrane, surrounded by close-packed neuropil profiles. The endoplasmic reticulum is dilated and the nuclear membrane is invaginated (←) (× 2720).

(C) *Changes in the Neuropil*

The neuropil shows remarkably few changes. The profiles retain their individuality. Even around the large pericapillary vesicles they constitute a continuous pattern of separate profiles. They are more clearly joined than immediately after a five-minute freezing lesion.

DISCUSSION AND CONCLUSIONS

The direct consequence of freezing and rewarming of brain tissue seems to be a disturbance of the astroglial membranes. The changes are most clearly visible around the neurons, oligodendrocytes and capillaries, where astrocytes or their processes are an important morphological component. The clear astroglial processes found around these cells and capillaries in control tissue disappear after freezing; the neuropil profiles are exposed and show rounding-off. These changes are not in evidence to the same degree throughout the tissue.

Freezing and rewarming gives rise to disintegration of cell membranes (Trump *et al.*, 1965). We are unable to explain why it is the astroglial membranes in particular that are disturbed.

The following manifestations are prominent 30 minutes after cold application. The neurons, oligodendrocytes and astrocytes show dilatation of rough and smooth endoplasmic reticulum (vacuolation) and clumping of nuclear chromatin. The disturbances of the astrocytes—cells which play an important role in water metabolism and govern transportation of salts and metabolites—can give rise to degenerative changes of this kind, according to Clasen and co-workers (1962) and Hager (1964). This effect is likely to be intensified by the disturbance in capillary circulation. The capillaries show swelling of the endothelial lining and sometimes partition of the basal membrane. Some homogeneous material accumulates around the capillaries. This material—exuding blood plasma—accumulates in membrane-enclosed structures. We have no conclusive explanation for the origin of these membranes. The exudation of plasma is probably caused by the disturbance of the capillary components and the astrocyte processes.

The neuropil remains a coherent system of profiles even around the pericapillary vesicles. There is no appreciable enlargement of the intercellular crevices between the neuropil profiles. The distinct neuropil profiles observed immediately after five-minute freezing are not observed after 30 minutes, when the neuropil profiles show marked swelling, especially around the capillaries. The swollen profiles are similar in appearance to the pericapillary vesicles, and the exuded plasma is probably taken up in these neuropil profiles.

SUMMARY

A temperature of −60°C was applied with a freezing-probe to the brain of the anaesthetized cat. Tissue was removed for electron microscope examination either immediately after the freezing period (five minutes) or after the tissue had been left *in situ* for 30 minutes. Freezing was found to cause a disturbance of the astroglial membranes, in particular. Tissue left *in situ* for 30 minutes showed dilatation of the endoplasmic reticulum and clumping of nuclear matter, with disturbances in capillary circulation and swollen neuropil profiles.

REFERENCES

Clasen, R. A., Cooke, P. M., Pandolfi, S., Boyd, B., and Raimondi, A. J. (1962). *J. Neuropath. exp. Neurol.*, **21**, 579–596.
Hager, H. (1964). *Die feinere Cytologie und Cytopathologie des Nerven Systems.* Stuttgart: Fischer.
Palade, G. E. (1952). *J. exp. Med.*, **95**, 285.
Palay, S. L., and Palade, G. E. (1955). *J. biophys. biochem. Cytol.*, **1**, 69.
Reynolds, E. S. (1963). *J. Cell Biol.*, **17**, 208.
Sabatini, D. D., Bensch, K., and Barnett, R. J. (1963). *J. Cell Biol.*, **17**, 19.
Trump, B. J., Young, D. E., Arnold, E. A., and Stowell, R. E. (1965). *Fedn Proc. Fedn Am. Socs exp. Biol.*, **24**, suppl. 15, 114.

DISCUSSION

Luyet: In my experience with freezing curves (Luyet, 1957; see, in particular, Fig. 6A), the supercooling depression and the freezing plateau could be obliterated when the cooling rate was very high, when the thermocouples were bulky, or when the recording instruments had appreciable inertia. Would those who studied this problem care to comment on it in the case of the cooling and freezing curves produced by the introduction of a probe at −60°C into a mass of living tissue?

Mazur: Meryman has shown (1966) that there is no supercooling and no freezing plateau when freezing is done with a probe in this way.

The surface of the probe Dr Walder used may be at −60°C but the periphery of the sphere of ice must be at the freezing point of the tissue, about −0·5°C. Initially the cooling rate is high near the probe and lower further away. Finally, a steady state is attained, the cooling rate becomes zero and there is a thermal gradient from −60°C to the melting point of the tissue at the periphery.

Meryman: The situation is that freezing is moving outwards from the centre, so that the surface to volume ratio is decreasing. In the usual system, where freezing is from the outside going inwards, there tends to be a plateau as the surface to volume ratio increases. Where the freezing is going outwards the plateau is not seen.

What proportion of the effects you described do you think could be attributed to anoxia rather than to freezing injury, Dr Walder?

Walder: Transplantation of artificial sarcomas into mice directly after the tumours were frozen gave a survival of about 30–35 per cent. But when the sarcomas which had been frozen to −60°C for five minutes were kept *in situ* for a week before being transplanted into other mice (recipients), the survival of the tumour cells was 2 per cent.

The electrical activity in the frozen brain tissue went down and after rewarming it returned to almost the normal EEG pattern seen in the cat brain. But when we inserted the same electrode at the same place after two or three days there was no more electrical activity. So I concluded that there were two mechanisms: first the destructive mechanism of direct freezing damage and then the hypoxaemia due to the vascular changes and to the oedema around the lesion (Walder, 1966).

Meryman: What do you think is the cause of the vascular stasis?

Walder: Some people say it is due to the leaking out of plasma or to so-called thermal shock.

Meryman: Have you observed in your sections anything that looked like plugging of the vessels by platelet or leucocyte plugs, as reported by Mundth, Long and Brown (1964)?

Walder: We often see capillaries with platelets all packed together and the fluid leaking out.

Mazur: How does the size of the lesion compare with the size of the ice sphere? Does a 10 mm ice sphere give a 10 mm lesion?

Walder: When we place the electrodes at different distances from the cryoprobe, we use a double control. First we make a plastic reprint of the position of the thermocouples after they are taken out of the brain. Then after a week or two we section the brain, and after allowing for the shrinkage due to fixation we note the locations of the thermocouples again. The boundary of the lesion is at about 0°C. However the method is not too accurate; we place the thermocouples at a minimum distance of 1·5 mm from each other, because otherwise there may be interference between the two thermocouples.

Mazur: It is difficult because there is such a steep thermal gradient.

Huggins: Dr Walder has raised a number of very important points. Morphology of cells does not necessarily correlate well with the degree of freezing damage. Variations in sensitivity of different types of cells during freezing and thawing have both advantages and disadvantages. Lastly, and most important, there is the remarkable ability of the body to repair sublethally injured cells. Without *in vivo* repair I doubt that any cells could be preserved for transplantation.

Mazur: Does the body repair sublethally damaged cells or replace them by new ones?

Huggins: Both may occur. With preserved mammalian cells, however, the affected cells themselves must survive and function for successful transplantation. Dr Walder, would you comment on freezing-thawing and refreezing?

Walder: I did refreeze sometimes, but I haven't much experience with that. All my experiments were done just to see how freezing for five minutes affected what we are doing clinically. Our results from freezing *in vitro* are very limited but they must be rather different from freezing *in situ*, because the blood supply from the surrounding tissue plays an important role.

Huggins: In clinical frostbite—nature's cryosurgery—refreezing greatly increases the ultimate loss of tissue.

Walder: That is rather peculiar, because what I am working on now shows a clinical application of freezing vascular anomalies in brain. After a cryoprobe is placed in a cervical artery of the dog, we see a proliferation of the endothelial layer after three to six weeks' survival. This gives again a clinical application by reinforcement of the vascular wall in arteriovenous malformations. These experiments are at a preliminary stage.

Meryman: Many reliable reports (see Meryman, 1957) refer to tissues *in vivo* frozen to quite low temperatures which showed almost complete recovery of function after thawing under proper circumstances, the proper circumstances being the maintenance of circulation. In much work with dogs, rabbits or rats rapid thawing, alone (Meryman, 1957) or with the addition of low molecular weight dextran (Mundth, Long and Brown, 1964) which prevents the vascular stasis, can result in an almost complete recovery of tissues briefly frozen to quite a low temperature. We all tend to forget that cells can withstand freezing if the conditions of their environment after thawing are ideal.

Walder: That is why I think it is important to limit the time all the tissue is in ice to five minutes. If you freeze for a short time you can be sure that the cryoprobe is below 0°C. But 2 or 3 mm from that point the tissue will be at +10°C or something like that. You have to find the point at which it is stabilized, so that freezing is not continued for too long or too short a time.

REFERENCES

LUYET, B. (1957). *Biodynamica*, **7**, 293–336.
MERYMAN, H. T. (1957). *Physiol. Rev.*, **37**, 233–251.
MERYMAN, H. T. (1966). In *Cryobiology*, pp. 1–114. London & New York: Academic Press.
MUNDTH, E. D., LONG, D. M., and BROWN, R. B. (1964). *J. Trauma*, **2**, 246–257.
WALDER, H. A. D. (1966). *The application of cryogenic method in neurosurgery*. M.D. thesis. Amsterdam: Scheltema & Holkema.

ULTRASTRUCTURAL AND FUNCTIONAL CHANGES IN SMOOTH MUSCLE ASSOCIATED WITH FREEZING AND THAWING

C. A. WALTER

National Institute for Medical Research, Mill Hill, London

THERE have been few ultrastructural studies on frozen and thawed organized mammalian tissues; and the contribution that electron microscopy has made towards our understanding of freezing injury at the cellular level has been correspondingly small. This is not so surprising when considered in relation to the complexity of organization that exists between groups of cells having an overall function depending on various cell types, each with different susceptibilities to the damaging effects of freezing and thawing. Also there is the added difficulty of being able to test such tissues satisfactorily for viability *in vitro*, and the possibility that the procedures necessary in order to be able to examine the tissue electron microscopically may cause further damage to cells already made vulnerable by freezing and thawing. Electron microscopy of frozen and thawed tissues cannot therefore be considered as an end in itself, but must be considered in conjunction with other relevant experimental data.

A correlated biochemical and ultrastructural study by Trump and co-workers (1965) on mouse liver after different rates of freezing and thawing showed that alterations were evident in virtually all cytoplasmic organelles and subcellular systems. These alterations tended to be minimized after relatively slow rates of freezing followed by rapid rewarming, especially in the presence of protective agents such as glycerol, or dimethylsulphoxide (DMSO). Farrant, Walter and Armstrong (1967), in a combined structural and functional study of guinea-pig uterine smooth muscle after conventional slow freezing and rapid thawing, showed that the smooth muscle cells were severely damaged after freezing to −79°C in the presence of 10 per cent DMSO. This difference in results could be an expression of the known variable effects that the freezing and thawing process has on many different types of cells, or it could be due to differences in experimental techniques.

Most of our current understanding of the causes of freezing injury has come from work on microorganisms (Mazur, 1966) and especially the human red blood cell (Lovelock, 1953; Doebbler, Rowe and Rinfret, 1966; Meryman, 1968). The red blood cell because of its structural simplicity has definite limitations in relation to the effects of freezing and thawing on cells in general. For this reason, some attempt has been made to study the effects of freezing and thawing on a more representative mammalian cell, the smooth muscle cell, which despite its functional

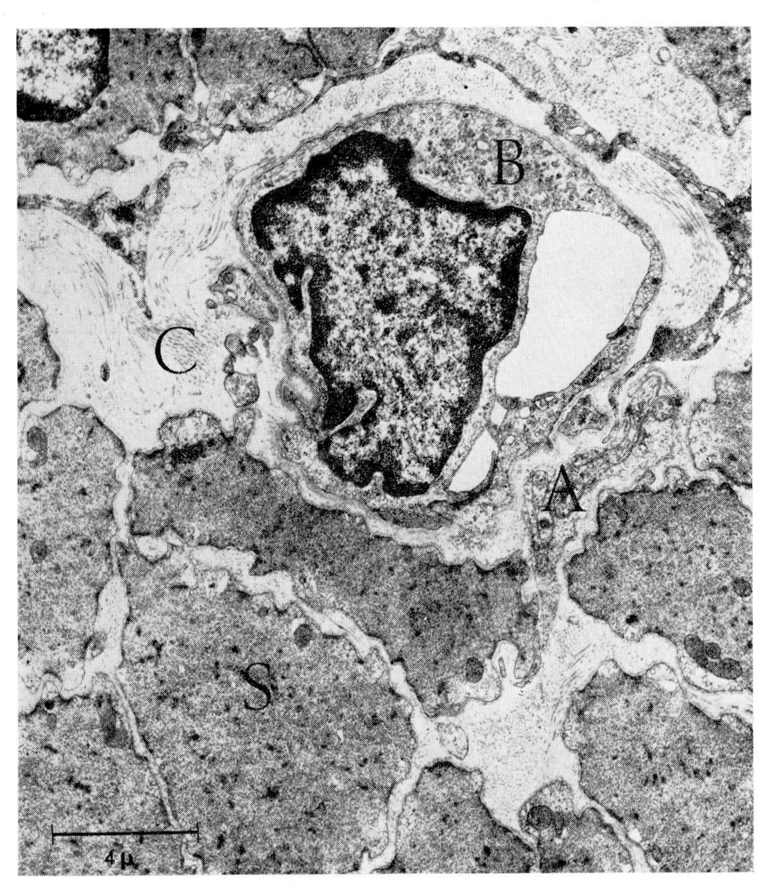

FIG. 1. Electron micrograph of a representative area of the taenia coli muscle from the guinea-pig caecum as seen in transverse section. The muscle was fixed immediately after removal in 1 per cent osmic acid in Krebs' solution (3 hours at 0°C), embedded in Epon and stained with alcoholic uranyl acetate followed by lead citrate. All the micrographs were taken from transverse sections of muscle prepared in this way after the various experimental treatments (except Fig. 14). S, smooth muscle cell. B, blood capillary. A, non-medullated nerve fibres. C, collagen.

specialization has the characteristic attributes of most cells. Smooth muscle offers distinct advantages for such a study in that it is easily tested for viability *in vitro*, and over the years has been extensively studied physiologically (see Eichna, 1962).

The present paper is concerned with some preliminary experiments directed towards establishing certain base-line information relating to functional and ultrastructural changes in smooth muscle associated with freezing and thawing. The smooth muscle used in this investigation was the longitudinal smooth muscle (taenia coli) from the guinea-pig caecum. Individual muscle cells are approximately 200 μm long and 2–4 μm in diameter in the relaxed state (Bennett and Rogers, 1967). The muscle cells are loosely arranged in bundles, being separated from each other by an extracellular connective tissue composed mainly of collagen. In the extracellular space ramifying blood capillaries, bundles of nerve fibres, and occasional fibroblast cells make up the general organization of the taenia coli (Fig. 1). Regarding the effects of freezing and thawing on the tissue as a whole, attention has been focused only on the smooth muscle cells although it is appreciated that damage to other component cells may affect the overall functional recovery.

As a routine procedure, transverse sections of the taenia coli were examined electron microscopically at low magnifications (× 4000) in order to survey a large sample of cells; for the sake of uniformity, representative areas from transverse sections only are considered in this paper.

FREEZING BETWEEN 0°C AND −79°C

Organized tissues have not yet been successfully maintained in a viable state after freezing and thawing. In order to justify a correlation between structure and function it was necessary to know over what range of temperature, between 0°C and −79°C, smooth muscle could be frozen whilst retaining some measure of viability after thawing.

Lengths of smooth muscle were cooled slowly in a modified Krebs' solution without a protective agent at 1 to 4°C per minute, and individual pieces were maintained for periods of five to 30 minutes at 10°C intervals between 0° and −79°C. This was followed by rapid thawing, i.e. in less than two minutes. Before freezing, each muscle segment was briefly tested for contractility in an organ bath at +37°C, in Krebs' solution. Isometric contractile responses to a standard dose of histamine (1 to 4 μg) were recorded using a weight transducer coupled to a pen chart-recorder. After thawing each muscle was retested for contractility.

A surprising result was that some contractile responses were recorded from all muscles frozen and thawed between 0° and −79°C. Specimens frozen below −20°C gave impaired contractile responses after thawing. An accurate assessment of the contractile responses from frozen and thawed muscles compared with responses of the same muscles before freezing could not be made by direct comparison, because of variations in the way each muscle was tested. Rough comparisons showed that the contractile responses after freezing and thawing estimated as a percentage of the responses from unfrozen muscle were less than 5 per cent for muscle frozen to and thawed from −20°C, and that this low level of contractility was retained in muscles frozen down to −79°C (Fig. 2). Muscle frozen

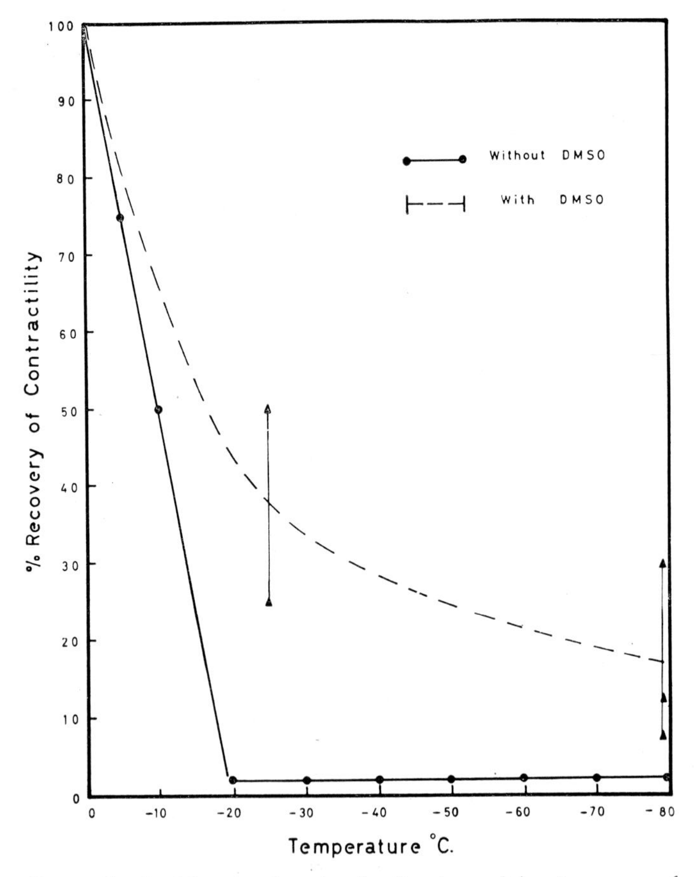

FIG. 2. Graph of functional results after freezing and thawing expressed as a percentage of the responses obtained before freezing. Segments of muscle were frozen either with or without 10 per cent w/v DMSO. In the absence of DMSO, minimal contractile responses were obtained after freezing to and thawing from temperatures at and below −20°C.

in the presence of 10 per cent w/v DMSO gave improved contractile responses compared with unprotected muscles.

Electron microscopy

Muscle frozen to −20°C or below, without DMSO, was severely damaged structurally. A high proportion of smooth muscle cells characteristically had disrupted plasma membranes, gross intracellular alterations to the contractile myofilaments, complete disruption of mitochondria and abnormally shrunken nuclei (Fig. 3). However, despite the general extent and severity of the damage by electron microscopic assessment, some foci of cells with remarkably good structural preservation were always present (Fig. 4). It is considered that in severely damaged muscles of this kind, capable of producing only minimal contractile responses, electron microscopic assessment of the minority of apparently intact muscle cells provides a reasonable basis for the correlation of morphology and function.

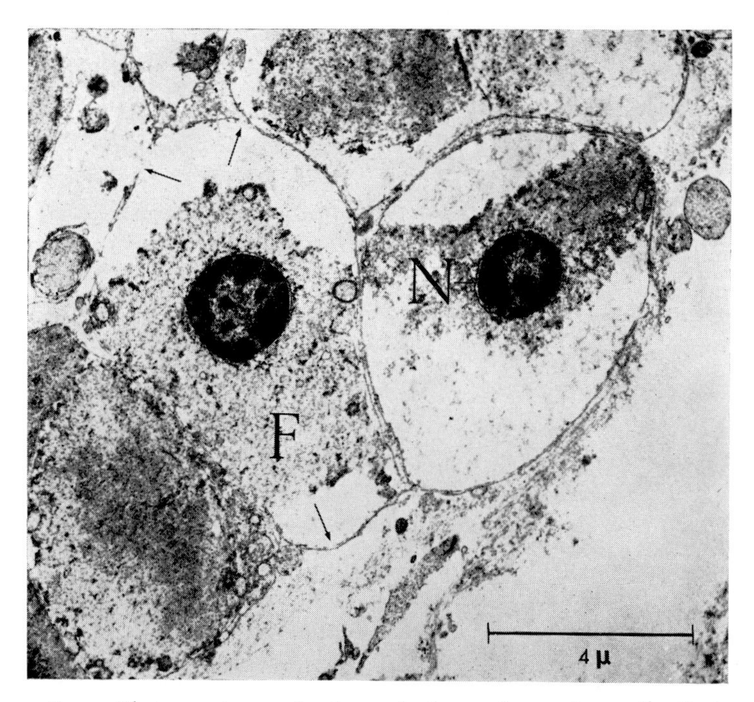

Fig. 3. Electron micrograph of severely damaged smooth muscle cells after freezing to and thawing from −60°C. Note the shrunken appearance of the nuclei (N), and gross changes in the organization of the myofilaments (F), together with extensive breaks in the continuity of the cell membrane (arrows).

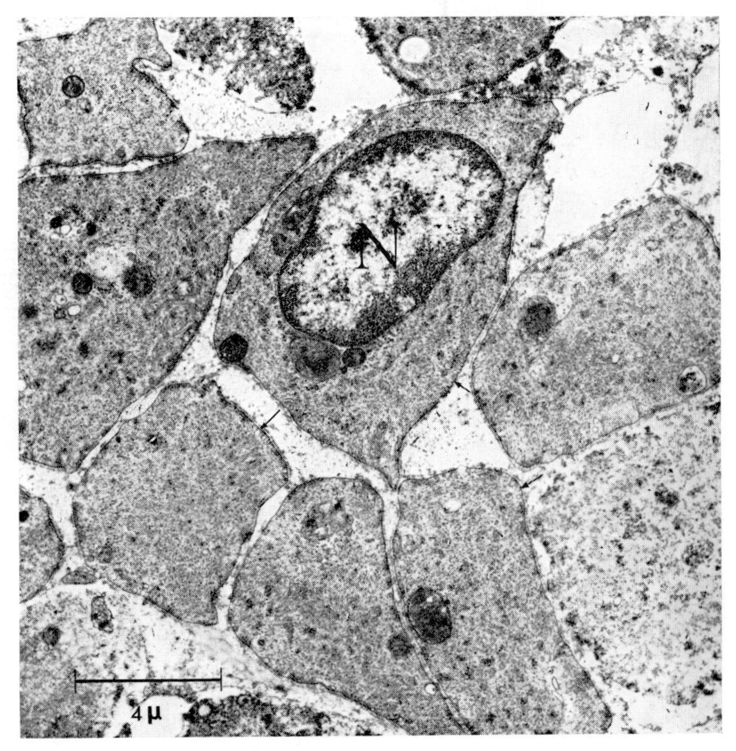

FIG. 4. Micrograph from same muscle as shown in Fig. 3, but from a different area. From the morphological appearance of these cells it is considered likely that they represent the smooth muscle cells that were capable of some contractile function even after being frozen to and thawed from −60°C without protection. N, nucleus. Pinocytotic vesicles (arrows).

FREEZING BETWEEN 0° AND −30° C

The results obtained with unprotected smooth muscle after cooling to temperatures between 0° and −79°C demonstrated clearly that the minimal recovery attained by −20°C represented a transitional zone of useful functional recovery. Further examination of muscles frozen to temperatures below −30°C was therefore considered to be unnecessary. There seemed however to be a real possibility, providing a more accurate method of estimating contractile responses before and after freezing could be devised, of detecting a sequence or pattern of structural and functional damage in smooth muscle frozen and thawed in the temperature range 0° to −30°C.

In order to make comparative assessments between muscle frozen to different temperatures a standard procedure was adopted throughout.

A 2 cm length of muscle was equilibrated for 30 minutes in modified Krebs' solution bubbled with O_2 (95 per cent) + CO_2 (5 per cent) at + 37°C in an organ bath before testing for contractility in response to a set dose sequence of histamine (2, 4, 8, and 16 μg calculated as base). Histamine was added to the organ bath every two minutes, with a contact time of 30 seconds before the histamine was washed out and replaced with fresh Krebs' solution. Each muscle was slowly frozen at approximately 1°C/min to the required temperature and left at that temperature for 15 minutes before rapid thawing. The muscle was then retested for

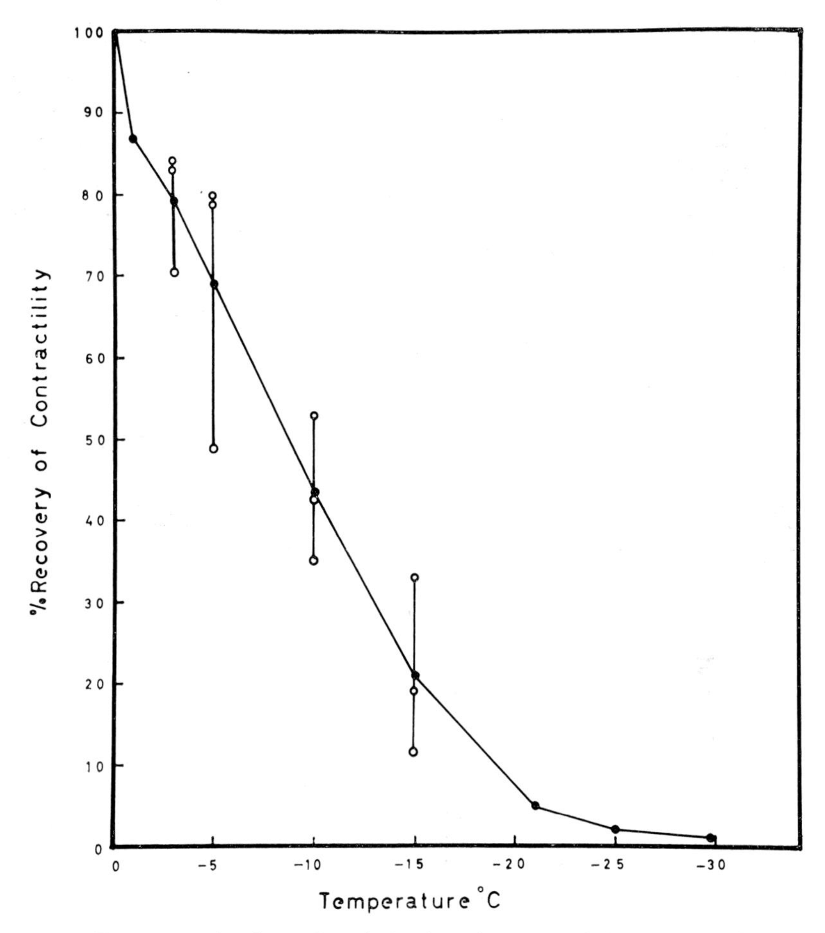

FIG. 5. Functional results obtained under controlled experimental conditions from muscles frozen within the temperature range 0° to −30°C. The curve passes through the mean of the results at each temperature. Note the variation (open circles) in results obtained where several muscles were frozen to and thawed from the same temperature.

contractility in exactly the same way as before freezing. Partial dose-response curves were plotted from the measured responses obtained before and after freezing. Percentage recovery of the frozen and thawed muscles was calculated from measurements of the respective areas under the two dose-response curves. Three experiments were performed at most of the chosen temperatures, and a curve representing percentage recovery of contractility was plotted through the mean of the results in each experiment (Fig. 5). The recovery-of-contractility curve is seen to follow the same general pattern as that obtained by less accurate methods, except that nil recovery was recorded from muscles frozen to and thawed from −30°C. This is a reflection of differences in the two methods of estimating recovery. At any chosen temperature there was a range in the calculated recoveries, with some overlap from one temperature to the next. Variation in the recoverable responses could be due to inherent muscle variability, as each muscle tested came from a separate animal; or due to inevitable minor variations in the freezing technique, despite attempts to keep the procedure uniform. A further possibility is that the pattern of freezing and thawing damage to the muscle as a whole is inconstant, which could be expected to result in differences in the overall recovery even if other factors were controlled.

In an attempt to answer some or all of these questions, four guinea pigs were selected from the same strain and of approximately the same weight and age. Two adjacent muscle preparations were removed from each animal and tested for contractility; they were then cooled to −10°C simultaneously and under identical conditions. The results of these experiments are shown in Table I.

TABLE I

THE PERCENTAGE RECOVERY OF CONTRACTILITY OF TWO MUSCLES, A AND B, FROZEN TO AND THAWED FROM −10°C, FROM EACH OF FOUR GUINEA PIGS

Guinea pig no.	*% Recovery*		
	Muscle A	*Muscle B*	*Mean*
1	25·0	37·5	31·2
2	44·9	23·4	34·1
3	37·6	37·1	37·3
4	37·3	31·1	34·2

Standard deviation (S.D.) 34·2 ± 2·6 per cent

As the freezing and thawing procedure was identical for the four groups of muscles, variation in results due to freezing technique can be excluded, as also can the possibility that differences are due to muscle variation. That the variation is due just to differences in the extent of the freezing and thawing damage sustained remains as a possible, though perhaps over-simplified, explanation. The average results for each pair of muscles show very close agreement, and with more quantitative data of

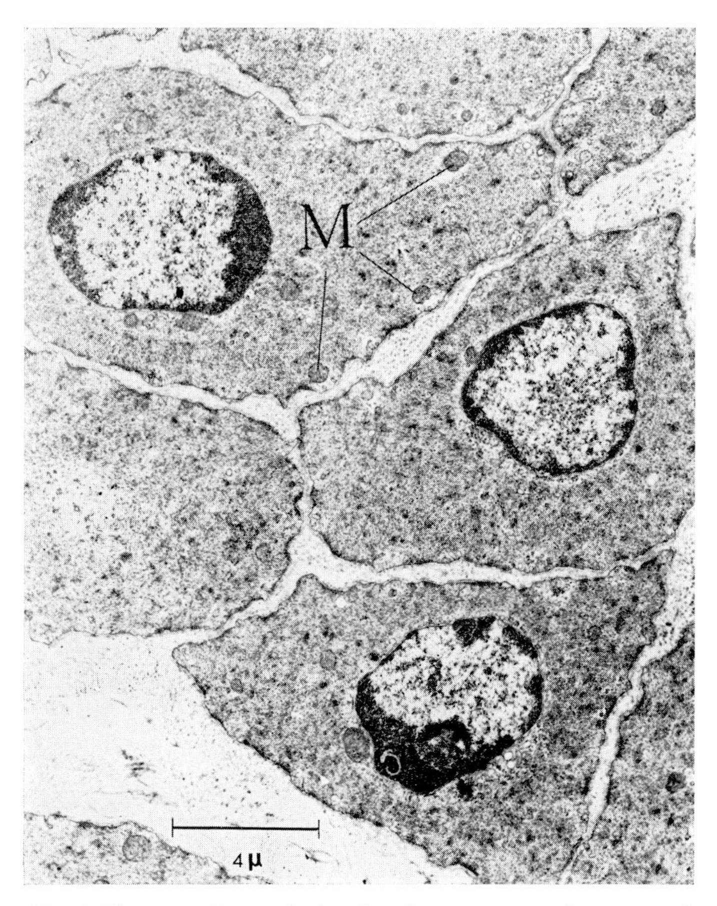

FIG. 6. Electron micrograph showing the appearance of a group of smooth muscle cells (compare with Fig. 1) from a segment of taenia coli kept for 24 hours in Krebs' solution at 0°C, which also gave unchanged contractile responses afterwards. M, mitochondria.

this kind from muscle frozen to other temperatures a really reliable recovery-of-contractility curve could undoubtedly be obtained.

Electron microscopy of muscle that had been kept at 0°C for one, three and 24 hours gave normal contractile responses at the end of these periods and showed little evidence of structural damage (Fig. 6). Similar results were obtained from muscle supercooled to −10°C for 15 minutes.

Muscle frozen to between 0° and −30°C showed that just as there was a progressive decrease in functional recovery (Fig. 7) so there was also a progressive increase in structural damage to the smooth muscle cells over the same temperature range. A characteristic feature of muscle frozen below −2·5°C was that severely damaged smooth muscle cells with obviously disrupted plasma membranes differed strikingly from the other, apparently intact cells (Fig. 8). Because of this distinction "damage" has been assessed here purely on the basis of whether or not the smooth

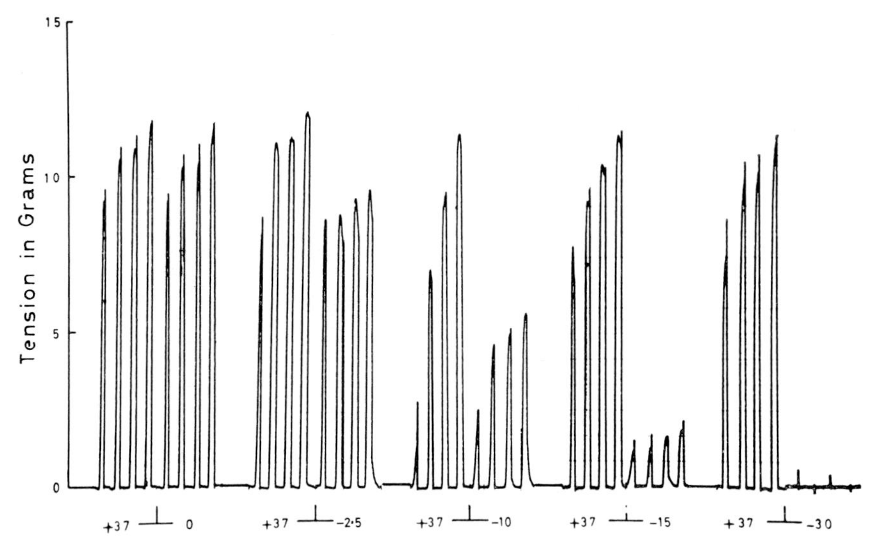

Fig. 7. The last sequence of four contractions obtained before (+37°C) and after freezing and thawing between the temperature range 0° and −30°C (i.e. at 0°, −2·5°, −10°, −15° and −30°C). Note the decrease in contractile responses after freezing to and thawing from progressively lower temperatures.

muscle cells appeared to have intact plasma membranes. However, cells that appear electron microscopically to have intact membranes are not, of course, necessarily to be considered as having complete structural integrity or normal functional capabilities.

The ratio of "damaged" to "undamaged" cells was high in muscles that had been frozen to −21°C (Fig. 9); whereas at the other end of the scale, "damaged" cells were infrequent in muscle frozen to −2·5°C (Fig. 10). At the same time a high proportion of the "undamaged" cells showed marked intracellular abnormalities, including many membrane-bound cytoplasmic vacuoles. Nuclear changes were also evident, in that the nucleoli often had an unusually condensed appearance.

An interesting feature of the pattern of damage observed was that the topographical relationship between intact and disrupted cells was completely random. Smooth muscle cells near the periphery of the taenia coli were not more likely to be damaged than cells further in;

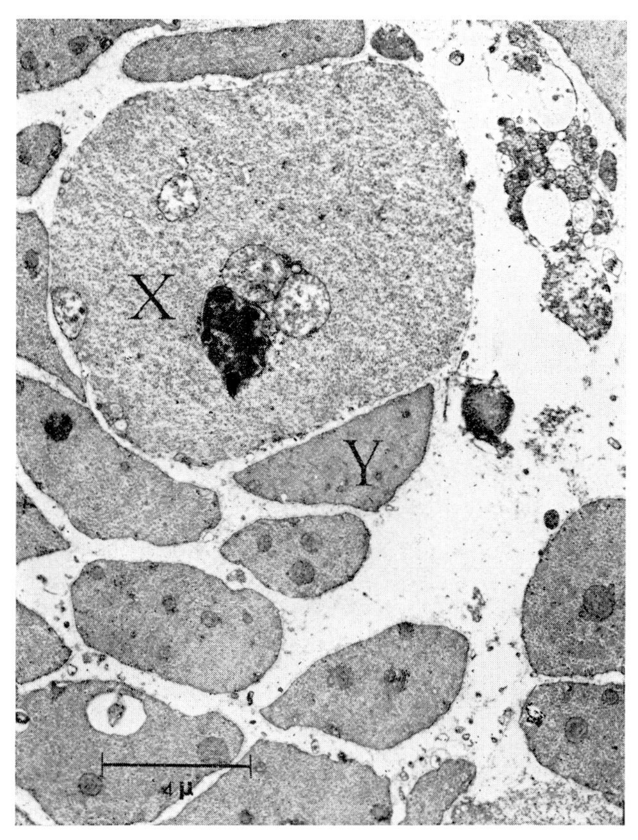

Fig. 8. Electron micrograph of frozen (−10°C) and thawed smooth muscle. Note the distinction between an obviously damaged cell (X) with a disrupted cell membrane and apparently intact (Y) smooth muscle cells.

nor were the smooth muscle cells adjacent to the wider extracellular spaces any more prone to damage than those cells situated more centrally within a bundle group.

In general, the electron microscope findings have shown that the proportion of cells "damaged" increases in parallel with the decrease in functional recovery of the frozen and thawed muscles over the very interesting temperature range 0° to −30°C. Although a direct correlation has not been made to determine whether in fact a real relationship exists between the proportions of disrupted cells with decrease in functional recovery, it can be argued even on the scant data available that some correlation is possible between fine structure and function.

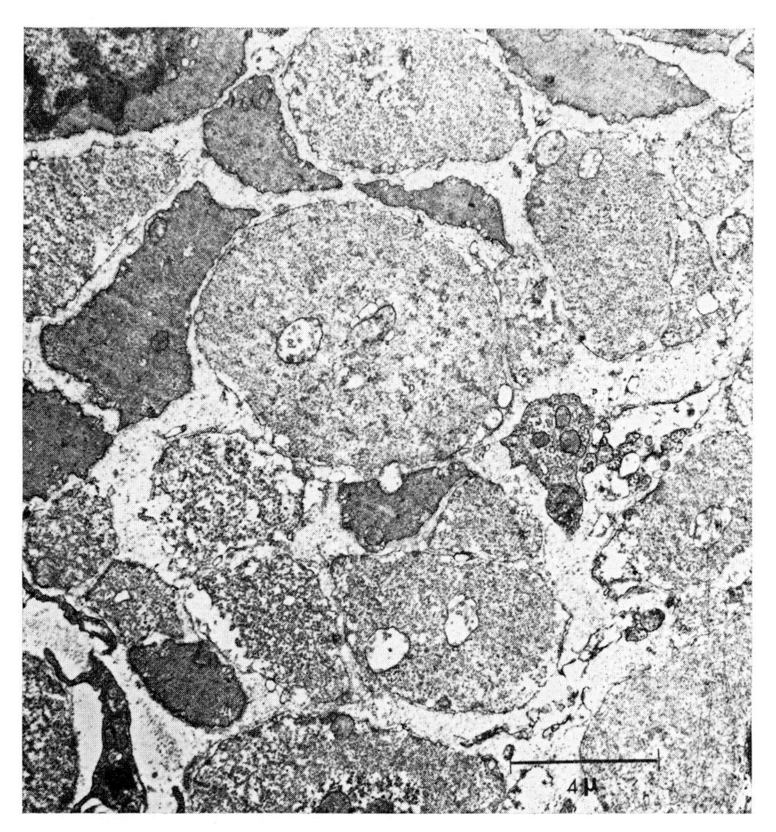

FIG. 9. Electron micrograph of frozen (−21°C) and thawed smooth muscle. The proportion of damaged smooth muscle cells was greater in muscle that had been frozen to and thawed from −21°C compared with muscle frozen to −10°C. However, the distinction between "damaged" and "intact" cells was still apparent.

EFFECTS OF HYPERTONIC SOLUTIONS

With the establishment of a reliable survival curve for smooth muscle after freezing and thawing under controlled conditions, certain possibilities can become a reality. The various freezing parameters can be altered and permutated to test their effects in relation to the results already obtained. Also, instead of being confined to observation of the effects of freezing and thawing in smooth muscle cells, these investigations can be extended to a more general consideration of the cause or causes of freezing injury.

One possible cause of damage to cells during freezing and thawing has been identified in the raised electrolyte concentrations that develop as ice

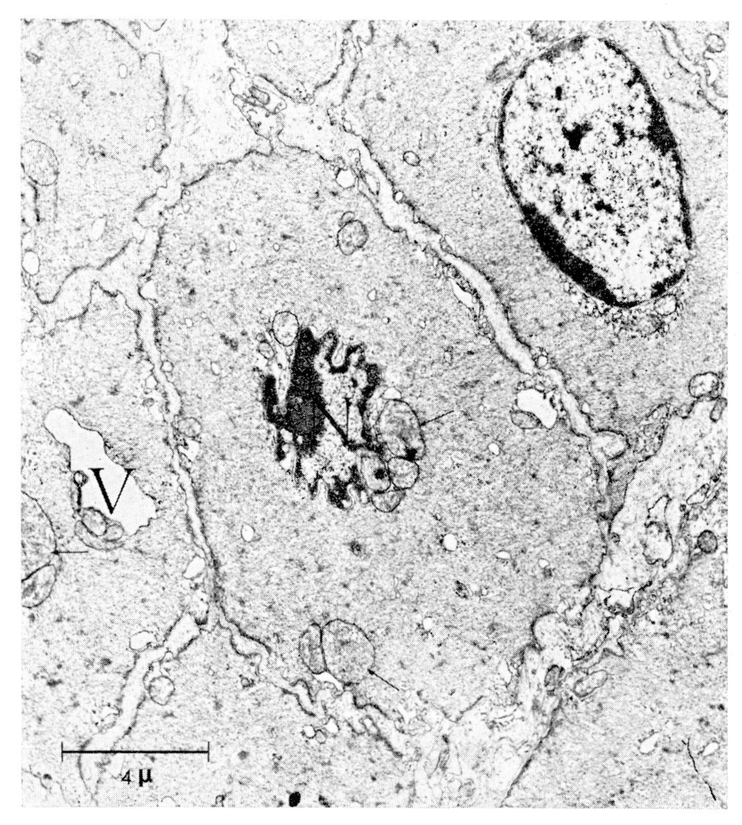

FIG. 10. Electron micrograph of frozen (−2·5°C) and thawed smooth muscle. Severely damaged smooth muscle cells with disrupted cell membranes were infrequently seen, but many of the cells showed marked intracellular alterations. Note the vacuolations (V) and the abnormal appearance of the nucleus (N); also the swollen mitochondria (arrows).

separates from the liquid phase during the freezing process (Lovelock, 1953; Farrant, 1965). Segments of muscle were subjected to hypertonic salt solutions at 0°C followed by return to isotonic conditions to determine whether there is any similarity between the effects of hypertonic salt solutions and the effects of freezing and thawing on the fine structure and function of smooth muscle. The hypertonic solutions tested were made up from the basic salt constituents as a modified Krebs' solution of the following composition (mM): NaCl, 121; $CaCl_2$, 2·5; $MgCl_2$, 1·2; KCl, 5·4; NaH_2PO_4, 1·21; dextrose, 11·5; and $NaHCO_3$, 15·0. The hypertonic solutions were as follows:

(*a*) Modified Krebs' solution increased to five times isotonic.
(*b*) Modified Krebs' solution increased to 12 times isotonic.
(*c*) Modified Krebs' solution with additional NaCl (total NaCl, 1·45 M).
(*d*) Modified Krebs' solution with additional NaCl (total NaCl, 3·4 M).
(*e*) NaCl only (1·85 M).

Smooth muscle appears to be remarkably tolerant of the effects of osmotic stress. Muscle immersed in Krebs' solution made five times isotonic (Soln *a*) and left in this concentration for 30 minutes or 50 minutes before dilution to isotonic conditions showed only slight impairment of contractile function and minimal structural alterations. In contrast muscle subjected to Krebs' solution made 12 times isotonic (Soln *b*) showed marked impairment of functional recovery, and in this respect was very similar to muscle immersed in isotonic Krebs' with only the sodium chloride level increased by a factor of 12 (Soln *c*). Furthermore, muscle after treatment with sodium chloride only (Soln *e*) gave similarly impaired functional results. These results seemed to indicate that functional impairment was in some way linked to the total sodium chloride level, irrespective of the other salts present.

To determine whether the same relationship exists for muscle that has been frozen and thawed, the results obtained from the freezing experiments were replotted in terms of the sodium chloride concentration that would be attained in the Krebs' solution at the various temperatures to which the muscles were cooled (Fig. 11); the percentage recovery of contractile responses of muscles after treatment with the hypertonic solutions is superimposed on the same curve. It is apparent that some correlation exists between the effects of freezing and thawing and of hypertonic salt solutions, although more data are needed to establish the relationship firmly.

The ultrastructural appearance of the muscle cells was in many respects very similar after exposure to hypertonic solutions to that of frozen and thawed specimens (Fig. 12). The same clear distinction was evident between "damaged" and apparently intact cells, and in morphological terms it was usually not possible to distinguish the effects of the two

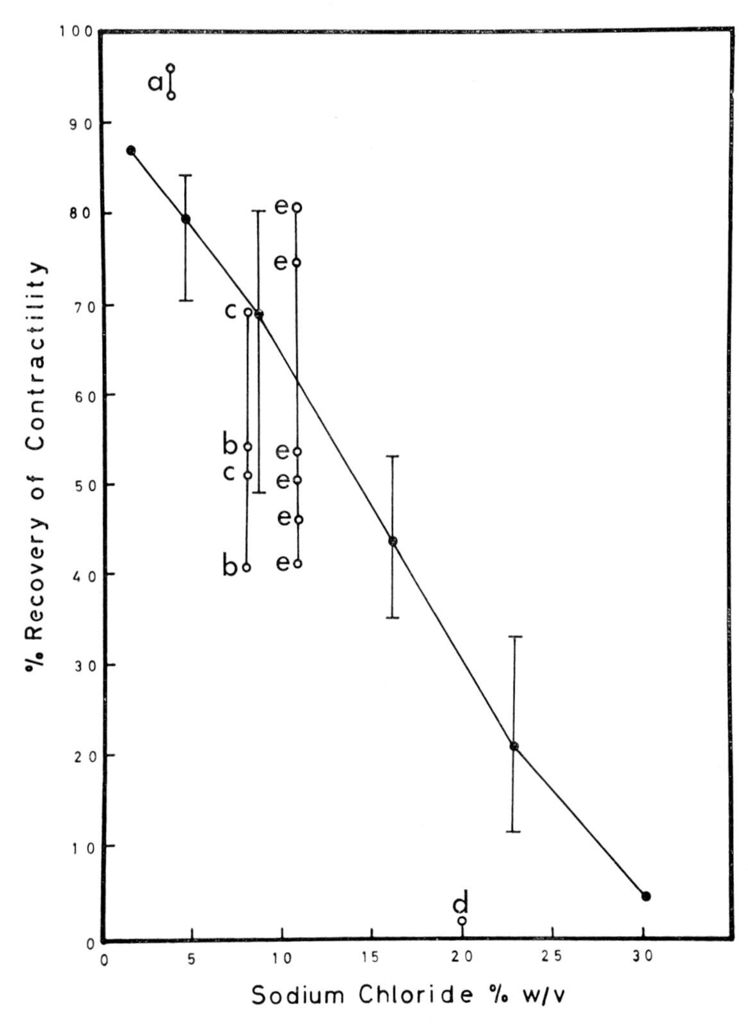

FIG. 11. Percentage recovery-of-contractility-curve (see Fig. 2) obtained from muscles frozen and thawed in Krebs' solution, replotted in terms of the sodium chloride concentrations attained at the different freezing temperatures, irrespective of other electrolytes present. The functional results from muscles subjected to various hypertonic salt solutions (*a*, *b*, *c*, *d* and *e*: see text for composition) have been superimposed on the same curve.

forms of stress. However, in the muscle immersed in isotonic Krebs' solution with a total NaCl concentration of 3·4 M, cellular damage far exceeded that seen in muscle frozen to the nearest equivalent freezing point (i.e. −10°C), and muscle exposed to that concentration of sodium chloride was also functionally dead (Fig. 13).

Lovelock (1953), working with human red blood cells, found that exposure of red cells to sodium chloride at concentrations less than 3·0 M (equivalent to freezing conditions down to −10°C) was itself insufficient to account for the damage that occurred during freezing and

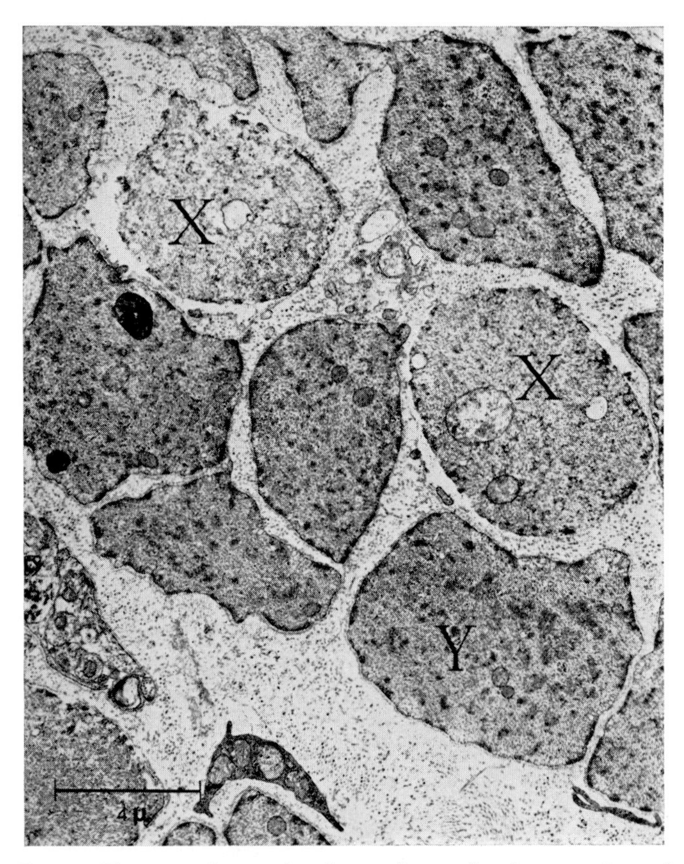

FIG. 12. Electron micrograph of smooth muscle after experimental treatment with increasing sodium chloride solutions at 0°C (final concentration 1·85 M) before return to isotonic conditions. Note the same distinction between disrupted cells (X) and apparently intact cells (Y) as was found in frozen and thawed muscles.

thawing, unless the cells were subsequently returned to isotonic conditions. Lovelock further proposed that a possible cause of damage to red cells exposed to sodium chloride concentrations greater than 3·0 M was a denaturation effect involving the lipoprotein complex of the cell membrane. My results, so far, have shown that with smooth muscle cells there is a possible correlation between the injurious effects of freezing and thawing and the rising concentration of sodium chloride during cooling, as far as $-10°C$. It has not been possible, however, to extend the connexion further: the marked severity of cell damage found in muscle treated with 3·4 M-sodium chloride at 0°C in comparison with that seen

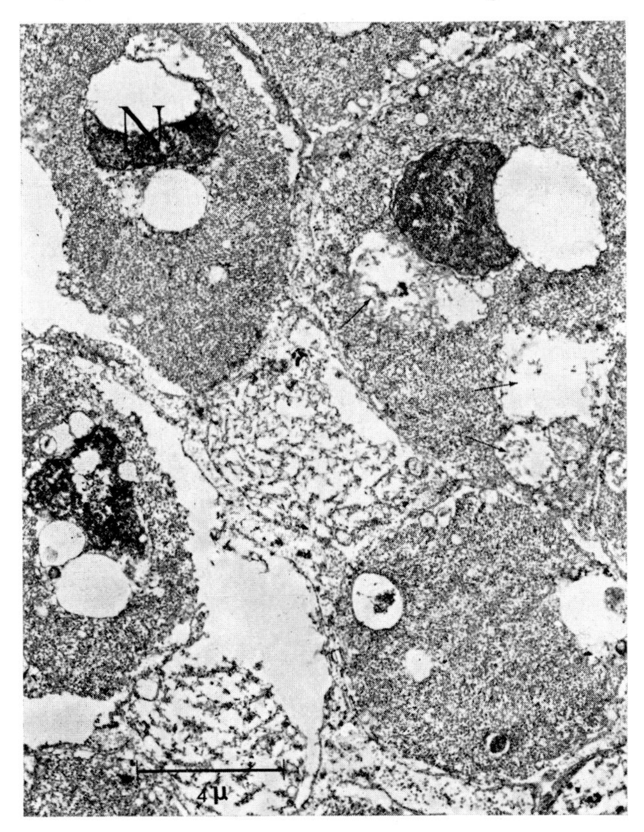

FIG. 13. Electron micrograph of smooth muscle subjected to increasing hypertonic sodium chloride solutions at 0°C (final concentration 3·4 M, with a freezing point of $-13°C$) and then returned to isotonic conditions. Cellular damage was more extensive than was seen in muscle frozen to and thawed from comparable freezing temperatures. N, nucleus. Mitochondria (arrows).

in muscle frozen even to temperatures as low as $-79°C$, followed by thawing, suggests that the lyotropic properties of concentrated salt solution (Lovelock, 1957) may have less significance below $-10°C$.

EFFECTS OF FIXATION

The electron microscopic results described so far in this paper were obtained from taenia coli muscle that had been fixed, for three hours at

0°C, with 1 per cent osmium tetroxide in isotonic Krebs' solution. Although probably not the ideal fixative for smooth muscle, this had the advantage of avoiding the further stress involved in transferring the injured tissue into another, more conventional buffer system. All specimens were dehydrated with acetone and embedded in Epikote 812 epoxy resin. Muscles for electron microscopy, whether frozen and thawed or exposed to hypertonic salt solutions, were in every case tested for contractility before fixation. Thus, a delay of not less than two hours at +37°C preceded fixation. Nevertheless, the ultrastructural findings need not, on that account, be dismissed as unrelated to the immediate effects of freezing and thawing or of hypertonic salt solutions. Almost certainly the injurious effects of these various experimental procedures were accentuated by progressive autolytic and osmotic phenomena in the damaged cells; but this may only have served to emphasize the changes in already injured cells, and could account for the clear differences that always distinguished "damaged" from "undamaged" muscle cells.

With the establishment of a reliable survival curve for smooth muscle of the taenia coli it will be unnecessary, in future, to subject all experimental muscles to a lengthy functional test before fixation. This will enable the muscle to be frozen and thawed, under specified conditions, and promptly fixed for electron microscopy using a variety of different fixation procedures in order to obtain a more precise understanding of freeze–thaw injury as it affects the smooth muscle cell. Preliminary experiments on muscle frozen to and thawed from −6°C, and then fixed immediately in a variety of solutions, with osmolarities ranging from 156 to 582 m-osmoles, have produced some interesting results. For example, muscle placed in an isotonic fixative solution immediately on thawing showed the same two cell types ("damaged" and "undamaged") as seen previously; but cellular disruption was less severe than in specimens that had been functionally tested after thawing. Moreover, better structural preservation was obtained in frozen and thawed muscles fixed in solutions of higher osmolarity than after fixation in the so-called isotonic solutions. With isotonic fixatives (300 m-osmoles) disrupted cells were numerous, and most of the cells had characteristically swollen mitochondria; but after hypertonic fixation disruption of cell membranes was less frequently seen, and it was possible to observe various degrees of intracellular damage (Fig. 14).

To conclude, it has now been established that in smooth muscle of the guinea-pig taenia coli slow freezing to temperatures not lower than −20°C, followed by thawing, results in severe injury to only a certain

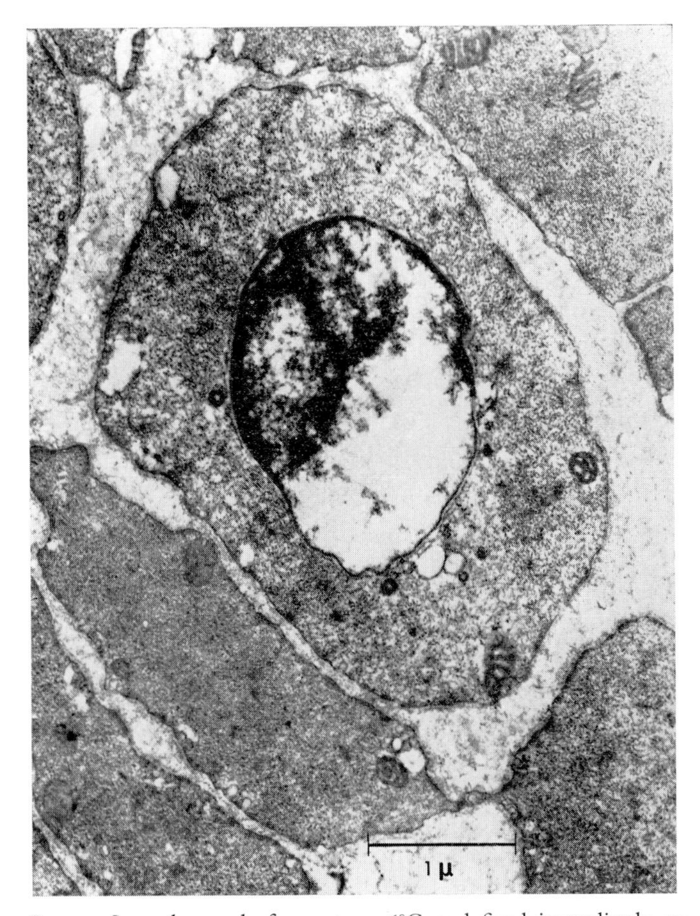

FIG. 14. Smooth muscle frozen to −6°C and fixed immediately on thawing in 1·25 per cent glutaraldehyde in Krebs' solution (412 m-osmoles). Very few smooth muscle cells with disrupted cell membranes were seen, although severely damaged cells with marked intracellular alterations were evident. Disrupted cells were commonly seen in muscle frozen to the same temperature but fixed on thawing in a near isotonic fixative (317 m-osmoles).

percentage of the muscle cells. It seems reasonable to hope that further electron microscopic study of these particular cells, using optimal methods of tissue fixation, will throw further light on the precise nature of freezing injury at the cellular level.

SUMMARY

Preliminary studies have been made of the structure and function of guinea-pig taenia coli smooth muscle after freezing and thawing. The findings were as follows:

(1) Contractile responses were obtained from unprotected smooth muscle after freezing and thawing between 0° and −79°C. Estimations of the contractile responses after freezing and thawing, compared with contractile responses obtained before freezing, showed that functional capability reached a minimal level by −20°C which was maintained by muscles frozen down to −79°C.

(2) A further study of muscle frozen to temperatures between 0° and −30°C established that a percentage recovery-of-contractile-response curve can be produced under standard conditions of freezing and thawing.

(3) Electron microscopically it was possible to distinguish between apparently intact cells and obviously damaged cells after freezing and thawing. Some correlation exists between functional impairment and the proportion of severely damaged cells in muscles frozen to temperatures between 0° and −30°C.

(4) Some correlation exists between the damaging effects of exposure to hypertonic salt solutions followed by return to isotonic conditions, and the effects of freezing and thawing. Sodium chloride has been implicated as the most likely electrolyte causing damage to smooth muscle during freezing and thawing and also after immersion in hypertonic salt solutions.

(5) In muscles fixed immediately after thawing, in a variety of fixative solutions with total osmolarities ranging from 156–582 m-osmoles, better structural preservation was obtained using fixative solutions of higher osmolarity than with so-called isotonic fixatives (300 m-osmoles).

(6) Although more information is needed the preliminary results show that smooth muscle is potentially a suitable tissue for investigating the effects of freezing and thawing on mammalian cells.

REFERENCES

BENNETT, M. R., and ROGERS, D. C. (1967). *J. Cell Biol.*, **33**, 573–596.

DOEBBLER, G. F., ROWE, A. W., and RINFRET, A. P. (1966). In *Cryobiology*, pp. 407–450, ed. Meryman, H. T. London & New York: Academic Press.

EICHNA, L. W. (ed.) (1962). *Physiol. Rev.*, **42**, suppl. 5.

FARRANT, J. (1965). *Nature, Lond.*, **205**, 1284–1287.

FARRANT, J., WALTER, C. A., and ARMSTRONG, J. A. (1967). *Proc. R. Soc. B*, **168**, 293–310.

LOVELOCK, J. E. (1953). *Biochim. biophys. Acta*, **11**, 28–36.

LOVELOCK, J. E. (1957). *Proc. R. Soc. B*, **147**, 427–433.

MAZUR, P. (1966). In *Cryobiology*, pp. 214–316, ed. Meryman, H. T. London & New York: Academic Press.

MERYMAN, H. T. (1968). *Nature, Lond.*, **218**, 333–336.

TRUMP, B. F., YOUNG, D. E., ARNOLD, E. A., and STOWELL, R. E. (1965). *Fedn Proc. Fedn Am. Socs exp. Biol.*, **24**, suppl. 15, 144–168.

DISCUSSION

Mazur: In the hypertonic solution only the sodium chloride was concentrated, whereas with freezing all the salts in the balanced salt solution were concentrated.

Walter: Nevertheless, there does seem to be some indication that the damaging effects of both freezing and hypertonic salt solutions are in some way related to sodium chloride concentrations, rather than to the total electrolyte concentrations present in Krebs' solution.

Mazur: How long was the time interval between thawing and fixation?

Walter: In those cases where I wanted to know what the percentage recovery of contractility was after freezing and thawing, it was never less than two hours after thawing the muscle. However, despite this delay it would not have been possible to correlate the electron microscope findings with the functional assessment had the muscles been fixed immediately on thawing, because then there is not such a clear distinction between damaged and intact smooth muscle cells.

Mazur: There is a very good correlation in yeast between cells that would be classified as living or dead on a cytological basis and the percentage of survival (Mazur, 1961).

MacKenzie: What happens if you leave them longer than two hours after thawing? Was this two-hour period before fixation used to develop the injury?

Walter: I have not tried leaving the muscles longer than two hours before fixation, as this two-hour period represented the time over which the muscle was tested for contractility after freezing and thawing. Although damage to some smooth muscle cells was undoubtedly being accentuated, I was more concerned with keeping to the same régime of testing the muscle as I had followed before freezing.

MacKenzie: To explain the fact that after this two-hour period some cells look normal and others are disrupted, but perhaps disrupted to somewhat different degrees, you could postulate the existence of a repair mechanism with a certain maximum working efficiency.

Walter: I am not prepared to say whether or not smooth muscle cells are capable of repair, especially in an *in vitro* system, but would prefer to consider that the cells that appeared to be intact were so because they were not severely damaged during freezing and thawing.

MacKenzie: In any case, if the damaging event occurs so many times per cell, you have a frequency of damaging events, and if the repair mechanism can only take care of a damaging event when it occurs below a certain frequency, then all cells which are repaired will be totally repaired and all cells which are damaged will be damaged in proportion to the frequency of occurrence of the damaging event. We could in this way explain why we see intact cells and variously damaged cells.

Mazur: Why do you need to postulate that? There could simply be a threshold level of stress. If that threshold is not exceeded the cell is not damaged; if it is exceeded, the cell is irreversibly damaged.

MacKenzie: That, I believe, is what I was trying to say. If any level of damage were to destroy the cell, one would not expect to see a certain proportion of cells each looking equally good. Cells exposed to conditions of least damage should all look good; they should look worse and worse, without distinction, one from another, after exposure to conditions less favourable to survival.

Meryman: I don't agree with Dr MacKenzie that one would expect to see a smooth gradation of injury. Certainly with the red cell injury appears to be an all-or-none phenomenon: a cell either succumbs to stress or it doesn't. In Mr Walter's pictures there are cells in which the membrane has obviously broken down. It is interesting that where the outer membrane has gone everything inside is destroyed; where the outer membrane is intact the nuclei and the mitochondria appear undamaged. It is as though the outer membrane were the major barrier. If this breaks down then everything else breaks down, perhaps due to an osmotic shock on return to an isotonic solution.

Mazur: In yeast too survival is an all-or-none phenomenon, but in some bacteria this is not true—the percentage recovery depends on the plating media on which they are grown (Mazur, 1966).

Elford: How hypertonic were the solutions in which the cells were immersed, before fixation in a hypertonic fixative?

Walter: It may sound paradoxical but the muscles were immersed in isotonic Krebs' solution before fixation in hypertonic fixative. This is because the muscles were frozen and thawed in Krebs' solution. But if one follows the drained wet-weight changes of a frozen and thawed muscle it takes at least 30 minutes before the muscle returns to equilibrium weight conditions. The muscles in fact weigh less immediately on thawing than they do before freezing. If it is presumed that the loss in weight is due to the removal of intracellular water from the cells during freezing, then on thawing, even though the muscle cells are in an isotonic medium, the cells are hypertonic to it intracellularly. Alternatively, the cells could have become more permeable to the increasing concentrations of extracellular electrolytes, particularly sodium chloride, during freezing. If this happens then one would expect the muscle to show a gain in wet weight rather than a loss. It is also possible that there is both a loss due to removal of intracellular water, and a gain due to altered electrolyte permeabilities. For example, after being frozen to −2·5°C the loss in wet weight (21 per cent) is greater than in muscles that had been frozen to and thawed from −21°C (11 per cent). However, the net result is that the smooth muscle cells are intracellularly hypertonic to the bathing solution, and this could explain why hypertonic fixative solutions are better for fixation of frozen and thawed muscle cells if fixation is attempted immediately on thawing.

Walder: Your survival percentage of 35 per cent is the same as I get with freezing and direct transplantation of artificial sarcomas into mice. But when I leave the tumour *in situ* I find a survival of 2 per cent. The suggestion is that much damage is caused by the freezing and rewarming process, and the secondary damage has another cause such as anoxaemia due to vascular changes and oedema (see also p. 267).

REFERENCES

MAZUR, P. (1961). *Biophys. J.*, **1**, 247.
MAZUR, P. (1966). In *Cryobiology*, pp. 214–315, ed. Meryman, H. T. London & New York: Academic Press.

GENERAL DISCUSSION

OSMOTIC DAMAGE

Mazur: Several speakers have implied that osmotic shock of some sort is the chief cause of damage in frozen and thawed cells. I would like to raise some objections. First, I know of no microorganism in which slow thawing is less damaging than rapid thawing, but there are a number of instances in which slow thawing is more damaging, even to cells that have been cooled slowly enough to have avoided freezing intracellularly. This is also true in hamster tissue culture cells (see Fig. 7C, p. 79), yeast (Mazur and Schmidt, 1968) and chloroplast membranes (but it is not true in higher plants). If osmotic shock were the cause of injury, slow thawing ought to be protective or at least no more damaging than rapid thawing.

Levitt: If it is true in the chloroplast membrane and not in plants as a whole, are the different kinds of freezing injuries the same?

Heber: I guess so, only the plant cells are much more complicated. There are very large vacuoles and something else could complicate the situation.

Levitt: There is no ice in the vacuoles.

Heber: No, but the whole thing has shrunk during freezing and it has to re-expand, so osmotic injury could be part of the picture. To avoid a whole set of factors which can lead to injury in plants we are working with grana, where dehydration produced by freezing causes damage.

Levitt: In other words the kind of protection that you are getting for your chloroplast may not protect the plant because it is a different kind of injury.

Heber: I think we should distinguish two different things. To preserve a cell we have first to protect it against dehydration, or salt accumulation, which may be the same thing. However, if a cell is protected it can still be killed by thawing. Then we have to adjust conditions during thawing so that it can survive thawing, and these latter conditions may be different in animal and plant cells. By supplying sufficient amounts of protective agent we can protect against the effects of dehydration and thereby against freezing damage.

Levitt: Your lamellar system therefore corresponds to one part of the damage to the whole cell.

Mazur: Recently Quatrano (1968) reported that when plant cells in tissue culture were frozen under the usual conditions used for animal cells (i.e. suspended in DMSO, frozen slowly and thawed rapidly), some 14 per cent of the cells were viable. Furthermore, as Dr Leibo mentioned earlier, Lovelock (1953) found that red cell haemolysis was higher after slow thawing than after rapid thawing. The fact that slow thawing is usually more damaging than rapid makes an osmotic shock explanation of freezing injury difficult to accept as a general mechanism.

Farrant: If slow warming is not as harmful in plants what are the rates of rewarming in plants?

Mazur: Very slow (0·01 to 0·1°C/min)—as slow as the rates of freezing.

Farrant: At these very slow rates it is possible that one might also find increased survival with nucleated mammalian cells, similar to the results that Dr Meryman reported with his glycerol experiments on red cells when rewarming was done at 0·3°C/min (Meryman, 1967).

Meryman: I don't think one would find increased survival in red cells. The red cell can swell to approximately twice its original volume, which means that one should be able to introduce into it approximately twice its normal solute concentration without the cell haemolysing in isotonic solution. So in frozen packed cells, where the extracellular solution is only 3 or 4 per cent of the total volume, even if one forced all of the solute into the cell, and if there were no other damage to the cell, it would only be slightly larger on thawing, regardless of how one thawed it, and there should be no injury. In fact, one gets 100 per cent haemolysis of the cells. I would conclude from that and other evidence I have presented here that the injury from hypertonic solute is an injury to the membrane and not simply the reversible passage of solute into the cell leading to an osmotic shock on thawing.

Mazur: I thought you were implying that osmotic shock is itself the major cause of irreversible injury.

Meryman: The stress from the concentration of extracellular solute leads to a membrane injury. What causes the cell to fall apart subsequently and to haemolyse may be the osmotic shock on thawing, but the membrane has already been irreversibly damaged; if this were not so one should be able to freeze packed cells without any haemolysis.

Leibo: Why, in fact, is slow warming of red cells slowly frozen in 1 M-glycerol less damaging than rapid warming, Dr Meryman (Meryman, 1967)?

Meryman: As the erythrocytes freeze only some cells haemolyse, as Professor Nei can confirm. We believe that these cells haemolyse as a result of thermal shock. The remaining cells are subjected to hypertonic stress and a proportion of them will break down and leak solute. When these cells are thawed and returned to isotonic solution, they will haemolyse from hypotonic stress, because of the increased amount of intracellular solute. If they are thawed very very slowly, perhaps the excess intracellular solute can leak back out again, and the cells will not haemolyse. Other than this possibility, I have no good answer.

Mazur: Could the improved survival you got when you rewarmed red cells at 0·3°C/min rather than at 1°C/min be accounted for by an osmotic shock from the glycerol itself? That is, during very slow freezing glycerol might diffuse into the cell, producing an osmotic shock on thawing.

Meryman: There hasn't been any shift of glycerol. There is no need for glycerol to go in and out of the cell to maintain osmotic equilibrium. The diffusion rate of glycerol across the cell membrane at those temperatures is negligible. The benefits of slow thawing may have been related to the redistribution of water from thawing ice.

Levitt: What are your rates of thawing, Dr Heber?

Heber: The fast rates are pretty fast, of the order of a few hundred degrees per minute. Small frozen samples are immersed in a water bath at about +25°C. The slow rates are obtained by just transferring the samples to room temperature in air. They are still of the order of some degrees per minute.

I would like to propose an answer to the question, why is slow thawing usually damaging and not protective? In our membrane system, salt damage can be viewed as a chemical reaction. It can be followed kinetically, and it is temperature-dependent. So if one thaws slowly, one exposes the membrane to the damaging effect of salt for a long time.

Mazur: That seems to me a valid argument, because in your case, in Lovelock's case with red cells, and in *E. coli* (unpublished data) the total damage appears proportional to the total time the cell is exposed to concentrated unfrozen solution.

Heber: We can easily follow the course of salt damage. Depending on the conditions, it may take hours.

Levitt: Are you specifying that this slow thawing is after intracellular freezing?

Mazur: If you cool fast enough to get intracellular freezing, then there are enormous effects from slow thawing, probably because of recrystallization, but I am talking now about the time spent during freezing and thawing, when freezing is too slow for intracellular ice. The plant is certainly an interesting exception to the idea that damage is proportional to total time of exposure. In that case, osmotic shock may be operating.

Farrant: It depends on the rate of thawing, which is different in the plant and animal cell experiments. Has anybody thawed animal cells extremely slowly?

Meryman: We thawed red cells as slowly as 0·1°C/hour, and it doesn't seem to make any difference.

Heber: In plant cell thawing, in most instances a huge vacuole and two large continuous membranes bordering the protoplasm on either side are involved. One of the membranes is in contact with a cell wall, which is non-existent in animal cells. During the process of fast thawing, different rates of water entry into the different phases could easily create disorders which would lead to injury.

Levitt: I think it is a cell wall phenomenon.

Meryman: In the grana exposed to hypertonic salt, there is a big time factor. If they are exposed to, say, 2000 m-osmolal of salt, injury develops slowly over a period of an hour. So if a suspension is taken back to a lower concentration at any point, there will be a proportional reduction in total activity. The rate of thawing therefore could add to the injury if it were very slow.

Farrant: Zade-Oppen (1968) showed that when red blood cells were diluted after exposure to hypertonic conditions there was more haemolysis the further one diluted.

Leibo: What about the dilution rate?

Farrant: Rate was not investigated, just the concentration steps.

Leibo: My point is that slow dilution would be less damaging than rapid dilution.

Meryman: In red cells if one dilutes very slowly and carefully in salt, potassium leakage can occur without haemolysis. In sugars on the other hand one can dilute by instant mixing and get instant relief of osmotic stress by the loss of potassium without any haemolysis. The potassium must be leaving the cell as rapidly as the water is coming in, otherwise the cell would rupture.

Mazur: How slow is the slow dilution?

Meryman: We haven't investigated different rates, but by diluting at the rate of say 10 m-osmolal/min or less, a lot of the cell potassium will come out without haemolysis, in the presence of salt. In the presence of sugar one can simply drop red cells into a hypotonic solution and lose 90 per cent of the cell potassium without any haemolysis.

Mazur: But red cells won't survive slow thawing, which is tantamount to slow dilution.

Meryman: There is a very substantial increase in the recovery, though not complete protection. If one freezes red cells in an inadequate concentration of glycerol, say 1·0 M, with the osmotic support provided by sugar, the recovery of cells is about twice what it would have been with salt. I don't know precisely what the mechanism is.

Farrant: Valdivieso and Hunter (1961) increased hypertonic conditions with sucrose. They used twice isotonic sucrose in the presence of isotonic salt, and then they resuspended the red cells in three times isotonic salt and got haemolysis. If they resuspended in twice isotonic salt they got more haemolysis. This indicates that sucrose may be getting in, yet there is still this relative obstacle to sucrose diffusing out. This would produce haemolysis through swelling.

Meryman: I have not seen sucrose getting in when salt is at least isotonic. I have seen no changes in the cells and no exchange of cations.

Farrant: What is your measurement of whether the sucrose is getting in?

Meryman: Cell volume and a check against intracellular sodium and potassium to make sure that they are not exchanging.

Elford: Dr Meryman, you are saying that all or most of the damage is done when the red blood cell is dehydrated, yet on the other hand you found that in salt solutions at eight and ten times the isotonic concentration there was only 2 per cent haemolysis, so there isn't gross damage to the cell membrane under these conditions.

Meryman: But the membranes were apparently completely leaky to low molecular weight compounds, and over a period of two hours all the cell potassium leaked out.

Elford: You got swelling during reimmersion in isotonic conditions, simply because you rendered the membrane leaky to cations.

Meryman: No. They have been filled with excess solute because they have been made leaky in a hypertonic solution.

Elford: Cell membranes may have been made leaky, but if you re-dilute extremely slowly so that salts and water have time to maintain osmotic equilibrium then no damage should be seen under these conditions.

Meryman: It is possible, on slowly returning cells to isotonic solution, to prevent most of the haemolysis, but one just has a leaky bag from which the haemoglobin has not yet escaped. The cells appear to be irreversibly damaged.

Elford: When you did a very slow redilution of red blood cells which had been exposed to hypertonic salt solution above 0°C, did you get results which were the same as with very slow thawing of red blood cells after contact with ice and hypertonic salt solution in the remaining liquid phase?

Meryman: We get what I consider to be a false indication of recovery of red cells by a very slow thawing or very slow redilution—we get a cell which still contains haemoglobin but which is leaky.

Mazur: You are saying the red cell may look normal and have an intact membrane, but that it is still essentially a dead cell.

Meryman: Yes. It just isn't leaky enough to leak haemoglobin but it leaks everything else.

Farrant: If that were so, when you went back to isotonic conditions you wouldn't expect swelling due to any discrepancy between the diffusion of electrolytes and water.

Meryman: The rate at which it will leak is influenced by the nature of the solute. In sugar the solute will leak out as fast as water comes in. In salt the rate of leakage is much slower. I haven't the remotest idea what the mechanism is.

Woolgar: In view of the possible strong dependence of the passive permeability to cations on such factors as pH or ionic strength, what evidence have you that the red cell membrane becomes *irreversibly* permeable to cations, Dr Meryman?

Meryman: The only evidence I have is that while the cells are in the hypertonic solution they continue to leak cations or small molecules in and out until their internal constitution is the same as the suspending medium. This may take two to four hours. Presumably the osmotic stress has long since been relieved by the inward leak of the extracellular solute; the cell size has come back to something near normal, yet there continues to be a diffusion of cation across the cell membrane until the inside and outside come to equilibrium.

Woolgar: At low ionic strength, a change in pH alone can cause a drastic change in the permeability to cations (Passow, 1964). Without information on the factors involved in the very hypertonic situation, it is difficult to decide whether permeability changes are due to direct membrane damage or to changes in the physical environment.

MacKenzie: Rapatz and I stored frozen red cells in various suspending media at −60, −55 and −50°C for periods of up to two weeks. We generally found an increase in haemolysis over this two-week period. Haemolysis after thawing was something like 5 or 10 per cent in the controls, and it rose to something like 70 to 80 per cent after one or two weeks at these temperatures. This deterioration was more or less independent of the suspending medium so long as it contained a cryoprotective substance in 10 or 20 per cent concentration, w/w, together with Hank's salts. We couldn't correlate the increased haemolysis with any measurable physical property of the suspending medium (MacKenzie and Rapatz, 1968).

MUST ADDITIVES PERMEATE TO PROTECT?

Mazur: My second general question concerns permeation again. Must additives permeate to protect? Dr Leibo, Dr Farrant and I have shown (pp. 69–85) that sucrose did not permeate the hamster cell at 0°C. Furthermore, because of its high molecular weight, PVP probably didn't permeate either. Yet both additives protected. Dr Meryman apparently feels that the protective effect of sucrose on the red cell is due to the fact that it does permeate during freezing. Didn't you say that sucrose prevented the cell from shrinking below its critical volume because just before that volume is attained the cell becomes permeable to sucrose?

Meryman: The sucrose effect simply lowers the threshold for the penetration of low molecular weight molecules across the membrane, under an osmotic gradient. It seems to make the membrane more plastic; the cell can swell to a slightly larger size and when it is under an osmotic stress it will leak small molecules readily, rather than be overstressed.

Elford: Do you include sucrose among the small molecules that will leak into red blood cells?

Meryman: Yes. The sugars, PVP, and some of the proteins we talked about will permit leak under stress in the red cell. This possibility alters the things that can happen during freezing but does not itself provide protection. In the red cell the only way we have ever seen protection, ignoring rapid freezing for the moment, is by getting into the cell a high concentration of a penetrating solute that protects on a colligative basis.

Mazur: But sucrose does protect hamster cells, even though it cannot permeate them at 0°C.

Levitt: In those cases where sucrose protects without getting in have you ever kept them frozen for a really long time and still obtained protection?

Mazur: We held the cells at −196°C for times up to 16 hours but we haven't tested the effect of time at higher temperatures.

Levitt: But has external freezing ever been as successful for maintaining cell life over a long period of time as very rapid internal freezing, that is in cases where a protective solution has penetrated the cell?

Mazur: I don't know. I would expect that it wouldn't be so successful. However, survival should not be affected by the length of time at −196°C.

Farrant: Dr Meryman, how do you explain the results I got, where PVP would protect the red blood cell at a freezing temperature of about −6°C to −10°C? Does PVP itself act like sucrose and lower the threshold for penetration of these low molecular weight substances? Is PVP itself going to penetrate into the cell in these experiments?

Meryman: I would like to look at the intracellular sodium and potassium and the cell volume. Whether PVP is going to go in or not I don't know, but there are other solutes outside that could. The first thing to check is whether PVP permits the reversible influx of extracellular solute.

Farrant: But like the lower molecular weight protective substances PVP lowers the gradient of electrolytes across the cell membrane during freezing, because it produces a lower concentration outside the cell. So a gradient is produced which would lead to less influx during the hypertonic phase.

Meryman: PVP is not going to reduce the osmolality of the solution or the hypertonic stress that might develop. So if hypertonic stress is the cause of injury, one would expect to get it regardless of what mixture of solutes there is outside.

Farrant: I get protection with PVP, though.

Meryman: Then you have to assure yourself that this protection is not an illusion in which the cell fails to haemolyse because of the transfer of solutes. This is an important question—how does one define injury to the cell? If the membrane is irreversibly permeable to small molecules, whether the cell haemolyses or not, I would say that it had been injured.

Farrant: But when you use PVP as Dr Rinfret has done you recover clinically viable cells.

Meryman: That is rapid freezing.

Farrant: Yes, but the cells are still exposed to the same system. Although the conditions are different one can get cells which can be used and which are in good shape.

Meryman: When the technique is not ideal one recovers cells which are very seriously depleted of potassium. In the early experiments with glucose as a protective agent haemolysis was very low in some samples, yet there was a loss of perhaps 35 per cent of intracellular potassium.

Rinfret: As I recall our studies on lactose and glucose as cryoprotective additives, there was always increased potassium in the suspending medium after thawing. Some part of this would have been due to gross cell disruption. With the PVP, at least with rapidly frozen preparations, our analyses indicated extracellular concentrations up to 20 m-equiv./l after thawing, less than the shifts reported by Chaplin, Schmidt and Steinfeld (1957) in glycerol-treated blood stored at −20°C and more than those observed by Haynes and co-workers (1962) when such blood was stored at −80 to −120°C. These latter authors reported essentially no change in cellular potassium.

Mazur: I must emphasize again that protection by sucrose and PVP in marrow and hamster cells cannot be illusory, because we are in fact assaying the cells after growth and replication.

Rinfret: The body of evidence on the *in vivo* survival of red cells preserved in the frozen state would support that statement.

Luyet: On the question of whether or not the penetration of cryoprotective substances into the cells is a requirement for protection, Malinin, Fontana and Braungart (1969) have reported that, as judged by radioactive tracing, dimethylsulphoxide (DMSO) would not penetrate the cells.

Elford: I haven't read that paper but I have also done similar experiments on DMSO distribution. The water space in smooth muscle is identical to the space measured using ^{35}S-labelled DMSO. At +37°C a transient loss in the wet weight of the guinea-pig smooth muscle, taenia coli, is produced by immersing strips of muscle in Krebs' solutions containing 20 per cent DMSO. But after an hour the muscle returns to its normal resting wet weight *in vitro*. The [^{35}S]DMSO space under these conditions is very close to the water content of the cells, and this water content is identical to that of muscles which haven't been immersed in 20 per cent DMSO. Henderson, Bickis and Edwards (1967) also looked at the exchange of [^{35}S]DMSO in perfused dog kidney, and concluded that equilibration occurs within about 20 minutes at room temperature, even though less than 100 per cent of the available water in the kidney was available to dissolve DMSO. Yet at Mill Hill J. Farrant and D. E. Pegg have shown that all the water in the perfused rabbit kidney is available to [^{35}S]DMSO (personal communication). It would appear that the criterion of Henderson, Bickis and Edwards for deciding on the extent of equilibration was in fact invalid, as one cannot look at the concentration of [^{35}S]DMSO in the effluent from a perfused kidney and come to the conclusion that equilibration has been reached between rates of exchange of water and DMSO within the organ.

Mazur: The packed cell volume data that I presented on the hamster cells are difficult to explain unless one assumes that DMSO permeated the cells. How good is the evidence that glycerol permeates cells other than human erythrocytes under the conditions used in freezing?

Meryman: The influx of glycerol is very temperature-dependent, but the efflux is almost not temperature-dependent at all. This would imply that there is an active process involved in the influx but that the efflux is largely passive diffusion.

Mazur: Bickis and co-workers (1967) found that the rate of permeation of glycerol into ascites cells is very slow and is practically zero at +2°C.

Elford: Yes, but they did say that equilibration seemed to be complete in that all the water was available to dissolve glycerol. However, permeability coefficients seemed to be markedly dependent on the temperature. They gave values for the half-time of equilibration for DMSO and glycerol of 90 seconds and about 20 minutes respectively, at +37°C; the corresponding values at room temperature were about 5 and 60 minutes.

Mazur: It was much slower at 0°C. In fact little glycerol permeated even after 50 minutes.

Meryman: We have measurements of both DMSO and glycerol transfer into mouse ascites cells. At room temperature the equilibration with 1 M-DMSO is complete in about 12 seconds. For glycerol it is a matter of over a minute. At +4°C the equilibration of DMSO takes seven or eight minutes, while that of glycerol is too slow to measure.

Mazur: Glycerol generally has a very high temperature coefficient for permeation.

Elford: The final volume of your hamster cells in glycerol was only about 70 per cent of the initial volume, yet after contact with DMSO the cells returned to 110 per cent of their initial volume. How do you explain this surprising result?

Mazur: There are at least three possible explanations. (1) Only a proportion of each cell was permeable to glycerol. (2) Part of the population was permeable and part was not permeable. (3) Glycerol permeates by facilitated diffusion, under conditions in which the membrane carriers saturate at very low levels. Facilitated diffusion is a form of active transport in the sense of being enzyme-mediated. But it differs from true active transport in that the solute cannot be moved against its concentration gradient. LeFevre (1954) showed intermediate volumes of red cells exposed to hyperosmotic concentrations of sugars that permeate by facilitated diffusion.

"SALT BUFFERING" AND PROTECTION

Mazur: My third general question is, can protection by non-permeating additives be satisfactorily explained on the basis of "salt buffering" as Dr Farrant has suggested for PVP? Can the action of Dr Heber's protein be explained on that basis?

Heber: No; if it could, other proteins of equal molecular weight should have the same effectiveness, and another protein, let's say of twice the molecular weight, should have half the effectiveness. But this is not so.

Mazur: As Dr Farrant pointed out, the ability of a macromolecule to be a "salt buffer" depends on its osmotic coefficient. However, since the mole ratio of Heber's protein to salt is only 1:2000, it seems unlikely the protein could act in that way.

Farrant: It depends on how the macromolecule affects the activity of water in high concentration. We should not rule out the possibility that Dr Heber's protein has an effect on the concentrations of salts, just because of its low concentration. It depends on the relative concentration of the protein and salt. In fact Dr Heber has quite similar concentrations of protein and salt on a weight basis in his system.

Mazur: They are similar on a weight basis, not on a molar basis.

Farrant: It also depends on the eutectic characteristics and the freezing characteristics of the three-component system—water, protein and salt, with all the non-ideal behaviour these have. This would have to be determined and until that is done one can't prove it one way or the other.

Elford: You would have to postulate a depression of freezing point, caused by Dr Heber's protein, to support your idea of a colligative mechanism.

Farrant: Yes, it is being concentrated during freezing. The extent of the concentration depends on the temperature and on where you start on the eutectic diagram.

Mazur: What about the molecular weight of PVP? Dr Rinfret, you said earlier that 10 000 molecular weight PVP is not very protective.

Rinfret: That statement has to be taken in the context of a desire to preserve blood in a frozen state in a form suitable for transfusion. The smaller molecular weight PVPs do not afford the protection that PVP of a higher molecular weight (40 000) does. However, all PVP is protective to some extent. With 10 000 molecular weight PVP you might get recovery of 80 to 85 per cent of the red blood cells, rather than the 98 per cent we aim at.

Mazur: What about protection with very high molecular weight PVP?

Rinfret: We did not use very high molecular weight PVP in our studies because such material is unacceptable for human transfusion.

Huggins: Nash has found *N*-methylpyrrolidone to be an excellent cryophylactic agent. This finding may be very significant since this compound, biologically, is the monomer of PVP.

Levitt: Dr Heber, your chloroplast system has potassium chloride but no sodium chloride in it.

Heber: Potassium doesn't make much difference when compared to sodium. The anions have very drastic effects compared to the cations (Santarius, 1969). Iodide is much more destructive than fluoride. Bromide and chloride are in between. For other anions, inactivation follows in principle the Hofmeister series (Hofmeister, 1888), but in reversed order. Different univalent cations do not make much difference.

Levitt: But most of the salt is in the vacuole.

Mazur: When one multiplies the concentrations of the salts by their osmotic coefficients I wonder if the different anions would behave so differently?

Heber: If one just plots ionic strength against inactivation there are large differences. From this one can see the specific effects of differently charged anions.

Mazur: But ionic strength doesn't take osmotic coefficients into account.

Levitt: The proteins of mammals might show a greater reversible denaturation at the lower temperatures than those of plants, because these are temperatures at which plants grow normally (around 0°C). Therefore the animal cell may be more in danger of injury, and this could be why one has to thaw cells through that temperature range rapidly.

Mazur: The fact that most animal cells can be supercooled without damage argues against that.

Levitt: When the animal cell is supercooled at that low temperature the chemical changes don't take place. I am talking about a denaturation that would be followed by aggregation, for instance.

Brandts: Our guess is that the aggregation would tend to take place to a greater extent in frozen solutions, since the local protein concentrations presumably are much higher. But one also changes the salt concentration and other things when freezing cells.

* * *

Rinfret: The role of temperature is an important one in providing, or failing to provide, protection. If the temperature is high enough, even though well below the point at which ice formation is initiated, changes can and do take place with relative rapidity. For example, Mollison and his co-workers stored blood at temperatures we now think of as very high, about −20 to −40°C. They observed unmistakable evidence of change and cellular degradation with time (Mollison, Sloviter and Chaplin, 1952; Jones, Mollison and Robinson, 1957). These changes were taking place over periods measured in days, weeks or months at these temperatures.

The microscopic examination of blood frozen in thin films emphasizes the rapidity with which destructive reactions can occur in this state. One need not thaw to achieve total cellular distintegration. In some of our experiments thin films were supercooled to −8°C and seeded. No protective additive was present. The temperature was maintained at −8°C. The erythrocytes, surrounded by ice crystals, were in contact with clear (colourless) channels of residual unfrozen fluids. Within seconds after freezing the cells were crenating and within two minutes haemoglobin was observed to be diffusing along the channels from those points. Two minutes later the cells collapse, two or three in any given region combining to form an amorphous pool of haemoglobin. All this takes place in what would appear grossly in a larger volume to be a solidly frozen mass. At this temperature Professor Luyet's data would, I believe, indicate that 93 to 95 per cent of the freezable water has been converted to ice. The intracellular contents, however, have not frozen, as indicated by the progression of the pigmented haemoglobin along the channels.

Mazur: This confirms what Dr Nei said (pp. 131–142) about getting appreciable haemolysis in the frozen state before thawing. This is again a major difference from Lovelock's findings in unfrozen solutions, in which case appreciable haemolysis became evident only after dilution.

Rinfret: I can't help but think that the blood stored by Mollison and Jones was haemolysing at least in part at the temperatures they used.

Luyet: The question of the time required after thawing for the injury to show up is a question of interest for its own sake. To become injured and to die are processes which involve vital activities. Like all other life processes, the development of injury and the progress towards death are slow at low temperatures and are accelerated by a rise in temperature. This was well illustrated in some of our studies on leucocytes. After freezing and thawing, some cells recovered, others deteriorated further and died. In both cases, the processes involved took time and were accelerated at higher temperatures.

Mazur: There are advantages to measuring viability on the basis of colony formation, for colony formation requires a cell to live perhaps eight or ten days, and also requires it to undergo a number of replications and divisions. Whatever latent injuries were present immediately after thawing should be expressed by then.

Davies: Has anyone found any change in biochemical activity, or antigenic changes in cells frozen to low temperatures or alternatively by using hypertonic salts? This would provide evidence of damage to the membrane at a sublethal level.

Mazur: Some bacteria, including *E. coli*, show different percentages of survival depending on the plating medium. An enriched medium gives a higher recovery than a minimal medium. This is not true for yeast.

Davies: I was thinking particularly of mammalian cells, where the dye exclusion test has not been a good assay of viability. This could possibly be linked with a modification in the protein structure without a gross alteration in the overall membrane structure.

Mazur: I know of no evidence of this in mammalian cells.

Heber: In plant cells several reports have shown the accumulation of proline after sublethal freezing (Heber, 1958). This is presumed to be the consequence of the breakdown of damaged proteins which occurs before resynthesis.

★ ★ ★

Mazur: My closing remarks are going to be very brief. Although I don't think we have reached many definite answers, we have sharpened a number of questions, and perhaps that is equally important. It is useful to have a clear idea not only of questions but also of possible experimental approaches in order to get definitive answers. There seems to be a certain lack of numerical data relevant to some of the problems. We all have our own hypotheses to interpret the facts. Hopefully critical experiments can be designed to support or rule out some of these hypotheses. Still a good chance exists that no single hypothesis will explain all freezing injury. It is likely that different factors operate in different cells under various circumstances.

REFERENCES

BICKIS, I. J., KARAKS, K., FINN, J. J., and HENDERSON, I. W. D. (1967). *Cryobiology*, **4**, 1.

CHAPLIN, H., SCHMIDT, P. J., and STEINFELD, J. L. (1957). *Clin. Sci.*, **16**, 651.

HAYNES, L. L., TURVILLE, W. C., SPROUL, M. T., ZEMP, J. W., and TULLIS, J. L. (1962). *J. Trauma*, **2**, 2.

HEBER, U. (1958). *Planta*, **52**, 431–446.

HENDERSON, I. W. D., BICKIS, I. J., and EDWARDS, P. (1967). *Cryobiology*, **3**, 373.

Hofmeister, F. (1888). *Arch. exp. Path. Pharmak.*, **24**, 247–260.
Jones, N. C. H., Mollison, P. W., and Robinson, M. A. (1957). *Proc. R. Soc. B*, **147**, 476.
LeFevre, P. G. (1954). *Symp. Soc. exp. Biol.*, **8**, 118.
Lovelock, J. (1953). *Biochim. biophys. Acta*, **10**, 414–426.
MacKenzie, A. P., and Rapatz, G. L. (1968). *Fedn Proc. Fedn Am. Socs exp. Biol.*, **27**, 700.
Malinin, G. I., Fontana, D. J., and Braungart, D. C. (1969). *Cryobiology*, **5**, 328–335.
Mazur, P., and Schmidt, J. J. (1968). *Cryobiology*, **5**, 1–17.
Meryman, H. T. (1967). In *Cellular Injury and Resistance in Freezing Organisms*, pp. 231–244, ed. Asahina, E. Sapporo: Institute of Low Temperature Science, Hokkaido University.
Mollison, P. L., Sloviter, H. A., and Chaplin, H. (1952). *Lancet*, **2**, 501.
Passow, H. (1964). In *The Red Blood Cell*, pp. 71–145, ed. Bishop, C., and Surgenor, D. M. London: Academic Press.
Quatrano, R. S. (1968). *Pl. Physiol., Lancaster*, **43**, 2057.
Santarius, K. A. (1969). Submitted to *Planta*.
Valdivieso, D., and Hunter, F. R. (1961). *J. appl. Physiol.*, **16**, 665.
Zade-Oppen, A. M. M. (1968). *Acta physiol. scand.*, **73**, 341–364.

INDEX OF AUTHORS*

Numbers in bold type indicate a contribution in the form of a paper; numbers in plain type refer to contributions to the discussions.

* Author and Subject Indexes compiled by Mr. William Hill.

INDEX OF SUBJECTS

Printed by Spottiswoode, Ballantyne & Co. Ltd., London and Colchester.